77 Structure and Bonding

Chemistry, Spectroscopy and Applications of Sol–Gel Glasses

Editors: R. Reisfeld and C. K. Jørgensen

With contributions by
A. Emmerling, J. Fricke, M. Henry,
C. K. Jørgensen, J. P. Jolivet, J. Livage,
R. C. Mehrotra, R. Reisfeld, S. Sakka,
H. Schmidt, T. Yoko

With 132 Figures and 16 Tables

Springer-Verlag
Berlin Heidelberg GmbH

Guest Editor

Professor Renata Reisfeld, Department of Inorganic and Analytical Chemistry,
The Hebrew University of Jerusalem, Jerusalem 91904, Israel

ISBN 978-3-662-14977-5 ISBN 978-3-540-47588-0 (eBook)
DOI 10.1007/978-3-540-47588-0

Originally published by Springer-Verlag Berlin Heidelberg New York in 1992
Softcover reprint of the hardcover 1st edition 1992

Typesetting: Macmillan India Ltd, Bangalore-25;

51/3020-5 4 3 2 1 0 – Printed on acid-free paper

Editorial Board

Foreword

For about 3,000 years, nearly all glasses have been made at temperatures between 500 °C and 1200 °C using a method invented by the Phoenicians. The major constituents of these conventional glasses are the non-stoichiometric sodium and calcium silicates, although even the compounds lead silicate, SiO_2 and polymeric $NaPO_3$, can be made into glass.

Only recently, did low temperature preparation of glass become possible by the sol–gel method. This process has acquired technological importance for a variety of reasons. This volume of *Structure and Bonding* treats aspects of this process in six reviews as previously done for "Solar Energy Materials" in volume 49. The first is by R. C. Mehrotra, a pioneer working for many years on tetrahedral silicon alkoxides and oligomeric alkoxides of other elements, the typical sol–gel precursors. J. Fricke and A. Emmerling write about the paradoxical properties of aerogels (having a bulk density as low as 2% of that of silica) used in buildings. S. Sakka and T. Yoko review the coating of conventional glass objects with thin films prepared by the sol–gel technique. H. Schmidt discusses composite materials, the chemical processes leading to gelation, and aspects of thin films. M. Henry, J. P. Jolivet and J. Livage treat the formation of monomeric and polymeric hydroxo complexes in aqueous solution (with the cation electronegativity as a significant parameter) and their consecutive polycondensation to oxo-bridged structures, the major step in gel formation.

Finally, we discuss the optical properties of inorganic and organic colored and/or luminescent materials introduced at small concentrations at accessible low to moderate temperatures. Tunable solid-state lasers in the visible range become possible with composite sol–gel glasses. These materials are also suitable for non-linear optics when they include either finely dispersed nanometer-size semiconductors, or single organic molecules.

The number of publications on each of the subjects presented in this volume, has increased exponentially during the last few years. We are grateful to Dr. R. Stumpe and to the reviewers for the opportunity to bring the breakthroughs to a wide spectrum of interested readers.

Renata Reisfeld (Guest Editor)
Christian K. Jørgensen

Table of Contents

Present Status and Future Potential of the Sol–Gel Process

R. C. Mehrotra

Department of Chemistry, University of Rajasthan, Jaipur 302004, India

The development and the current status of the Solution–Sol–Gel or the Sol–Gel (SG) process for preparation of different ceramic materials in various (i.e., bulk, powder, wire, thin film, aerogel, etc.) forms and shapes have been outlined in this article.

In view of the much greater attention having been paid to oxide-ceramics, the chemistry of the main precursors being employed for them, i.e., metal alkoxides, is briefly described followed by an indication of the efforts being made to elucidate (by the latest physico-chemical techniques) the mechanism of the different steps involved, e.g. mixing of solutions; conversion from solution to sol and then to gel and finally sintering the gel to the desired ceramic material, in the Sol–Gel Procedure. A brief account is also presented of the efforts being made to extend the applications of the technique to new demands such as those of super-conducting materials.

The much lower temperature(s) of operation involved in the process facilitate the applications of the SG technique to the *OR*ganically *MO*dified *CER*amics (Ormocers) and other materials suitable for applications in areas like non-linear optics and bio-systems. The SG technique is now being rapidly extended to many other types of materials such as nitrides and sulphides.

Finally, an attempt has been made to peep into the future potential of the fastly developing SG processes.

1 Introduction

Although noticed [1] as early as 1846 and known to play a part in natural processes like the formation of end products such as opal [2], it is mainly during the post-World War II period that the Solution–Sol–Gel (S–S–G) commonly known as the Sol–Gel (S–G) process has been increasingly exploited for the preparation of glasses and other ceramic materials.

Glasses and ceramics have been prepared for thousands of years by heating together at high temperatures, say between 1000–2000 °C, mixtures of finely grained solid oxides, e.g. SiO_2, Al_2O_3, CaO (or their decomposable compounds such as carbonates) for different durations (hours to days) of time. In fact, the word 'ceramics' itself is derived from the Greek word '*keramikos*' (meaning pottery) or the Sanskrit root '*shrapaka*' (meaning to fire/heat on fire).

During the last decade or so, glasses and ceramics have been transformed gradually from 'stone age' to 'space age' materials [3, 4]. Research work on them has advanced mainly by evolving new procedures for creation of novel materials to meet the challenges of newer applications. Elucidation of the mechanism of the new procedures by physico-chemical techniques has led from a 'continuum modelling' to a 'model informed' empiricism [5], resulting in refinements which enhance the capabilities of the procedures for synthesis of ever-developing novel materials. Out of the new procedures [6] evolved during the past 3–4 decades, MOCVD (Metalo-Organic Chemical Vapour Deposition) [7, 8] and SG (Sol–Gel) [9–13] processes have assumed special significance.

The S–G process (Fig. 1) consists of (i) preparing a homogeneous solution of easily purifiable precursor(s) generally in an organic solvent miscible with water or the reagent used in the next step; (ii) converting the solution to the 'sol' form by treatment with a suitable reagent, e.g. water with HCl for oxide ceramics; (iii) allowing/inducing the sol to change into a 'gel' by polycondensation; (iv) shaping the gel (or viscous sol) to the finally desired forms or shape [14] such as thin film [15], fibre [16, 17], spheres or grains [18] and (v) finally converting (sintering) the shaped gel to the desired ceramic material [19] at temperatures generally much (~ 500 °C) lower than those required in the conventional procedure of melting oxides together:

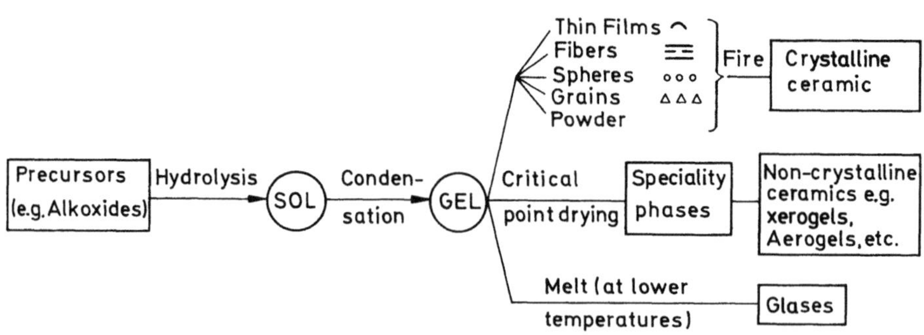

Fig. 1. Steps in the Sol-Gel process for ceramic materials

Although the S–G technique has also been extended to other types of materials more recently, the preparation of oxide-ceramics has till recently received the major attention so far. Oxide-ceramics will therefore be dealt with initially, followed by a brief account of some other materials, e.g. nitrides, sulphides, etc.

Out of a large number of precursors (e.g. metal nitrates or acetates [20], monodispersed metal hydrous oxides [21], oxides dissolved in alcohols [20], oxide alkoxides [22], alkoxides [20]), metal alkoxides have from the beginning [23] attracted considerable attention as suitable starting materials for oxide ceramics, in view of the ease of their purification (generally by distillation and sometimes by crystallization, e.g. $Zr(OPr^i)_4 \cdot (Pr^iOH)$, solubility in organic solvents (generally parent alcohols which are miscible with water) and facile hydrolysability. On hydrolysis, they are converted into hydroxy derivatives in a 'sol' form which can be converted into a 'gel' by polycondensation. Fortunately, the alkoxide chemistry [24] of elements has also grown almost simultaneously albeit independently as the sol–gel process for ceramics during the same period (i.e. 4 decades). It may, however, be pointed out that there was hardly any interaction between the material scientists developing the sol–gel process and metallo-organic chemists working in the field of metal alkoxides. This can be illustrated by the absence of even a mention of such applications of metal alkoxides in the book 'Industrial Applications of the Organometallic Compounds' by Harwood (1963) [25]. Even the book 'Metal Alkoxides' by Bradley, Mehrotra and Gaur (1978) [24] deals with the topic in a single paragraph mentioning only a few patents by I. M. Thomas (1972–75), who was earlier in the research group of Prof. D. C. Bradley. These facts are being emphasized as the future potential of the sol–gel process in the preparation of novel materials by overcoming the higher costs of alkoxide precursors would depend largely upon a better understanding of their chemistry and the physico-chemical changes they undergo during the sol–gel process, as emphasized by Ulrich [11]. Taking into account the exceptionally homogeneous nature of multicomponent oxide glasses obtained from a solution of alkoxides, Dislich [26] (1971) conjectured about the 'strong tendency of alkoxides of various elements to react with one another. These reactions lead to 'alkoxo-salts', which are often very complicated ......'. Thus in addition a gain factor of $\sim 10^4$–10^5 in the intimacy (at the molecular level of ~ 0.5 nm scale) of mixing of the components in the precursor solution for the S–G process over that (size of particles of ~ 50 μm) in the conventional process (involving melting of few metal oxide particles), the extraordinary homogeneity of the final ceramic led Dislich to the conjecture that new chemical bonds must be formed between alkoxides of different elements in the original precursor solution. On the other hand, besides a large number of research articles on bi- and heterometallic alkoxides in well-known international journals and treatises [24, 27] and several invited lectures at the International Conferences on Coordination Chemistry in different parts of the world (Sao Paulo, 1977 [28]; Toulouse, 1980 [29]; and Athens, 1986 [30]), Mehrotra et al. [31] almost simultaneously (1971) with Dislich's conjecture (1971) published a review article on double alkoxides indicating the tendency of formation of such species through coordination bonds between different metals.

However, it was only in 1987, when Mehrotra [32] delivered a lecture on 'Synthesis and Reactions of Metal Alkoxides' at the IV International Workshop on Glass-Ceramics at Kyoto that the two groups of workers became aware of each other's work on almost parallel lines. This glaring example of lack of communication indicates an urgent need for active collaboration of scientists of different disciplines (materials science, chemistry, physics, engineering and even biochemistry, etc.) for rapid synergetic advances in the interdisciplinary field of the Sol–Gel technique. The essential requirement of "an application of chemical principles unprecedented in the history of ceramics" has been re-emphasized recently [13] in view of the fastly developing concept of molecular manipulation of the processing of ceramics, glasses and composites.

It is a happy sign that bi- and hetero-metal alkoxides are beginning to draw attention as precursors [33–36] in the sol–gel process. Special mention may also be made of recent publications on (i) "sol–gel chemistry of transition metal oxides" [37], (ii) "chemistry of alkoxide precursors" [38] and (iii) "a new route to alkoxy-silanes and -siloxanes (from cements and minerals) of use for the preparation of ceramics by the sol–gel technique" [39].

One of the most convenient (though it may not be fully realistic in an applied field such as ceramics) indicators of the development of any topic could be the number of publications in succeeding years (Fig. 2), although counting of papers could again be rather approximate due to intrinsic difficulties in locating them in a large variety of journals belonging to different disciplines. In spite of the uncertainties involved, the graph (Fig. 2) shows a dramatic increase in the number of publications from the early seventies:

The fastly accelerating interest in the sol–gel technique is also reflected in the number of international conferences (with steeply rising numbers of participants) such as International Workshops on Glasses and Glass-Ceramics from Gels; Material Research Society Symposia on Better Ceramics through Chemistry and the International Conferences on Ultrastructure processing of Ceramics,

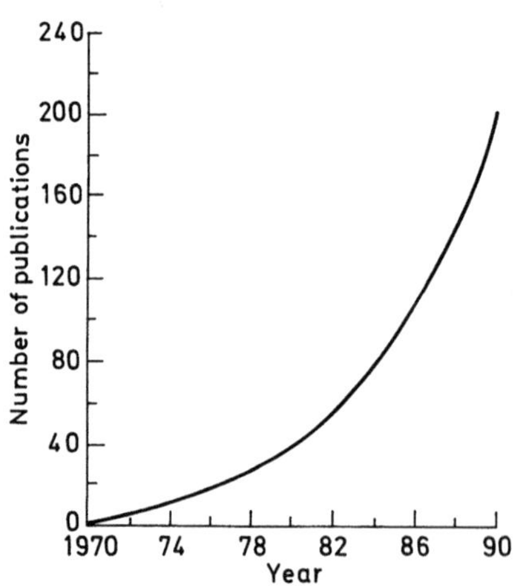

Fig. 2. Yearly publications on the sol-gel process

Glasses and Composites. In addition to the above, many new conferences such as the 'First International Conference on Advanced Ceramics (ICAC-I)-Molecular design of Ceramics by Sol–Gel Processing (Kyoto, November 1990)' are appearing on the scene.

The proceedings of these conferences give a consolidated picture of the fast progress made in the different facets of Sol–Gel Technology [18]. A recent book [14] presents in its seventeen chapters by established workers a vivid account of the advances made particularly in the fields of thin films, fibres, preforms, electronics and speciality shapes.

As regards commercialisation, progress in the field of multi-component sol–gel products has been rather slow. The bulk glass products which began to be manufactured in 1970 by Owen-Illinois, Inc. had to be discontinued due to much higher costs involved. In fact, the feeling in general is that the S–S–G process may not be an industrially productive route for large bulk quantities except for materials involving components like ZrO_2 [40–44], which are extremely high melting.

The S–S–G process appears to be highly attractive for coatings [45] and fibres [46]. On a square foot basis, raw material and processing costs are rather economical for coatings by the S–S–G process. Schott Glasswerke has been utilising the process for thin (0.1 to 0.5 μm) coatings on large surfaces of glass for architectural (e.g., IROX) and other purposes such as windows and rear view mirrors for vehicles. Dislich [47] has recently mentioned the success achieved at Schott in developing contrast enhancement filters for monitors; anti-reflective windows; multicomponent oxide layers and cermet coatings. Special optical coatings such as a cheap SiO_2–TiO_2 anti-reflective coating for silicon solar cells developed by Yoldas and O'Keefe [48] appear to be highly attractive for exploitation by the S–S–G process.

In the area of different types of fibres [16, 17], the S–S–G process appears to offer really distinct advantages. These fibers whether spun or drawn may be continuous or woven for use in refractories, composite reinforcements and thermal insulation. New products like silica fibres from Asahi Glass Company Ltd., solid silica glass rods, tubes and plates, may be cited in these directions. In a recent article [49], Matsuzaki and coworkers have described a new industrial technology (Fig. 3) for manufacture of continuous silica glass fibre at temperatures below 1000 °C. According to them the sol–gel derived fibres contain very few impurities, which make them very attractive as electronic and thermal insulation materials, particularly at higher temperatures.

In his '1988 Arthur L. Friedberg Lecture', Sowman [16] emphasized the emergence of 'a new era in Ceramic Fibres via Sol–Gel Technology', beginning with the development of Nextel 312 (Al_2O_3–B_2O_3–SiO_2) heat resistant fabric [50] by the 3M Co. (USA), used in gasketing and other heat shielding applications in space shuttles (e.g. Discovery and Atlantis). Blending it with silica fibres at high temperatures results in 'unique rigid thermally insulating composites', with attractive applications, such as 'filter bags' for high temperature refinery operations. Another variety, Nextel 440 ceramic (Al_2O_3–SiO_2–

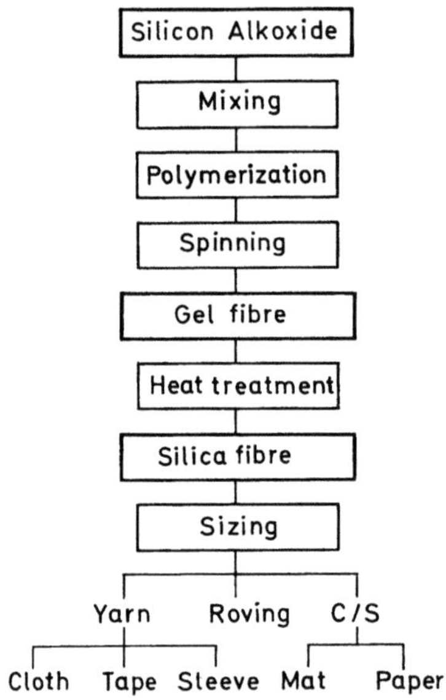

Fig. 3. A schematic diagram of continuous glass fibre production (S-G process)

$2\%B_2O_3$) fibres have found yet another interesting application as ceramic fibre-protected 'booms' for collection and burning of petroleum from spills on water.

Properties, such as colour, index of refraction, modulus of elasticity, chemical resistance, etc. of oxide fibres of Al_2O_3–B_2O_3–SiO_2 system can be modified by a variety of additives like nickel oxide upto sometimes about 23% nickel corresponding to the formation of the 'spinel' species, $NiAl_2O_4$.

A distinct advantage of the S–S–G process over the conventional molten phase technique can be illustrated by the preparation of SiO_2–TiO_2 fibres from polytitanosiloxanes [51] and also of low content SiO_2 fibres, such as those of TiO_2 from $Ti(OPr^i)_4$.

Exemplifying the bright potential of the S–G process in preparation of special ceramic materials like potassium-tantalate-niobate (KTN), Mackenzie [52] has pleaded for a more penetrating study of the basic science involved in the different steps involved in it. Similarly, interdisciplinary understanding and inputs have been strongly emphasized by Schmidt [53], while presenting an excellent account of his efforts to incorporate the attractive features of 'polymers' in a series of excitingly novel materials. Ulrich [54], as well as Hench and West [13], have also argued that ultrastructural control of materials through S–G processes offers significant promise for the highest achievements for reliable performance of glass, glass-ceramics, ceramics and composites.

Another area, which may be emphasized, encompasses special applications such as preforms, microballoons and electronics. In areas like the last one, cost is no bar to the use of expensive precursors like the pure metal alkoxides, as highest standards (ppb) of chemical purity and control of stoichiometry within strict limits command the highest premium in such applications. A large number of developments, e.g. a series of metal silicate solutions for doping silicon wafers with arsenic, antimony, boron or phosphorus for semi-conductor applications [55], may be cited for this purpose. Modification of silica aerogels by addition of infra-red spacifiers like soot or titanium dioxide allows the application of these highly porous materials for CFC-blown insulating polyurethane foams in refrigerators and neat storage systems [56]. Finally, serious attempts [57] are being made for the preparation of super-conducting oxide-ceramics by the Sol–Gel process, which offers an attractive possibility of obtaining them in wire form [17].

2 Precursors for Oxide-Ceramics by the S–G Process

A reference [2, 55] has already been made to the interest aroused amongst colloid chemists in the use of 'oxide' sols in the S–G process. Luth and Ingamells [58] re-emphasized the use of 'silica' sols instead of tetraethyl orthosilicate which might tend to be lost by volatilization during the process. More recently, Rabinovich et al. [59] have used SiO_2–B_2O_3 precursor compositions from aqueous colloidal silica and boric acid solutions. Such applications are obviously limited to water soluble reagents for the other component, but their potential has been shown by casting such mixtures into moulds for obtaining large sinterable bodies. Similarly, the use of Al_2O_3 and ZrO_2 sols has also been attempted and it has been shown that in all these cases, the final products isolated at higher temperatures were identical with those obtained at comparatively lower temperatures from colloidal oxides or metal organics as starting materials.

In a recent article entitled 'Monodispersed Metal (Hydrous) Oxides—A Fascinating Field of Colloid Science', Matijevic [21] has summarized the effects of different parameters, e.g. temperature; pH of the medium; and the anions/complexing agents present, on the nature of hydrated oxide particles obtained. Amongst the systems studied by his research group, mention may be made of a few: e.g. nature of TiO_2 & Al_2O_3 by hydrolysis of droplets of an aerosol of $Ti(OEt)_4$ [60] and $Al(OBu^{sec})_3$ [61]; preparation of ferrites of Co [62], Co/Ni [63] and Cr [64] as well as titanates of Ba/Sr [65] through colloidal species. Matijevic [66] has also presented an excellent review entitled "Monodispersed Colloids: Art and Science" on the topic.

Besides their application in sol form, metal oxides and hydroxides, which dissolve in alcohols forming alkoxides (at least partially) have been used in the SG process. The reactions of these oxides with alcohols are reversible and solubility of the oxides can be generally increased by removal of water [24]:

$$MO_n + 2x\,ROH \longrightarrow MO_{n-x}(OR)_{2x} + x\,H_2O$$

$$M(OH)_n + x\,ROH \longrightarrow M(OH)_{n-x}(OR)_x + x/2\,H_2O$$

For example, oxides and hydroxides of alkali metals, boric acid and oxide, phosphoric acid and oxide, lead monoxide and vanadium pentaoxide have been employed in this manner. In view of its availability in pure form and solubility in alcohols, boric acid is probably the most common material used in this manner. Phosphoric acid is generally used as a source of phosphorus, as phosphorus alkoxides (or alkyl orthophosphates) are rather resistant to hydrolysis; phosphoric acid, however, reacts readily with silicon alkoxides in particular to give useful phosphorosiloxane intermediates [67].

The reactions of boric acid (or trialkyl borates) with tetraalkyl orthosilicates and their hydrolysis mixtures have been monitored [68] by NMR and IR spectroscopy during all the stages, from initial alkoxide solution to final heat-treated product. Only a small number of borosiloxane (=B–O–Si=) bonds could be detected at the time of gelation and most of the boron remained in the form of boric acid, which appears to be condensed into borosiloxane bonds only upon heat treatment of the dried gel. For higher boron compositions, a small number of =B–O–B= bonds are also formed and it has been found that these do not get converted into =B–O–Si= bonds upon heating to 500 °C.

Studies [69] of phosphosilicate sols from hydrolysis of $Si(OEt)_4$ and $O{=}P(OEt)_3$ by ^1H, ^{13}C and ^{31}P NMR spectroscopic studies indicated that the hydrolysis of $Si(OEt)_4$ is much faster than that of $O{=}P(OEt)_3$. ^{31}P NMR spectra show that there are no P–O–P and P–O–Si bonds in the sols 9 h after the beginning of the reaction, but such bonds are found in the gels.

In cases where suitable alkoxide precursors are not available, a solution is sometimes prepared with the other components and the odd one is added as an oxide in a fine powder form followed by vigorous stirring to facilitate the coating of these oxide particles with the solution of other precursors. With the limited homogeneity thus achieved, Thomas [70] was able to prepare lead (or zinc) alumino- as well as boro-silicates by suspending lead (or zinc) oxide in a solution of other components followed by gelling and sintering in the usual manner; the advantage claimed for this procedure was that it was easier to free the product from the carbonaceous residue than in the process employing lead (or zinc) acetate as precursor.

In the case of a few metals, it might not be convenient to use alkoxides due to their unavailability and/or difficulties in synthesis of soluble derivatives, and alternative precursors may have to be employed. For example, alkoxides of Groups I & II metals are non-volatile solids and often display low solubility in organic solvents. In their cases, therefore, one of the advantages of easy purification either by volatilization or crystallization of the alkoxide derivative is lost. In view of importance of barium (a component of $YBa_2Cu_3O_{7.5}$ superconductor), attempts have been made to prepare a soluble alkoxide by using a chelating alcohol, e.g. methoxyethanol for soluble alkoxides of barium [71], copper [72] and also of yttrium [73].

In the case of the above types, metal salts may sometimes provide a viable alternative, but care has to be exercised to choose those precursors which can be converted easily to the oxide by thermal or oxidative decomposition; their solubility in organic solvents preferably from the initial stage or during the sol–gel processing would be a further advantage.

Among the inorganic salts, metal nitrates are probably the best candidates [20] as other salts like the sulphates or chlorides are thermally more stable and in their cases, it may be difficult to remove the anionic portion effectively from the final ceramic products.

Amongst the salts of organic acids, acetates have been most commonly employed. Formates, citrates, tartrates, etc. also have been utilised as precursors in addition to acetates. Metal silicates [74] and complex derivatives like alumino-silicates [75], silyloxides [76] and polymetalloxanes [77] have also been used as precursors more recently.

The earlier sol–gel preparations described by Roy [78] employed tetraethyl orthosilicate and solutions of nitrates of Group I & II metals as well as those of aluminium, lead, iron, lanthanum, titanium, zirconium, thorium, nickel and gallium. A major drawback [79] in the use of metal nitrate precursors is their tendency to disturb the homogeneity of the system by the crystallization of one or more nitrates during dehydration. This disadvantage was the most severe in the cases of sodium, lead and barium nitrates, but it could be circumvented to some extent by evaporating a large fraction of the water at comparatively lower pH before the gelling actually sets in. It may not be out of place to emphasize that nitrates are strong oxidising agents and the simultaneous presence of large quantities of oxidisable organic material(s) may sometimes lead to uncontrollable exotherms and even explosions during the drying stage.

In their bonding and solubility characteristics, metal acetates are closest to the nitrates. In view of the nitrates and acetates being the commonest metal salts employed in the sol–gel process, a brief comparison between the functionality and the applicability of the two is presented. In some cases, like those of sodium, potassium and barium, the acetates are more soluble than the corresponding nitrates and the crystallization problem(s) encountered in the use of nitrates are to a large extent avoided. The explosion hazards with large quantities of nitrates in the presence of organic matter are also avoided by the use of acetates. Disadvantages with the use of acetates are that they do not degrade thermally as cleanly as nitrates and might have a tendency to leave carbonaceous residues. Further, in view of the weakly acid character of the parent acid, solutions of the acetates of many metals tend to be basic in character, but this can be at least partly overcome by buffering the solution with acetic acid.

Sol–gel preparations using mixtures of metal alkoxides and salts tend to be more complex due to the more facile hydrolysability of the alkoxides. This, as well as the disadvantages of carbonaceous residues from acetates, can be lowered by reacting the metal alkoxides with acetates together when a reaction of the following type has been observed in many cases [30, 80–82]:

$$M(OR)_n + M'(O_2CCH_3)_x \longrightarrow (RO)_{n-1}M\text{–}O\text{–}M'(O_2CCH_3)_{x-1} + CH_3COOR\uparrow$$

The above reaction continues with further interaction of acetate and alkoxide groups leading to higher molecular weight complex products. If the reaction mixture containing more alkoxide than acetate groups is heated together in the absence of solvent, almost all (90–95%) the ester can be distilled out. This partially hydrolyzed alkoxide derivative is in many cases soluble in organic solvents or its solubility can be enhanced to some extent by refluxing the same with an alcohol. The changes in the coordination state (3–6) of aluminium in such interactions has been investigated [82, 83] in detail and these should be taken in consideration when employing aloxane materials as precursors.

In addition to extensive work in the author's laboratories, Thomas [70] used the technique for $CaO–MgO–Al_2O_3–SiO_2$ and $PbO–B_2O_3–SiO_2$ glass compositions. Gurkovich and Blum [84] used it for preparing monolithic lead titanate. Reduction of acetate content appears to be particularly advantageous in the presence of less basic metallic species like lead, as these are more prone to leaving carbonaceous impurities in the final ceramic product.

Hydroxy carboxylic acid like glycolic, lactic, tartaric acids appear to offer advantages [85–87] in some cases as the hydroxy groups also in these tend to react with orthosilicate and other systems with similar characteristics.

Mention may also be made of the use of β-diketone and β-ketoester [32, 88, 89] derivatives of titanium [90, 91], aluminium [92] and other metals as precursors for the preparation of stable photosensitive colloids via the hydrolysis-condensation mechanism.

For low thermal expansion high refractive index $SiO_2–TiO_2$ glasses, polytitanosiloxane gels were used which were obtained [93] by condensation polymerisation of silicic acid and titanium chelates (bis(acetylacetonato) titanium diisopropoxide) ATP; or (bis(ethyl acetoacetato)titanium diisopropoxide), ETP. Silicic acid precursor was extracted with tetrahydrofuran from an aqueous sodium metasilicate solution neutralized with hydrochloric acid. Monolithic gels with TiO_2 content of less than 20 mol % were obtained by the reaction of silicic acid with ATP, whereas the reaction of ETP provided transparent monolithic $SiO_2–TiO_2$ gels of 57–94 mol % TiO_2.

Recourse has been made [94] to the use of such complex materials as $[Co(CO)_3L]_2$ and $Rh(CO)ClL_2$ (where $L = Ph_2PCH_2CH_2Si(OEt)_3$) for polycondensation with $Si(OEt)_4$ for increased surface area in the product.

In spite of quite straightforward methods available for synthesis of metal alkoxides [24, 32], work is limited to only commercially available alkoxides in most of the laboratories. Even in cases like tin tetra-alkoxides the direct synthesis of which is rather difficult and requires [24] the treatment of easily synthesised bimetallic alkoxides with required quantities of HCl or CH_3COCl,

$$SnCl_4 + 4ROH \xrightarrow[+\,4NaOR]{NH_3} \text{Impure } Sn(OR)_4$$

$$2SnCl_4 + 9NaOR \xrightarrow[ROH]{} NaSn_2(OR)_9 + 8\,NaCl\downarrow$$

$$NaSn_2(OR)_9 + HCl \longrightarrow 2Sn(OR)_4 + NaCl + ROH$$

Mehrotra [95] pointed out recently the possibility of the use of $NaSn_2(OR)_9$ which is soluble in organic solvents and can be purified by distillation as a much more convenient precursor than the tetra-alkoxides of tin.

It has been recently realized that heterometallic alkoxides, the formation of which was conjectured by Dislich [24] and a large variety of which have been actually synthesized and characterized since 1971 in the research school of Mehrotra [24, 27], are rather attractive as precursors [34] in the Sol–Gel process. For example, spinel type ceramic $MgAl_2O_4$ has been synthesized by hydrolysis of the bimetallic alkoxide $[Mg\{Al(OR)_4\}_2]$ in the presence of triethanolamine. It has also been shown by a careful ^{27}Al NMR study [33] of the system that the bimetallic alkoxide does not break down to its constituents during the hydrolysis, which can therefore be represented by the following type of mechanism [34, 96, 97]:

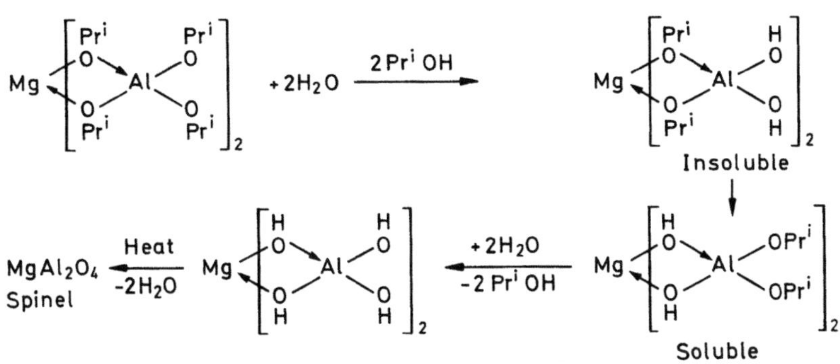

Similarly, crystalline lithium niobate (a well-known ferroelectric and piezoelectric material with a high Curie temperature) fibres [35] and potassium niobate discs [35a] have been prepared by sol–gel processing of the corresponding alkali hexaethoxyniobate $[Li/KNb(OEt)_6]$ [98, 98a]. The preparation of crystalline $LiNb_xTa_{1-x}O_3 (0 < x \leqslant 1)$ films has been described [36] recently by hydrolysis of a mixture of $LiNb(OEt)_6$ and $LiTa(OEt)_6$.

It may be expected that a large number of heterometallic alkoxide systems described during the last two decades by the research school of Mehrotra [99, 100] will find increasing applications for the preparation of ultra-homogeneous oxide-ceramic materials similarly by the sol–gel process in view of their solubility in organic solvents, but also by the CVD process in which their volatility should be an asset for the consistency of the vapour undergoing discomposition.

With the advent of superconducting materials such as $Y_1Ba_2Cu_3O_{7-x}$, frantic efforts are being made, as mentioned earlier, to prepare soluble alkoxides of bivalent metals barium and copper, as their ordinary alkoxides are not sufficiently soluble due to their extensive polymerization. Success has been achieved in this direction by using chelating ligands like methoxyethanol $CH_3OC_2H_4OH$ [72] or ethanoldimethylamine $HOC_2H_4NMe_2$ [27] and their applications [57, 101] in the preparation of the $YBa_2Cu_3O_7$ thin films or bulk powders. Another precursor which has also been used for preparing films of the above superconductor is $Ba_2Cu_2(OR)_4(acac)_4 2HOR$ ($R = CH_2CH_2OCH_3$) [102]. The yttrium derivative of this alcohol has been characterized as a hydrocarbon soluble decamer, $[Y(OC_2H_4OMe)_3]_{10}$ [73]. Khan et al. [103] have studied the rheology of aqueous solutions of copper acetate, barium acetate and yttrium hydroxide for the same 1,2,3 super-conducting material. Sakka [17] has reviewed the literature on the applications of the sol–gel process for the superconductor fibres and films and has tried to demonstrate the superiority of the technique over the conventional methods. With the development of superconducting materials involving bismuth, alkoxides of bismuth are also receiving renewed attention [104, 105].

In fact, with their increasing use as precursors, there appears to be an unprecedented spurt in research activity on the chemistry and structural features of metal alkoxides and efforts are being made to correlate the basic data with a view to explaining their role in the sol–gel process [32, 34]. A few of these findings out of a much larger number of publications during 1988–89 are mentioned as illustrative examples of this exciting trend relevant to sol–gel chemistry: synthesis of (i) methoxides of niobium/tantalum(M) clusters, $[M_6Cl_{12}](OMe)_2$ and $[K_2Ta_6Cl_{12}](OMe)_6 \cdot 6MeOH$ [106]; (ii) bimetallic methoxides of niobium (IV), $[Na(MeOH)_6][Nb_2(OMe)_9]$ and $[Mg(MeOH)_6]_2[Nb_2(OMe)_9]Cl_3$ [107]; and

(iv) $(Bu^tO)_4U(\mu\text{-}O)(\mu\text{-}OBu^t)(UO_2)(\mu\text{-}OBu^t)(\mu\text{-}O)U(OBu^t)_4$ [108]. Evans et al. [109, 110] have reported the formation of chloride products $Y_3(\mu_3\text{-}OBu^t)(\mu_3\text{-}Cl)(\mu\text{-}OBu^t)(OBu^t)(THF)$ and $Y_3(\mu_3\text{-}OBu^t)(\mu_3\text{-}Cl)(\mu\text{-}OBu^t)(OBu^t)_3Cl(THF)_2$ in the reactions of YCl_3 with 3 and 2 moles of $NaOBu^t$ respectively. The formation of a (μ_4-O)oxo product, $Y_{14}(OBu^t)_{28}Cl_{10}O_2(THF)_4$ with the interesting structure, $[Y_7(\mu_3\text{-}OBu^t)_2(\mu_3\text{-}Cl)(\mu_4\text{-}O)(\mu\text{-}OBu^t)_7(OBu^t)_5(\mu\text{-}Cl)_4(THF_2)_2$ by leaving $Y_3(\mu_3\text{-}OBu^t)(\mu_3\text{-}Cl)(\mu\text{-}OBu^t)_3(OBu^t)_3Cl(THF)_2$ in toluene for two weeks has been cited as an example of the type of mechanism which may be involved in the conversion of metal alkoxides into oxide-materials. A new class of cationic yttrium *tert*-butoxide complexes, e.g., $[Y_3(\mu_3\text{-}OBu^t)(\mu_3\text{-}Cl)(OBu^t)_3(\mu\text{-}OBu^t)_3(THF)_3]^+[BPh_4]^-$ has been synthesized [111] by the reaction of $[Y_3(OBu^t)_7Cl_2(THF)_2]$ with $Ag[BPh_4]$.

In another interesting publication [112], the crystal structure of a product (~ 75%) from the reaction of yttrium with isopropanol has been shown to have the structure, $Y_5(\mu_5\text{-}OPr^i)_4(\mu_2\text{-}OPr^i)_4(OPr^i)_5$. It has been suggested that this (μ_5-O) crystalline product has been formed from the rearrangement of $Y(OPr^i)_3$ itself, the presence of which in the crude product (solution) has been shown by distillation of the bimetallic product $Y\{Al(OPr^i)_4\}_3$ [28] itself after addition of $Al(OPr^i)_3$ to the crude reaction mixture; it was also shown that the final crystalline product $Y_5(\mu_5\text{-}O)(\mu_3\text{-}OPr^i)_4(\mu_2\text{-}OPr^i)_4(OPr^i)_5$ does not form the above bimetallic tetraisopropoxy aluminate with $Al(OPr^i)_3$. In view of the above, it would be interesting to reinvestigate the reaction of $YCl_3 \cdot 3Pr^iOH$ and also other $LnCl_3 \cdot 3Pr^iOH$ with 3 moles of $NaOPr^i$, which appear to yield the pure tris-products $Y(OPr^i)_3$ and $Ln(OPr^i)_3$.

Further, an interesting report [113] on the formation of $In_5O(OPr^i)_{13}$ appears to contradict earlier results [114] on the preparation of $In(OPr^i)_3$, which has been converted into a bimetallic alkoxide $Sn\{In(OPr^i)_4\}_4$ by Turova et al. [115].

Livage et al. [116] have reported interesting XANES and EXAFS studies of titanium alkoxides and their investigation on the sol–gel chemistry [117] of titanium alkoxides as well as titanium triisopropoxide monoacetylacetone [118]. Livage et al. [119] have also followed the hydrolysis of niobium pentaethoxide at the niobium K-edge. In the original alkoxide, only one Nb–Nb distance is observed which is typical of edge sharing octahedra. The niobium coordination does not change significantly upon hydrolysis but a new Nb–Nb distance appears corresponding to corner-sharing octahedra. The small number of niobium neighbours suggests a very open structured gel net-work. In a more recent study, Livage and coworkers [120] have followed, by $^{47,49}Ti$ and ^{51}V NMR spectroscopic studies, the changes which occur when their isopropoxides, $Ti(OPr^i)_4$ and $VO(OPr^i)_3$ are dissolved in tertiary amyl alcohol, followed by hydrolysis in the SG process.

A novel method of synthesizing ceric tetra-*tert*-butoxide, $Ce(OBu^t)_4$ and nitrate tertiary butoxide adducts, $Ce(NO_3)_x(OBu^t)_{4-x}$. 2THF has been described by the reactions of $(NH_4)_2\,Ce(NO_3)_6$ with $NaOCMe_3$ in different molar ratios in THF [121]. The reactions of $(NH_4)_xLn(NO_3)_{3+x}$ (Ln = Y, Ce, Nd, Pr) with (3 + x) equivalents of $NaOSiPh_3$ in THF have been reported [122] to yield $[Ln(OSiPh_3)_3(THF)_3THF]$ complexes. Following the characterization [123] of $Y[OC_6H_3(Bu^t)_2\text{-}2,6]_3$, Evans et al. [124] have been able to synthesise $[Y\,OC_6H_3(Me)_2\text{-}2,6\ _3(THF)_x]$ complexes with accessible coordination sites.

3 Physico-chemical Studies on the Steps in the Sol–Gel Process

The Sol–Gel process basically involves the synthesis of an inorganic network by a chemical reaction in solution at low (generally ambient) temperatures and involves the transition from solution to colloidal sol to a multiphasic gel form.

Any precursor, which is able to form an inorganic network can be used in the S–G technique. Even finely divided silica or metal oxide particles can be peptized and used for preparation of sols. In such cases, it is necessary to generate appropriate surface charges in order to provide stability to the 'sol' form.

As mentioned earlier, alkoxides are generally preferred as precursors. Tetraalkyl orthosilicates ($Si(OR)_4$ abbreviated as TMOS or TEOS when R = Me or Et respectively) are monomeric, but simple alkoxides of other metals (except bivalent ones) generally form low oligomers [24] and are soluble in organic solvents, e.g. alcohols which provide a convenient medium miscible with water used in the next step of the procedure. The hydrolysis, dealcoholation and condensation steps, resulting in the M–O–M (metallometalloxane) linkages, can be represented by the following equations:

$$\text{M–OR} + H_2O \rightarrow \text{M–OH} + \text{ROH} \quad (1)$$

$$\text{M–OR} + \text{M–OH} \rightarrow \text{M–OM} + \text{ROH} \quad (2)$$

$$\text{M–OH} + \text{M–OH} \rightarrow \text{M–O–M} + H_2O \quad (3)$$

The above reactions occur simultaneously in most cases. However, an attractive advantage of the alkoxide route is the possibility to control the rate of formation of M–O–M networks by chemical means (e.g. by addition of auxiliary bidentate ligands such as acetyl acetone) and hence, colloid chemistry procedures such as regulating the surface charges or species adsorbed on the colloidal moieties are not generally required in their cases.

In view of water being produced as a by-product in the condensation reactions of ≡SiOH moieties, complete hydrolysis of $Si(OR)_4$ to SiO_2 could be expected to occur when the molar ratio of H_2O to $Si(OR)_4$ is 2:1

$$Si(OR)_4 + 2H_2O \rightarrow SiO_2 + 4ROH$$

However, even in the presence of excess (2 moles) water, the hydrolysis reactions do not proceed to completion in the case of alkyl orthosilicates as well as other metal alkoxides [24]. Klemperer and Ramamurthi [125, 126] have explored the molecular growth pathways in silica sol–gel polymerization by employing a capillary gas chromatographic technique and showed that the species which can be detected in the hydrolysis of TMOS, $Si(OMe)_4$, can be represented by the formulae $[Si_nO_m](OMe)_{4n-2m}$; in which for simplicity each oxygen and silicon atom in the skeleton, $[Si_nO_m]$ is represented by a dot and a line respectively:

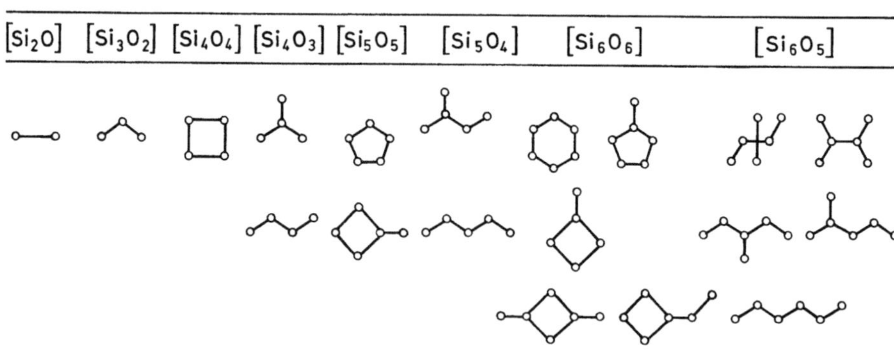

Fig. 4. Formulas and possible structures of the hydrolysis products of $Si(OMe)_4$

Although most of the work has been centred around silicon [13], there has been considerable research work on the chemistry of various steps involved in the SG process, e.g.

(i) investigation [120] on the interaction of a metal alkoxide, e.g. $Ti(OPr^i)_4$ and $VO(OPr^i)_3$, with the solvent, e.g. Pr^iOH and Am^tOH by $^{47,49}Ti$ and ^{51}V NMR spectroscopy respectively;
(ii) interaction of alkoxides of different metals and metalloids in a suitable solvent [24, 26–37];
(iii) controlling/catalyzing the hydrolysis of alkoxides by varying amounts of added water [24, 127, 128]; acid (e.g. HCl) or alkali (e.g. NH_4OH) [129]; addition of a drying control chemical additive (DCCA [11]) like formamide [130] and chelating agents like β-diketones [131] or β-ketoesters [92];
(iv) kinetics and mechanism of hydrolysis and condensation reactions [132–135];
(v) properties (e.g. syneresis [136] and critical stress [137]) of gels; and
(vi) sintering of gels into the final ceramic products [138–141].

Physico-chemical techniques employed in the above investigations vary from the conventional ones such as gas chromatography [125, 142]; DTA, TGA, IR [143]; viscosity [144]; rheology [145–147]; light scattering [148]; spectroscopic—Xanes and EXAFS [149]; NMR– (^1H, ^{17}O and ^{29}Si [150, 151], magic angle spinning MAS [152, 153] and solid state ^{29}Si [154] using both single-pulse (FT) as well as Si–H cross-polarization (CP) techniques) SAXS and SANS (small angle X-ray and neutron scattering) [141, 155]; CRAMPS (combined rotation and multiple phase spectroscopy) [156] techniques) differential scanning calorimetry (DSC) [13]; and image ellipsometry [15].

In addition to sophisticated experimental elucidation of the chemistry of the hydrolysis of orthosilicates, a considerable break-through has occurred in the theoretical aspects of this problem [86, 157, 158]. Out of the two possible (tri- and penta-) transition states, recent work [159] using high pressure Raman spectroscopy has confirmed the involvement of a pentacoordinated transition state:

$$\mathrm{HO^-} + \equiv\mathrm{SiOCH_3} \rightarrow \equiv\underset{\substack{\uparrow\\ \mathrm{O}\\ \mathrm{H}}}{\mathrm{Si}}-\mathrm{OCH_3^-} \rightarrow \equiv\mathrm{Si{-}OH} + \mathrm{OCH_3^-}$$

$$\mathrm{H_3O^+} + \equiv\mathrm{SiOCH_3} \rightarrow \equiv\mathrm{Si}-\overset{\substack{\mathrm{H}\\ |}}{\mathrm{O}}\mathrm{CH_3^+} \rightarrow \equiv\mathrm{SiOH} + \mathrm{CH_3OH} + \mathrm{H^+}$$

It is presumed that the OH^- has enough high nucleophilic power to attack the silicon atom directly, as the Si atom in the orthosilicate molecule carries the highest induced positive charge. In acidic hydrolysis, the proton is probably attracted by the oxygen atom of the OCH_3 group, inducing a higher positive charge on the silicon atom which may then be able to coordinate to a water molecule. This experimental verification [143] of the pentacoordinated silicon transition state has corroborated a considerable volume of theoretical quantum mechanical calculations [160, 161].

4 *Or*ganically *Mo*dified *Cer*amics (ORMOCERS)

In spite of their versatility in finding ever-increasing application, organic polymers generally suffer from the inherent drawbacks of organic compounds, i.e. instability to heat and tendency of natural degradation on ageing. Considerable attention has, therefore, been focussed on 'inorganic polymeric materials' [165] for the past few decades. Although really substantial industrial success has been limited to only a few instance like 'silicones', extensive work [166] on heterometalosiloxanes with metal-oxygen-silicon backbones has resulted in characterization of a large number of novel derivatives with attractive physico-chemical properties.

In addition to the above, the development of composite materials has elicited unprecedented interest as in many cases the properties of basic component materials are unable to cope individually with the special requirements of their ever-increasing applications such as those in "luminiscent solar concentrators" [167], "laminated wind screens", "fibre reinforced polymers", "corrosion resistant surfaces" [168], "non-linear optics", "ultraviolet photonics" [11] and "photochromic information recording materials" [169].

With a few exceptions such as those of the preparation of glass ceramics by formation of crystals in glasses, or of gas deposited films on suitable substrates, the traditional techniques of obtaining 'composites' involved combination of two (or more) basic types of materials by mechanical means either before or during the final processing/moulding stage. Most of these composites prepared by such mechanical mixing tend to attain 'macroscopic' rather than 'microscopic' homogeneity.

The homogeneity of mixing at molecular level attained in the initial solution and the lower temperatures of operations should, therefore, make the solution–sol–gel technique much more suited for the preparation of 'composites' and 'ormocers'. For example, Sakka et al. [170] have shown that fibres can be drawn from the gel obtained from hydrolysis of $MeSi(OEt)_3$, but hydrolysis of $Me_2Si(OEt)_2$ did not yield a gel and the product was non-spinnable.

Glaser et al. [171] have made a detailed study of orthosilicate based multifunctional sol–gel materials by the MAS ^{29}Si NMR solid state spectra (using both Fourier transform, FT and Si–H cross-polarization, CP techniques) of the hydrolysis and condensation reactions of TEOS$\{Si(OEt)_4\}$, MTEOS$\{MeSi(OEt)_3\}$ and DMDEOS$\{Me_2Si(OEt)_2\}$ systems, used in the preparation of 'hybrid' materials modified [172] with polymers like PDMS {poly(dimethylsiloxane)} and PTMO(polytetramethylene oxide).

The incorporation of fine carbon particles into oxide glasses has been known to bring about greater mechanical strength, electrical conductivity in addition to imparting a black colour to the glass. Carbon/silica glass composites have been prepared by carbonization of organic compounds such as furfuryl resins impregnated into the micropores of porous silica glass [173]. As far as the SG process is concerned, the small amount of carbon particles remaining in the silica glass on sintering of materials due to the incomplete removal of some remanent unhydrolyzed organic groups is generally considered as one of the draw-backs of the SG process. However, silica glass with higher (~ 20 wt %) carbon contents can be prepared [174, 175] by hydrolysis of $MeSi(OR)_3$. For example, in the following operation:

$$MeSi(OMe)_3 \xrightarrow[\text{stirred (60–80 °C) 4 h}]{EtOH + H_2O \text{ at pH } 3} MeSi(OH)_3$$

$$\downarrow \; NH_4OH(pH = 8) \text{ gelation (48 h), followed by drying the gel (70–80 °C) for 7 d}$$

$$\text{Black Silica glass } (CSiO_{3/2}) \xleftarrow[\text{Ar}]{\text{Sinter at 900 °C}} \text{Stiff Transparent Xero-gel}$$

One carbon atom attached to silicon appears to be retained (C content of the ideal composition $CSiO_{3/2}$ would be about 19%).

Schmidt [176, 177] has pointed out multifaceted possibilities of introducing organics into inorganic components by procedures such as:

(i) penetrating porous glass/gels by organic monomers, followed by their polymerization [178],
(ii) linking organic groups to an inorganic backbone through chemical bonds of the types:

(a) Covalent:
(b) Ionic:
(c) Coordinating:

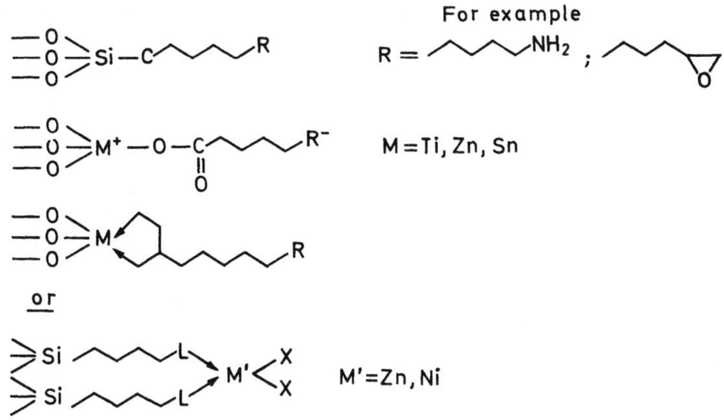

By examining the properties of a number of organic–inorganic materials, Schmidt [180] concluded that these generally tend to behave more like inorganic glasses rather than organic polymers. Schmidt has already demonstrated successfully the applications of ormocers in the scratch resistant coatings for eye-glass lenses with intrinsic hydrophilicity and oxygen permeability [181], biochemical enzyme immobilisation [182], reinforced polymers, sensors [177], and film formation [183]. In a recent publication, Schmidt et al. [184] have described the preparation of a hard coating material for polyethyleneterephthalate substrates by hydrolysis and condensation of γ-glycidyloxipropyltrimethoxysilane, $H_2C\overset{O}{—}CHCH_2O(CH_2)_3Si(OMe)_3$. Schmidt and Wolter [185] have also recently reviewed the applications of ormocers, particularly in the development of dielectric coatings for electronic materials.

Yamashita et al. [186] have shown that hydroxyl terminated siloxane polymers were reactive enough to be incorporated into silica networks, whereas hydroxyl terminated polyethers failed to react in sol–gel process involving TEOS. Some of the prepolymers used were prepared by the following types of reaction:

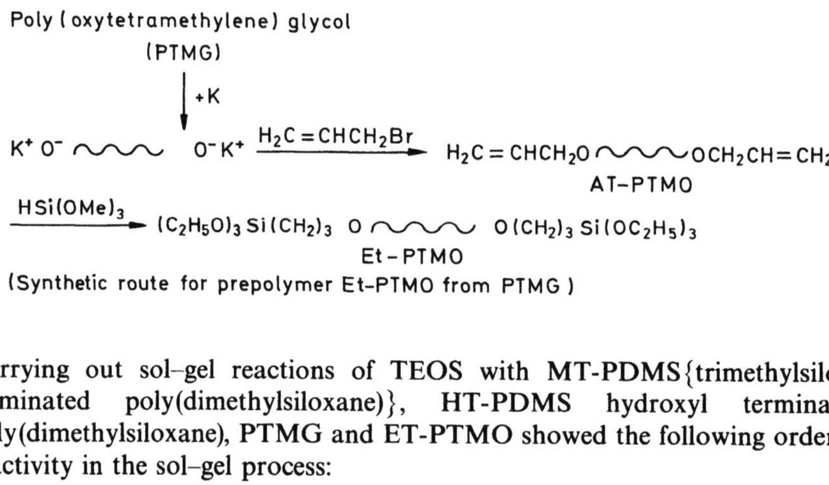

(Synthetic route for prepolymer Et-PTMO from PTMG)

Carrying out sol–gel reactions of TEOS with MT-PDMS{trimethylsiloxy terminated poly(dimethylsiloxane)}, HT-PDMS hydroxyl terminated poly(dimethylsiloxane), PTMG and ET-PTMO showed the following order of reactivity in the sol–gel process:

$$-Si(OEt)_3 > \equiv Si{-}OH \gg -CH_2OH$$

Following the development of ormocers or ormosils (*or*ganically *mo*dified *si*licates), novel types of solid state protonic material called 'aminosils' [187] have been synthesized, in which aminoorganotrialkoxysilanes, $R'Si(OR)_3$ (when R is Me or Et and R′ is an amino radical such as $-(CH_2)_3NH_2$, $-(CH_2)_4NH_2$, $-(CH_2)_3NMe_2$ or $-(CH_2)_3NH(CH_2)_2NH_2$) are used as precursors. Hydrolysis in the presence of an acid, followed by condensation results in materials which can be represented in the simplest form as $SiO_{3/2}R'(HX)_n$ (where $0 \leqslant n \leqslant 0.5$). The dried products are hard non-porous conducting materials (with a glass transition around $-30\,°C$) which are stable in air upto 180 °C.

5 Development of Optical Systems

The incorporation of organic molecules particularly, dyes results in the development of optical systems such as solar concentrators [188], dye lasers [189], phase conjugators [190] for non-linear response [11, 191], photochromic information recorders [169; 192] and indicators for environmental and biological impurities [193]. Instability to heat of organic moieties precludes their introduction into inorganic matrices, i.e. ceramics and glasses prepared by the conventional high temperature melting procedures. However, the lower temperatures required in the operations of the sol–gel process makes this conveniently possible [194, 195].

Reisfeld et al. have shown that while a dye like Rhodamine 6G, in spite of suitable spectroscopic properties [196], is not stable enough to be used in a practical device for luminiscent solar concentrators, much better results can be attained with the use of the photostable dye BASF-241 [197] in a polymer sol–gel system.

Compared to the well-known tunable laser based on crystals doped with chromium(III) [198], new lasers based on fluorescent glasses, and fluorol-555 and BASF-241 in a polymer material have some advantages, due to the singlet–singlet transitions between 400–700 nm in the dye systems [189].

Non-linear optics (NLO) deals with effects arising when a material is subjected to an intense light source such as a laser pulse. The third order NLO effects can be used to produce optical switching from a low to high optical transition state in a device of the directional couplet type used in telecommunications. This type of optical switching has extensive applications in the fast-emerging area of optical processing of information [191].

There is a great need for non-linear optical materials which can be utilised with low intensity light sources for several types of applications, e.g. in phase conjugation, image processing and optical switching. The phase conjugated waves are able to restore distortion in optical beams to their original unaberrated state under reflection from a non-linear medium [199].

Many non-linear materials are based on a variety of dyes incorporated in single crystals [200], polymers [201], liquid crystals [202], and doped solids [203]. An attempt [204] has been made to overcome the low optical quality of NLO polymers by a nanometer level conjugated polymer (poly-*p*-phenylene-vinylene) and silica glass in which the optical non-linearity of the polymer is coupled with the improved optical quality of the silica glass. Obviously such hybrid materials have a great future in the area of non-linear optics materials.

The extraordinary facility with which organic molecules can be trapped in inorganic oxide matrices [193] has led to synthesis of a new family of information recording materials by incorporation of photoactive molecules [205–209], many of which are capable of the hole-burning process [210, 211]. Following reports on the trapping of photochromic spiropyran derivatives in sol–gel matrices [208], Levy, Einhorn and Avnir [169] have shown recently that the glasses prepared by the polymerisation of $EtSi(OEt)_3$ and by copolymerisation of $Si(OEt)_4$ along with the oligomeric $[Me_2SiO]_n$ in the presence of a mild basic catalyst, sodium acetate, serve as much better photochromic information recording materials than those obtained from the hydrolysis of $Si(OEt)_4$ alone.

An interesting application of the oxazine-170 dye into sol–gel glasses [212] has been suggested recently as a reversible optical sensor for ammonia or acid vapours as environmental or biological traces.

In an important basic paper entitled "Processing of Gel-Silica Monoliths for Optics", Hench and Wilson [213] have emphasized the need for accurate drying data on various gel systems, specially the family of acid-catalyzed, alkoxide derived monolithic silica gels used to produce type V (fully dense with a very high level of purity and homogeneity) and type VI (optically transparent and chemically as well as thermally stable with 13–15% interconnected porosity, which can serve as a host matrix for organic polymers or metallic ions) gel-silicas [214, 215]. Phillips [216] has shown that uncoated etalons (which are optical elements with highly parallel surfaces used for frequency tuning in laser cavities) made from the alkoxide gel derived silica glass Type V is highly efficient in single-line to single-mode conversion.

Besides describing the tunable laser and optical gain applications of incorporating organic and organometallic molecules in silica and aluminosilicate gels, Dunn et al. [192] have shown that these molecules can be utilised as luminiscent probes to illuminate the chemical and structural aspects of the host gel itself.

6 Non-Oxide Ceramic Materials

Segal [217, 218] divided advanced ceramic materials into two broad groups. The members of the electroceramic group are generally oxides and have been used as insulators, semi-conductors and more recently as, super-conductors. To the structural group belong oxides, as well as a few others like carbide and nitride materials which have been known and used for a long time.

6.1 Nitride Ceramics

The conventional methods for preparing nitride and carbide materials involve, for example, high temperature reactions of the metal or its oxide (alone or mixed with a reducing agent) with nitrogen or carbon.

Significant achievements in the preparation of oxide-ceramics during the past few decades by a low temperature sol–gel process have led to a search for

Table 1

Structural Materials (Mechanical properties, e.g., tensile strength)	Electroceramics (Electrical and Magnetic Properties)
Al_2O_3	High Temp. Superconductors, e.g. $YBa_2Cu_3O_{7-x}$
ZrO_2	$BaTiO_3$ for capacitors
SiC	ZnO for varistors
Si_3N_4	Electro-optic
WC	Materials (e.g. $LiNbO_3$)
BN	Ferrites for magnetic Applications
AlN	Piezoelectrics
TiB_2	(e.g. Lead zirconium titanates)

similar routes for other materials. The sol–gel process for oxide ceramics involved hydrolysis of a solution of metal salts, but preferably alkoxides in a solvent miscible with water followed by condensation to the gel form.

Fortunately, the chemistry of metal amides has been investigated extensively and has been summarised in a recent book [219]. It may not be out of place to emphasize that a clear understanding of the following autoionisation equilibria in aqueous and liquid ammonia solutions has led to faster developments in the utilisation of amides for preparation of ceramic materials:

$$2H_2O \rightleftharpoons (H_3O)^+ + (OH)^-$$

$$2NH_3 \rightleftharpoons (NH_4)^+ + (NH_2)^-$$

Ammonolysis of titanium [220] and niobium [221] halides in liquid ammonia has been studied by Maya. Intermediates, isolated in these ammonolysis reactions, were converted to nitrides or to carbonitrides, on interaction with sodium acetylide.

Like the metal alkoxides, the metal dialkylamides with the general formula, $M(NR_2)_n$, are generally liquids with some volatility. These are highly reactive compounds which undergo facile transamination reactions with monoalkyl- and/or dialkylamines [219].

Brown and Maya [222] have studied the ammonolysis of $Ti(NMe_2)_4$, $Zr(NEt_2)_4$ and $Nb(NEt_2)_5$ in liquid ammonia solutions to examine the feasibility of obtaining high metal-content precursors which could be converted to nitride ceramics. Ammonolysis of the above dialkylamides yielded insoluble non-volatile solids, which corresponded in analysis to $Ti_3(NMe_2)\,(NH_2)_2(N)_3$; $Zr_5(NEt_2)\,(NH_2)_3\,(NH)_5\,(N)_2$; and $Nb_2(NEt_2)\,(NH)_3(N)$ respectively. On the basis of metal contents, titanium and zirconium compounds appear to be converted into nitride, whereas in the case of niobium, the final product is a mixture of nitride and carbonitride.

The progress of the reactions was checked by temperature programmed pyrolysis of the compounds upto 800 °C, which indicated a decomposition of the type:

$$Ti(NR_2)_4 \xrightarrow{\text{liquid } NH_3} Ti_3(NR_2)(NH_2)_2(N)_3$$

$$Ti_3(NR_2)(NH_2)_2(N)_3 \rightarrow Ti_3N_4 + NHR_2 + NH_3$$

$$Ti_3N_4 \rightarrow 3TiN + 1/2\,N_2$$

$$Ti_3N_4 + 3/2\,H_2 \rightarrow 3TiN + NH_3$$

The above findings are in conformity with earlier investigations [223, 224] on the decomposition reactions of metal dialkylamides at lower pressures.

Aluminium nitride (a III–V ceramic) which has attracted considerable attention because of its high thermal conductivity and high electrical resistivity is generally prepared by high temperature conventional routes such as direct combination of aluminium with nitrogen or ammonia [225], carbothermal reduction of alumina or by MOCVD of $(Me_2AlNH_2)_3$ [226], which has been obtained as a six membered ring compound by the reaction of Me_3Al and NH_3. Pyrolysis of a similar molecular reagent $[(Me_3Si)_2AlNH_2]_2$, obtained [227] by an equimolar reaction between $(Me_3Si)_3Al \cdot Et_2O$ and NH_3, leads [228] to the formation of AlN/SiC mixtures. Obviously, more intensive and controlled investigations of ammonolysis reactions of aluminium as well as silicon amides [229] in liquid ammonia are required for development of more convenient routes for nitrides of aluminium and silicon.

In a recent review, Paine and Narula [230] have summarized a large number of synthetic routes for the ceramic boron nitride and have finally described their efforts 'for preparing non-oxide preceramic polymers by methods related to sol–gel techniques....' . Out of a number of alternative routes, mention may be made [231, 232] of a fully B-cross-linked nitrogen and hydrogen rich polymer obtained in the gel form from a number of solvents:

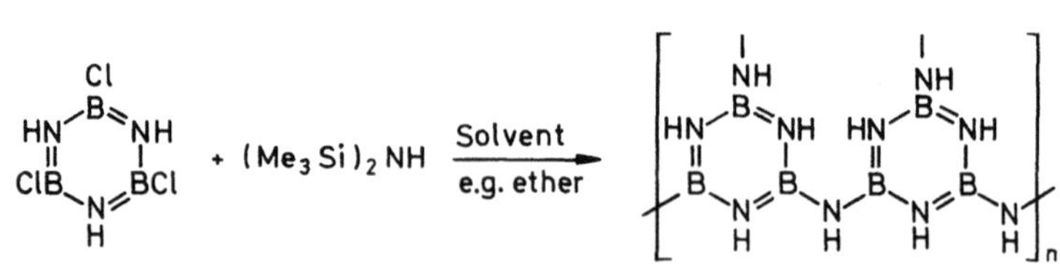

When bulk samples of the three-point gels are evaporated to dryness and the resulting glassy solids are polymerized under nitrogen at 1200 °C, relatively pure hexagonal h-BN powder is obtained with 55–60% ceramic yield. Small amounts of silicon may be retained from pyrolysis of residual methylsilyl end-capping groups. The silicon may be completely removed by pyrolysis of the gel under NH_3 at 1200 °C. Alternatively, the ceramic powder obtained from pyrolysis under nitrogen may be briefly heated in air, followed by washing with H_2F_2.

When samples of the above BN are exposed to liquid NH_3, the oligomer dissolves completely [233]. Subsequent evaporation of the NH_3 leaves a foamy white solid, which on being heated under argon shows a weight loss (42%) between 25 and 350 °C and no loss further upto 1100 °C. This solid and its solutions in NH_3 have quite interesting and useful properties: (i) it appears to 'melt' or 'flux' at about 1200 °C with the substrates (MgO, Al_2O_3, TiO_2) containing activated surface hydroxyl groups and upon cooling, the high quality adherent coatings of h-BN are obtained; (ii) the solution in NH_3 on being concentrated yields a heavy syrup from which green fibres of BN can be drawn and (iii) treatment of the gel by aerogel techniques yields a "low-density" boron nitride in the form of foam with potential applications in composites.

It has been further observed [234] that the aminolysis of $[(Me_2N)BNH]_3$ in toluene produces an insoluble white solid, which yields excellent-quality h-BN in good (50%) yield by pyrolysis of the solid in vacuo at 900 °C, followed by pyrolysis at 1200 °C in N_2 or NH_3.

It may not be out of place to mention about the investigations carried out in the author's laboratories on the reactions of alkanolamines (mono-, di- and tri-) with different metal alkoxides, showing a marked difference in the reactivity of the amino group. It has been shown that whereas alkoxides of silicon [235], and aluminium [236] exhibit reactivity towards the hydroxyl groups of the alkanolamines only with the formation of derivatives of the type, $(RO)_{n-x}M(OCH_2CH_2NH_2)_x$; alkoxides of germanium [237] and tin [238] react with both the hydroxyl as well as the amino groups of alkanolamines forming products of the type $(RO)_{n-2x}M(OCH_2CH_2\overset{H}{N})_x$.

6.2 *Sulphide Ceramics*

In recent years, chalcogenides have attracted research attention due to their infrared transmitting optical properties [239, 240] and their photoconductive electrical characteristics [241], which are being exploited in infrared optical fibres as well as transmitting windows, optical information storage media, xerography, and high resolution photolithography. At present, CVD produced ZnS is one of the most widely used infrared optical window materials but unfortunately it does not possess optimum mechanical properties for numerous other similar applications.

Although the sol–gel technique has been extensively explored and exploited for synthesis of oxide ceramics and glasses, very little attention appears to have been diverted to the analogous derivatives of chalcogen group VI elements, i.e. sulphur, selenium and tellurium.

In principle, the chemical synthesis of sulphide compounds could be achieved by methods similar to those employed in the metallo-organic sol/gel preparation of the oxides. The thiolysis/condensation reactions of organometallic sulphide precursors should be analogous to hydrolysis/condensation reactions of metal alkoxides:

Hydrolysis : $M(OR)_n + x\,H_2O \rightarrow M(OH)_x(OR)_{n-x} + x\,ROH$

Thiolysis : $M(SR)_n + x\,H_2S \rightarrow M(SH)_x(SR)_{n-x} + x\,RSH$

Condensation:

$$M(OH)_x(OR)_{n-x} \rightarrow MO_{n/2} + \left(\frac{2x-n}{2}\right)H_2O + (n-x)ROH$$

$$M(SH)_x(SR)_{n-x} \rightarrow MS_{n/2} + \left(\frac{2x-n}{2}\right)H_2S + (n-x)RSH$$

However, it is important to emphasize that there are salient differences in the chemical properties of oxygen and sulphur which would of course influence the above as well as other reactions with analogous oxygen and sulphur reagents. Sulphur is much less electronegative than oxygen, tending to make metal sulphides more covalent than the corresponding oxides. Due to lower electronegativity, sulphur has a much lower tendency to form hydrogen bonds, as reflected in the much greater volatility of H_2S compared to the associated $(H_2O)_n$. In addition to the above, the most important difference of far-reaching consequence is the extra-ordinary ability of sulphur to catenate and form polysulphide ions (S_n^{2-}) and polysulphanes (H_2S_n), etc.

The chemistry of thiols, RSH and metal thiolates, $M(SR)_n$ also presents some features quite distinct from the alcohols, ROH and metal alkoxides, $M(OR)_n$. Hydrogen sulphide H_2S and thiols, RSH are much stronger acids ($pK_a \sim 7$–8) compared to H_2O and alcohols, ROH($pK_a \sim 16$). A reference may be made to a review of the thiol and thio-β-diketonate derivatives of metals [242]. Metal thiolates of earlier transition metals (IUPAC groups III, IV and V of the periodic table) are hardly known in contrast to the extremely rich alkoxide chemistry of these transition elements [27]. For example, repeated attempts to isolate $Ti(SR)_4$ derivatives have been unsuccessful, but a derivative, $(C_5H_5)_2Ti(SR)_2$ appears to be quite stable [243]. This has been explained to be due to enhancement in the capacity of titanium to stabilise TIS bond by back-bonding arising from a higher electron density on the metal brought about by the donor C_5H_5 ligands. This highlights another essential difference between oxygen and sulphur, arising from the capacity of sulphur (with electronic configuration $1s^2 \cdot 2s^2p^6 \cdot 3s^2p^4$) to expand the octet $3s^2p^6$ in the valency shell by accommodating back-binding electrons in its '3d' orbital.

In the case of metal thiolates also, condensation polymerisation could be expected [244] to occur through the μ_2-bridges associated both with SH– and SR– species, viz.

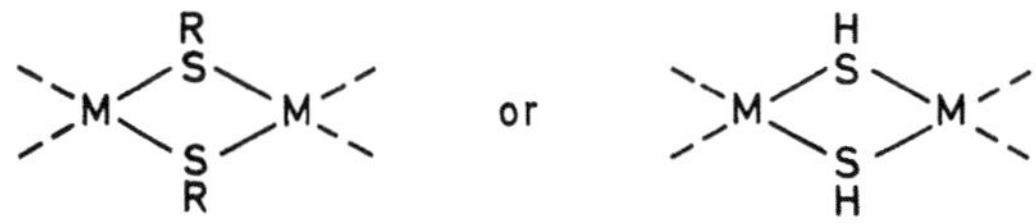

However, metal sulphide or thiolate ions have the capacity to form a wide range of other crosslinks, enhancing the tendency of condensation and precipitation of solids.

As stated earlier, the application of ZnS prepared by conventional methods as infrared optical windows tended to be offset by lack of mechanical strength. One of the successful techniques of enhancing the strength of a material is the formation of self-similar composites. For example, the fracture toughness of Al_2O_3 has been enhanced [244] three-fold by the preparation of Al_2O_3/Al_2O_3 fibre reinforced composites. For improving the strength of ZnS windows also, a similar attempt has been made to reinforce it with ZnS whiskers [245].

The methodology and the dimensions of the study [246] are represented by the following equations:

Reaction (1): $EtZnSR_x + xH_2S \rightarrow xZnS + xEtH + xRSH$

Reaction (2): $Et_2Zn + H_2S \rightarrow ZnS + 2\,EtH$

Reaction (3): $Zn(SR)_{2x} + xH_2S \rightarrow xZnS + 2x\,RSH$

Reaction (4): $Et_2Zn + (RS)_2S \rightarrow ZnS + 2EtSR$

The most promising precursor for synthesis of ZnS whiskers appears to be $\{EtZn(SBu^t)\}_5$ which on treatment with H_2S in pentane immediately yields EtH and Bu^tSH. As the product does not show any bands in its infrared spectra, the reactions occurring can be represented by the following equations:

$$\{EtZn(SBu^t)\}_5 + xH_2S \xrightarrow[\text{fast}]{} Et_{5-x}Zn(SH)_x(SBu^t)_5 + xEtH$$

$$Et_{5-x}Zn(SH)_x(SBu^t)_5 \longrightarrow Et_{5-x}ZnS(SBu^t)_{5-x} + xBu^tSH$$

It has been proposed that the morphology of the product is a direct consequence of the extent of thiolysis and condensation. Initially, as the polymeric pentamer's organic ligands are eliminated, the pentamer's structure consists of fibrous chains. As the thiolysis reaction proceeds, the extent of Zn–S–Zn crosslinking increases and the resulting product tends to be a fine nano-scale ZnS powder. A careful heat treatment of fibrous ZnS product under flowing H_2S at 500 °C led to the formation of single crystal α-ZnS whiskers, which are usually mixed with β-ZnS powder. During this treatment, the vapour phase transport of ZnS has been suggested, based on the observation that the whiskers are not formed when the flow rate is high as the (volatile) ZnS species are swept through the furnace.

Another route for the preparation of ZnS has been described as the reaction of diethylzinc, Et_2Zn with dibenzyl trisulphide, $(BzS)_2S$ in pentane or toluene at 0 °C. If the reaction conditions are stoichiometric (1:1), evacuation of the volatiles leads to a white precipitate. In the presence of excess (5:1) of $(BzS)_2S$, a transparent yellow viscous solution is obtained, which yields a gel on evacuation of the solvent. This gel yields a reddish orange solid after 48 hours:

$$Et_2Zn + (BzS)_2S \rightarrow \underset{\text{orange solid}}{(\text{–BzS–S–Zn–S–SBz–})} + 2EtSBz$$

$$\downarrow H_2S$$

$$\underset{\text{white solid}}{ZnS}$$

Another sulphide system leading to As_2S_3 fibres and films, which has been investigated [248] by the sol–gel process is that of As_2S_3-amine. It has been proposed that solutions of arsenic trisulphide in ethylene diamine consist of As_4S_6 rings which are interlinked through sulphur–sulphur bridges, which give rise to an intense band at 438 cm^{-1} in the infrared spectra of the solutions. This band is of the highest intensity in the most viscous gels, and its intensity diminishes on drying the gels. The intensity further decreases during heat treatment (under reduced pressure) of the fibres drawn directly from the gel.

On the basis of physico-chemical studies, the molecular complexation of As_2S_3 in amine solvents involves Lewis acid–base type of interaction, the ethylene diamine molecules acting as chelating ligands probably between the arsenic atoms in the same layer. The resulting soluble species can be, therefore, considered to be composed of As–S (with ethylene diamine ligand chelating between arsenic atoms) rings interlinked three-dimensionally through bridging sulphurs. The gelation process represents mainly to chain entanglements and produces no significant differences between the solution and gel infrared spectra.

7 Future Potential of the Sol–Gel Process

7.1 General

In the preceding pages a succint account of only a few of the rapidly enhancing areas of the sol–gel process has been presented, laying special emphasis on a few recent developments only in each of these directions. A reference has been made (in Sect. 1 of this article) to the extremely fast growth of the technique and its applications as evinced by the rate of annual publications. Fortunately, the emphasis has shifted from mere extensions of the already established applications to novel break-throughs in still untrodden directions, based on a better understanding of the basic chemical reactions involved. The importance of chemical knowledge has been repeatedly emphasized (11) particularly in the emerging ultra-structure processing achievements by more efficient manipulation at the molecular level and more stringent control of uniquely homogeneous structures.

In spite of determined efforts for maximum objectivity, any prediction about future potential tends to be subjective and dependent upon the background of the subject as well as that of the author. Fortunately, a number of recent attempts in these directions [4, 6, 9, 11, 13, 16, 54, 249] make the task of the present author easier. Further, as the future potentials of the different themes have already been indicated in the earlier sections of this article, attention here would be focussed mainly on points not fully covered earlier.

Tremendous success in the S–G technology has been achieved since the early 1980s, when in an article 'Glasses from Melts and Glasses from Gels—A Comparison', Mackenzie [250] discussed the advantages and deficiencies of the Sol-Gel process. Hench and West [13] in an article published (1990) in Chemical Reviews concluded that 'other engineering advantages of the lower temperature chemically based sol–gel processing such as net-shape casting, fibre-pulling, and film-coating have also reached economic potential'. These authors [13], however, emphasized that the understanding of the molecular reaction mechanisms, and the thermodynamics as well as kinetics, particularly of the several overlapping processing steps is rather meagre and requires concerted efforts for new break-throughs in sol–gel technique.

As emphasized earlier also, one of the main advantages of the sol–gel process, in addition to the requirements of lower temperatures of operations, is the homogeneity of the final ceramic product, brought about by mixing at molecular level in the original solution. Roy, who is one of the founders of the technique [23], indicated in a recent (1987) article (9) an interesting variation in the plan of his research group, which is to prepare 'maximally heterogeneous' materials by the S–G process. 'Maximally heterogeneous' in this reference has been quantified as the 'surface area of contact between the two (or more) phases involved. In view of the size of individual phases in the nanometer (100 to 1 nm) range, such 'hetero-materials' can be considered to be 'nano-composites'. It has been pointed out that nature abounds in such micro- and/or nano-composite structures in which the extraordinary toughness is provided by a ceramic-polymer combination; interesting examples of these include bones and teeth, in which 100 nm hydroxyapatite fibrils are topotactically related to the collagen phase. It is proposed to emphasize the high potential of the low-temperature sol–gel technique in bioprocessing towards the end of this article.

The success achieved in the field of multiphase composites can be illustrated by the synthesis of electron microscopically indistinguishable Gilsonite opals [251] and uvarovite garnets [252]. The structure of garnet presents a broad crystal chemistry with their general formula being $\{A\}_3[B]_2(C)_3O_{12}$ in which { }, [ ], and ( ) represent dodecahedral, octahedral and tetrahedral co-ordinations respectively. The extraordinary discovery of multiphase oxide superconductors has urged the sol–gel chemists to additional efforts [57] in these directions, which have been extended even to a new spurt in the alkoxide chemistry of concerned elements, e.g. yttrium [73, 108–111], barium [71], copper [72], bismuth [73], etc.

The possibility [26] of the formation of new chemical bonds in a solution of two or more alkoxides and these heterometal alkoxide systems being involved in the hydrolysis and condensation steps has given a new dimension to the work [34, 38, 100] on heterometal alkoxides being pursued in the research schools of the author of this article and his students. The possibilities of obtaining materials such as $NiAl_2O_4$ with the same composition but different geometries by a mild hydrolysis of octahedral $Ni\{Al(OMe)_4\}_2$ and tetrahedral $Ni\{Al(OBu^t)_4\}_2$ open up exciting possibilities [34].

In view of the extraordinary advantages in the use of metal alkoxides as precursors [20] in the sol–gel process, these investigations on alkoxide chemistry could be expected to have a synergetic effect on the progress of the sol–gel process.

Following an earlier (1988) article (54) on 'Prospects of the Sol–Gel Process', Ulrich [11] has more recently (1990) given a vivid account of the fastly emerging applications of the SG process for non-linear optics, ultraviolet photonics, bioprocessing and multifunctional materials.

The latter presentation [11] by Ulrich was made at the Fifth International Workshop on Glasses and Ceramics from Gels (Rio de Janeiro, 1989) and a number of other presentations at the above workshop on silica aerogels by Fricke [56], silica monoliths for optics by Hench and Wilson [213], thin film formation by Brinker et al. [253], spherical submicron oxide powders from oil/water emulsions of metal oxides by Messing et al. [254], insulating, electro-conducting and superconducting fibres by Sakka [17], ceramic superconductors by Kordas [57], spectroscopy and applications of molecules in glasses by Reisfeld [193], and organically modified ceramics by Schmidt and Wolter [185] have not only focussed on considerable accomplishments in the area concerned but have also pointed towards possibilities of novel developments in the future.

Attention has already been drawn to the exciting developments in the fields of fibre technology [16, 17, 49]. In addition to silica aerogels [56], Gesser and Goswami [255] have given a critical review of aerogels of many other elements (e.g. Ti, Ge, Zr, Sb, Al, Ce, Sn, V, N, V, Th, Sc) including their novel applications in catalysis and high energy physics. A novel process with a bright future has been described [256] for fabricating lightweight cellular ceramics by utilizing the rapid viscosity change during gelation to stabilize the structure of a foamed silica sol.

In the field of sol–gel processes for non-oxide ceramic materials, the work of Paine and Narula [230] has opened new dimensions which are likely to be followed with very promising results in the near future.

7.2 *Bioprocessing*

The sol–gel processes appear to offer new dimensions for bioprocessing and biomimicking. Besides such requirements arising from accidents, implantation of millions of artificial or prosthetic parts have to be carried out due to the natural deterioration of the human body on ageing. The challenge in the field of biomaterials [257, 258] arises from the basic fact that all implant devices replace living tissues evolved over millions of years. Hulbert et al. [259] cited some 38 different ceramic materials for such implantation devices. The types of ceramic materials for musculoskeletal prostheses may be divided [260] into 3 classes, according to their chemical reactivity with the environment: (i) completely resorbable, e.g. $Ca_3(PO_4)_2$; (ii) surface active bioglasses and bioceramics providing bonding sites for proteinaceous constituents of soft tissues and cell membranes; and (iii) inert compositions such as Al_2O_3.

Biological systems generally utilise the infiltration of one phase into another, as illustrated by teeth and bones, which consist of mineral particles imbedded in a polymer matrix [261]. Synthetic examples of in situ precipitation of sol–gel precursors within polymeric matrices leading to formation of ultrafine ceramic particles or composites with highly controlled microstructures have been provided amongst others by Aksay and coworkers [262]. These workers have also shown that silica is precipitated within polyacrylamide hydrogel matrices by diffusion and internal hydrolysis of tetraethoxysilane (TEOS). The hydrogels fill with ceramic as evinced by gain in weight which depends on the rate of diffusion of TEOS within the hydrogel and which can be controlled by pH and the pore size. These findings bring the field of sol–gel closer to bioprocessing, and similar techniques could be utilised for processing of piezoelectric, ferroelectric, electro-optic and superconducting thin films and powders.

The above concepts are being extended by Aksay et al. (cf. ref. 11) to the encapsulation of single crystal ceramic particles by a biomembrane and to the formation and crystallization of nanometer-sized magnetite particles in the freshwater magneto-tactic bacterium *Aquaspirillum magnetotacticum*. The bacterially produced magnetite particles form clusters which are more densely packed than the porous structures formed by nanometer-sized particles in synthetic ceramics. Another system of biosynthesized particles of colloidal alumina using lipid membranes in vesicle mediated synthesis has also been reported [263].

Considering the yet unresolved difficulties in producing sulphide compounds from gels, a promising route appears to be the use of porous sol–gel glasses as cages for bioprocessing reactions. In extending bioprocesses to sol–gel technique, the biomembrane appears to act as (i) a nanoscale container cage for the reaction with particle size control; (ii) an ion selection filter and valve regulating the rate of transport and (iii) a lubricant when packing particles. There are exciting possibilities of the membranes functioning, similar to their behaviour in nature, as a template for the nucleation of particles (cf., 11).

8 Conclusion

The primary goal of sol–gel processing, i.e. control of the structure of synthetic materials on a nanometer scale from the initial stage of processing, has been fairly well achieved for pure silica (in powder, fibre, and monolith forms), but the understanding of chemistry of the sol–gel process (particularly involving other metals and metalloids) is still at a rudimentary stage.

The phenomenal advances of the sol–gel technique in exciting fields like ormosils, aminosils, non-linear optics, ultraviolet-photonic and novel multifunctional materials (termed sometimes as 'smart' or/and 'intelligent') are also posing challenges to the experts for a much wider and penetrating grasp of the multidisciplinary principles involved.

Beginning with the medical applications of silatrane materials [264], the importance of silicon in biochemical fixation of nitrogen and carbon [265], metabolic pathways [266] and even biogenesis [267] is of considerable growing interest.

In spite of unprecedented advances in different disciplines of science, human ingenuity is still far behind the efficiency of the biological processes occurring all the time around in nature. With a clearer understanding of the physico-chemical principles involved in these natural phenomena of biological growth, industries based on biotechnology are emerging to the forefront. It is a happy sign that along with continued and more concerted efforts to probe into the physico-chemical intricacies of the fast advancing sol–gel process, attention has been focussed on bioprocessing and biomimicking from which much richer dividends can already be foreseen. It may, therefore, be appropriate to end this article on 'The Present Status and Future Potential of the Sol–Gel Process' with re-emphasis of the need of intimate interdisciplinary collaboration for not only sustained improvements in the existing procedures but for unforeseen break-throughs in novel directions. It may not be out of place to underline the synergetic effects of such multidisciplinary efforts. The author who has been one of the few workers around the world struggling during the past four decades in the field of metal and later heterometal alkoxide chemistry is happy to acknowledge the benefit from the unprecedented interest of many prominent schools of metalo-organic chemistry in these directions with highly exciting revelations. It is hoped that similar advantages accrue to the development of biological sciences, including probably biogenesis and biomedicine [267] by the growing interest of sol–gel scientists in bioprocessing.

9 References

1. Ebelman M (1846) Comptes Rend. de L. Acad. des Sciences 25: 854
2. Iler RK (1979) The chemistry of silica, John Wiley, New York
3. Suzuki H (1987) Phil Trans R Soc Lond A322: 465
4. Ashby MF (1987) Phil Trans R Soc Lond A 322: 393
5. Mackenzie JD (1988) J Non-Cryst Solids 100: 162
6. Segal D (1989) Chem in Britain pp 153–156, Chemical Synthesis of Advanced Ceramic Materials, CUP, Cambridge
7. Stinton DP, Besmann TM, Lowden RA (1988) Cer Bull 67: 350.
8. Bradley DC (1989) Chem Rev 89: 1317
9. Roy R (1987) Science 238: 1665

10. Mackenzie JD (1988) J Non-Cryst Solids 100: 162
11. Ulrich DR (1990) J Non-Cryst Solids 121: 465
12. Brinker CJ and Scherer GW (1989) Sol–gel science, Academic, New York
13. Hench LL, West JK (1990) Chem Rev 90: 33
14. Klein LC (ed) (1988) Sol–gel technology for thin films, fibres, preforms, electronics and speciality shapes, Noyes, New Jersey
15. Brinker CJ, Hurd AJ, Frye GC, Ward KJ and Ashley CS (1990) J Non-Cryst Solids 121: 294
16. Sowman HG (1988) Cer Bull 67: 1911
17. Sakka S (1990) J Non-cryst Solids 121: 417
18. Aegerter MA, Jafelicci Jr M, Souza DF, Zanotto ED (1989) Sol–gel–science and technology, World Scientific, Singapore
19. Scherer GW (1989) Ref 18 pp 221–257
20. Mehrotra RC (1989) Ref 18 pp 1–60
21. Matijevic E (1981) Acc Chem Res 14: 22
22. Kezuka K, Hiyashi Y and Yamaguchi T (1989) J Am Ceram Soc 72: 1660
23. Roy R and Osborn EF (1954) Am Mineral 39: 853
24. Bradley DC, Gaur DP and Mehrotra RC (1978) Metal Alkoxides, Academic Press, London
25. Harwood JH (1963) Industrial applications of organometallic compounds, Chapman and Hall, London
26. Dislich H (1971) Angew Chemie Int Ed (English) 10: 363
27. Mehrotra RC (1983) In: Emeleus HJ, Sharpe AG (eds) Advances in inorg and radiochemistry, Vol 26, Academic, New York
28. Mehrotra RC et al. (1980) Coord Chem Rev 31: 67
29. Mehrotra RC (1981) Coord Chem (IUPAC) 21: 113
30. Mehrotra RC (1986) XXIV Int Conf Coord Chem Abstracts p 27
31. Mehrotra RC and Mehrotra A (1971) Inorg Chim Acta Rev 5: 127
32. Mehrotra RC (1988) J Non-Cryst Solids 100: 1
33. Jones K, Davies TJ, Emblem HG and Parkes P (1986) Mat Res Soc Symp Proc 73: 111
34. Mehrotra RC (1988) Mat Res Soc Symp Proc 121: 81
35. Hirano S, Hayashi T, Nosaki K and Kato K (1989) J Am Ceram Soc 72: 707
35a. Nazeri-Eshghi A, Kuang AX, MacKenzie JD (1990) J Mat Sc 25: 3333
36. Hirano S and Kato K (1989) Bull Chem Soc Jpn 62: 429
37. Livage J, Henry M, Sanchez C (1988) Prog Solid State Chem 18: 259
38. Mehrotra RC (1990) J Non-Cryst Solids 121: 1
39. Goodwin GB and Kenney ME (1990) Inorg Chem 29: 1216
40. Kamiya K, Sakka S and Tatemichi Y (1980) J Mater Sc 15: 1765
41. Fegley B Jr, White P and Brown HK (1985) J Am Ceram Soc 68: C60
42. Kundu D, Biswas PK and Ganguli D (1989) J Non-Cryst Solids
43. Messing GL and Kumagai M (1989) J Am Ceram Soc 72: 40
44. Hirashima H, Onishi E and Nakagawa M (1990) J Non-Cryst Solids 121: 404
45. Farbes BD and Uhlmann DR (1990) J Am Ceram Soc 73: 978
46. Sakka S, Kamiya K and Yoko Y (1988) Sol–gel preparations and properties of fibres and coating films, ACS Symp Ser 360 Inorg and Organometal Polymers, Eds. Zeldin M, Wynne KJ and Allcock HR pp 343–353
47. Dislich H (Private Communication)
48. Yoldas BE and O'Keefe TW (1979) Appl Opt 18: 3133
49. Matsuzaki K, Arai D, Taneda N, Mukaiyama T and Ikemura M (1989) J Non-Cryst Solids 112: 487
50. Sowman HG (1988) Ref 14 pp 162–182
51. Gunji T, Nagao Y, Misono T and Abbe Y (1989) J Non-Cryst Solids 107: 149
52. Wu ET, Kuang AX and MacKenzie (1986) Int Symp on Applications of Ferromagnetics Proc.
53. Schmidt H (1988) J Non-Cryst Solids 100: 51
54. Ulrich DR (1988) J Non-Cryst Solids 100: 174
55. Ewell RS and Insler H (1973) J Res Nat Bur Stand 15: 173
56. Fricke J (1990) J Non-Cryst Solids 121: 188
57. Kordas G (1990) J Non-Cryst Solids 121: 436
58. Luth WC and Ingamells CO (1963) Am Mineralogist 50: 255
59. Rabinovich EM, Johnson DW, MacChesney JB and Vogel EM (1983) J Am Ceram Soc 66: 683
60. Visca M and Matijevic E (1979) J Coll Interface Sc 68: 308
61. Ingelrethsen BJ and Matijevic E (1980) 11: 271

62. Tamura H and Matijevic E (1982) J Coll Interface Sc 90: 100
63. Regazoni AE and Matijevic E (1983) Coll Surface 6: 189
64. Matijevic E, Simpson CM, Amin N and Arajs S (1986) Coll Surface 21: 101
65. Paola A and Matijevic E (1987) Eur Pat Appln EP 212, 935 (4 March) 1987
66. Matijevic E (1986) Langmuir 2: 12
67. Thomas I (1973) US Patent 3767 432
68. Irwin AD, Holmagreu JS, Zerda TW and Jones J (1987) J Non-Cryst Solids 89: 191
69. Tian F, Pan L and Wu X (1988) J Non-Cryst Solids 104: 129
70. Thomas I (1988) Ref 14 pp 7–8
71. Caulton KG, Chisholm MH, Drake SR and Huffman JC (1990) J Chem Soc Chem Commn 1498.
72. Goel SC, Kramer KS, Gibbons PC and Buhro WE (1989) Inorg Chem 28: 3619
73. Poncelet O, Hubert-Pfalzgraf LG, Darän J and Astier R (1989) J Chem Soc Chem Commun 1846
74. Hayashi A and Salto H (1980) J Mat Sc 15: 1971
75. Phallipou J, Prassas M and Zarzycki JJ (1982) J Non-Cryst Solids 48: 17
76. Pouxviel JC, Boilot JP, Poncelet O, Hubert-Pfalzgraf LG, Lecomlt A, Dauger A and Beloeib JC (1987) J Non-Cryst Solids 93: 277
77. Abe Y, Gunji T, Kimata Y, Kuramata M, Kasgoz A and Misono T (1990) J Non-Cryst Solids 121: 21
78. Roy R (1969) J Am Ceram Soc 52: 344
79. MaCarthy GJ and Roy R (1971) J Am Ceram Soc 54: 639
80. Mehrotra RC and Bohra R (1983) 'Metal Carboxylates', Academic Press, NY
81. Sharma HK and Kapoor PN (1988) Polyhedron 7: 1389
82. Mehrotra RC and Rai J (1990) unpublished results
83. Kumara Y, Tanimoto S, Yamane H and Kitao T (1990) Polyhedron 9: 371
84. Gurkovich SR and Blum J (1984) 'Ultrastructure processing of ceramics glasses and composites', Eds Hench LL and Ulrich DR, John Wiley, NY
85. Chen C, Ryder DF Jr and Spurgeon WA (1989) J Am Ceram Soc 72: 1495
86. Mehrotra RC (1966) Pure and Appl Chem (IUPAC) 13: 111
87. Mehrotra RC and Kapoor RN (1958) J Am Chem Soc 80: 3569
88. Mehrotra RC, Gaur DP and Bohra R (1978) 'Metal β-diketonates and allied derivatives', Academic Press, NY
89. Mehrotra, RC (1988) Pure and Appl Chem 60: 1349
90. Babonneau F, Sanchez C and Livage J (1988) J Non-Cryst Solids 106: 170
91. Babonneau F, Doeuff S, Leaustic A, Sanchez C, Carter C and Verdaguer M (1988) Inorg Chem 27: 3166
92. Babonneau F, Coury L and Livage J (1988) J Non-Cryst Solids 121: 153
93. Abe Y, Sugimoto N, Nagai Y and Misano T (1988) J Non-Cryst Solids 104: 164
94. Schubert U, Rose K and Schmidt H (1988) J Non-Cryst Solids 105: 165
95. Mehrotra, RC (1989) Plenary Lecture at the Ist Int Conf on Silicon and Tin, Kuala Lumpur, Abstracts pp 6, Oxford Univ Press (in print)
96. Govil S and Mehrotra RC (1975) Synth React Inorg Metal-Org Chem 5: 267
97. Mehrotra RC and Kapoor PN (1974) Coord Chem Rev 14: 1
98. Mehrotra RC and Agarwal MM (1968) J Chem Soc A 2673
98a. Eichorst DG, Payne DA, Wilson SR (1990) Inorg Chem 29: 1458
99. Mehrotra RC (1990) Chemtracts 2: 389
100. Mehrotra RC and Singh A (1991) Chem Rev 91 (in press)
101. Kordas G and Moore G (1989) Ref 16: 470
102. Sauer NN, Gracia E, Salazar KV, Ryan RR and Martin JA (1990) J Am Chem Soc 112: 1524
103. Khan SA, Barboux P, Bagley BC and Torres FE (1989) J Non-Cryst Solids 110: 142
104. Mehrotra RC and Rai AK (1966) Ind J Chem 4: 537
105. Matchett MA, Chiang MY and Buhro WE (1990) Inorg Chem 29: 360
106. Brnicevic N, Mustovic and McCarley RE (1988) Inorg Chem 27: 4532
107. Cotton FA et al. (1988) Inorg Chem 27: 3569
108. Burns CJ and Salteberger (1988) Inorg Chem 27: 3962
109. Evans WJ, Sollberger MS and Hanusa TP (1988) J Am Chem Soc 110: 1891
110. Evans WJ and Sollberger MS (1988) Inorg Chem 27: 4417
111. Evans WJ, Oloffson JM and Ziller JW (1990) J Am Chem Soc 112: 2308

112. Poncelet O et al. (1989) 29: 263
113. Bradley DC et al. (1988) J Chem Soc Chem Commun 1258
114. Chatterjee S, Bindal SR and Mehrotra RC (1976) J Ind Chem Soc 53: 867
115. Turova N Ya et al. (1986) Koord Chim 12: 1063; Chem Abstr (1986) 105: 182779
116. Livage J et al. (1988) Inorg Chem 27: 3166
117. Babonneau F, Sanchez C and Livage J (1988) J Non-Cryst Solids 106: 170
118. Babonneau F, Leaustic A and Livage J (1988) MRS Symp Proc 121–317
119. Vandenborre MT, Poumellec B, Alquier C and Livage J (1989) J Non-Cryst Solids 108: 333
120. Nabavi M, Doeuff S, Sanchez C and Livage J (1990) J Non-Cryst Solids 121: 31
121. Evans WJ, Denung TJ, Olofson JM and Ziller JW (1989) Inorg Chem 28: 4027
122. Gradeff PS, Yunlu K, Denung TJ, Olofson JM, Doeden RJ and Evans WJ (1990) Inorg Chem 29: 420
123. Hitchcock PB, Lappert MF and Singh A (1983) Chem Commun 1499; Hitchcock PB, Lappert MF and Smith RC (1987) Inorg Chim Acta 137: 183
124. Evans WJ, Olofson JM and Ziller JW (1989) Inorg Chem 28: 4308
125. Klemperer WG and Ramamurthi SD (1988) Mat Res Soc Symp Proc 121: 1
126. Klemperer WG and Ramamurthi SD (1990) J Non-Cryst Solids 121: 16
127. Bradley DC in Inorganic Polymers, eds Stone FGA and Graham WA (1962) Academic Press, New York
128. Chisholm MH and Rothwell IP (1987) in "Comprehensive Coordination Chemistry", eds, Wilkinson G, Gillard RD and McCleverty GAM, Vol II, Pergamon, London, pp 335–364
129. Song KC and Chung IJ (1989) J Non-Cryst Solids 108: 37
130. Boonstra AH, Bemards TNH and Smits JJT (1989) J Non-Cryst Solids 109: 141
131. Babonneau F, Leaustic A and Livage J (1988) Mat Res Soc Symp Proc 121: 317
132. Pouxviel JC, Boilot JP, Beloeil JC and Lallemand JY (1987) J Non-Cryst Solids 89: 345
133. Turner CW and Franklin KJ (1987) J Non-Cryst Solids 91: 412
134. Peltefer RF, Dupree R, Fernan I and Stenberg U (1988) J Non-Cryst Solids 106: 408
135. Huang M, Ding Z and Zhang F (1989) V Int workshop on glasses & ceramics from gels (Rio de Janeiro) Abstract A5
136. Scherer GW (1989) J Non-Cryst Solids 108: 18 (1990) ibid 121: 109
137. Zarzycki J (1988) J Non-Cryst Solids 100: 359 (1990) ibid 121: 110
138. Zelinski BJJ and Uhlmann DR (1984) J Phys Chem Solids 45: 1069
139. Rahaman MH, De Jonghe LD, Shinde SL and Tewari PH (1988) J Am Ceram Soc 71: C338
140. Lours L, Zarzycki J, Craievich AF, Aegerter MA and Santos DID (1988) J Non-Cryst Solids 106: 157
141. Scherer GW (1989) Ref 18 pp 221–256
142. Larbot A, Alary JA, Guizard C, Cot L and Gillot J (1988) J Non-Cryst Solids 104: 161
143. Bansal NP (1988) J Am Ceram Soc 71: 666
144. Sakka S, Kozuka H and Adachi T (1988) J Non-Cryst Solids 102: 263
145. Chang-Ching JY and Klein LC (1988) J Am Ceram Soc 71: 83, 86
146. Sakka S and Kozuka H (1988) J Non-Cryst Solids 100: 142
147. Filho OK and Aegerter MA (1988) J Non-Cryst Solids 105: 19
148. Boonstra AH, Meeuwsen TPM, Baken JME and Aben GVA (1989) J Non-Cryst Solids 109: 153
149. Babonneau F et al. (1988) Inorg Chem 27: 3166
150. Turner CW and Franklin KJ (1987) J Non-Cryst Solids 91: 402
151. Irwin ID et al. (1988) J Non-Cryst Solids 106: 359
152. Hartmann JS et al. (1989) J Non-Cryst Solids 108: 49
153. Meise-Gresch Ch. WK, Muller-Warmuth W, Beier W, Wellbrock U and Frischat GH (1990) J Non-Cryst Solids 116: 161
154. Glaser RH, Wilkes GL and Bronniman CE (1989) J Non-Cryst Solids 113: 73
155. Pouxviel JC, Boilot JP, Smaihi M and Dauger A (1988) J Non-Cryst Solids 106: 147
156. Marciel GE et al. (1988) J Am Chem Soc 110: 2023
157. Eaborn C (1960) Organosilicon compounds, Butterworth Scientific Publications
158. Ebsworth EAV (1963) Volatile silicon compounds, Pergamon Press
159. Zerda TW and Hoang GJ (1990) in Proc Fourth Int Conf on ultrastructure processing of ceramics, glasses and composites (Tuscan) Uhlmann D, Weinberg M, Ulrich DR (eds) Wiley, New York
160. Damrauer R, Burggraf LW, Davis LP and Gordon MS (1988) J Am Chem Soc 110: 6601

161. Tandura SN, Voronkov MG and Alekseev NV (1986) Top Curr Chem 131: 99
162. Watenpaugh K and Caughlan CN (1967) J Chem Soc Chem Commn 76
163. Bradley DC et al. (1968) J Chem Soc Chem Commn 1112
164. Morisin B (1981) Acta Crystallogr B33: 303
165. Lappert MF (1962) in 'Developments in inorganic polymer chemistry', eds Lappert MF and Leigh JJ, Elsevier, Amsterdam
166. Voronkov MG, Maletina EA and Roman VK (1988) Heterosiloxanes Volume 1: Derivatives of non-biogenic elements, Harwood Academic Publishers, GmbH, Switzerland
167. Reisfeld R (1989) Ref 18: pp 323–345
168. Schmidt H (1989) Ref 18: pp 432–469
169. Levy D, Einhorn S and Avnir D (1989) J Non-Cryst Solids 113: 137
170. Sakka S, Tanaka Y, Kokubo T (1986) J Non-Cryst Solids 82: 24
171. Glaser RH and Wilkes GL (1988) Polym Bull 19: 51
172. Huang HH, Glaser RH and Wilkes GL in Inorganic and Organometallic Polymers, Zeldin M, Wynne KJ and Allcock HR (eds) (1988) ACS Symp Series 360: 354 Washington DC
173. Elmer TH and Meissner HE (1976) J Am Ceram Soc 59: 206
174. Kamiya K, Yoko T, Sano T and Tanaka K (1990) J Non-Cryst Solids 119: 14
175. Zhang H and Pantano CG (1990) J Am Ceram Soc 73: 958
176. Schmidt H, Kaiser A (1981) Glastechn Ber 54: 338
177. Schmidt H (1989) Ref 16, pp 432–469
178. Schmidt H, Scholze H and Tunker G (1986) J Non-Cryst Solids 80: 557
179. Tunker G, Patzelt H, Schmidt H and Scholze H (1986) Glastechn Ber 59: 272
180. Schmidt H (1989) J Non-Cryst Solids 112: 419
181. Phillip G and Schmidt H (1984) J Non-Cryst Solids 63: 261
182. Kaisar A, Schmidt H, Hazenfratz-Schreier H and Kulbe KD (1984) Chem Ing-Tech 56: 652
183. Schmidt H, Rinn G, Nass R and Sporn D (1988) Mat Res Soc Symp Proc 121: 743
184. Nass R, Arpac E, Glaubitt W and Schmidt H (1990) J Non-Cryst Solids 121: 370
185. Schmidt H and Wolter H (1990) J Non-Cryst Solids 121: 428
186. Kohjiya S, Ochiai K and Yamashita S (1990) J Non-Cryst Solids 119: 132
187. Charbouillot Y, Ravaine D, Armond M and Poinsignon C (1988) J Non-Cryst Solids 103: 325
188. Reisfeld R and Jorgensen CK (1982) Structure & Bonding 49: 1
189. Reisfeld R (1987) J Physique Colloque 48: C7423
190. MacDonald KR, Tompkin WR and Boyd RW (1988) Opt Lett 13: 485
191. Reisfeld R (1989) Ref 18 pp 323–345
192. Dunn B, Knobbe E, McKiernan JM, Pouxviel JC and Zink JI (1988) Mat Res Soc Symp Proc 121: 331
193. Reisfeld R (1990) J Non-Cryst Solids 121: 254
194. Anvir D, Levy D and Reisfeld R (1984) J Non-Cryst Solids 88: 5956
195. Anvir D, Kaufman VR and Reisfeld R (1985) J Non-Cryst Solids 74: 395
196. Reisfeld R, Eyal M, Chernyak V and Zusman R (1988) Solar Energy Mater 17: 439
197. Reisfeld R, Brusilovsky, Eyal M, Miron E, Burshtein Z and Ivri J (1989) Chem Phys Lett 160: 43
198. Reisfeld R and Jorgensen CK (1988) Structure and Bonding 69: 63
199. Shkunov VV and Zel'dovich BYa (1985) Scientific American 253: 40
200. Carter GM, Thakur MK, Chen YJ and Hryniewicz (1985) Appl Phys Lett 47: 457
201. Kajar F, Messiar J, Zyss J and Ledoux I (1983) Opt Commun 45: 133
202. Wong KJ and Garito AF (1986) Phys Rev A 34: 5051
203. Kramer MA, Tompkin WR and Boyd RW (1986) Phys Rev A 34: 2026
204. Prasad PN (1990) Proc 4th Int Conf on ultrastructure processing of ceramics, glass and composites, Uhlmann DR and Ulrich DR (eds) Wiley-Interscience, New York
205. Pope EJA and MacKenzie JD (1987) MRS Bull March 17: 29
206. Reisfeld R, Eyal M and Brusilovsky D (1988) Chem Phys Lett 153: 211
207. Kobayashi Y, Imai Y and Kurokawa Y (1988) J Non-Cryst Solids 105: 198
208. Levy D and Anvir D (1988) J Phys Chem 92: 4734
209. Pouxviel JC, Dunn B and Zink JI (1989) J Phys Chem 93: 2143
210. Locher R, Renn A and Wild VP (1989) Chem Phys Lett 138: 405
211. Tanaka H, Takahashi J, Tsuchiya J, Kobayashi J and Kurokawa J (1989) J Non-Cryst Solids 109: 164
212. Chernyak V, Reisfeld R, Gvishi R and Venezky D (1990) Sensors and Materials
213. Hench LL and Wilson MJR (1990) J Non-Cryst Solids 121: 234

214. Hench LL, Wang SH and Nogues JL (1988) Multifunctional materials SPIE 878: 76
215. Hench LL, Wilson MJR, Balaban C and Nogue's JL (1990) in Proc 4th Int Conf on ultrastructure processing, Tuscon AZ, Uhlmann D, Weinberg M and Ulrich DR (eds) Wiley, New York
216. Phillips NJ (1990) in Proc Symp on Multi-functionals, Ulrich DR, Karasz FE, Buckley AJ and Gallaghar-Daggit G (eds) Mat Res Soc, Pittsburgh
217. Segal D (1989) Chem in Britain pp 153–156
218. Segal D (1989) Chemical Synthesis of Advanced Ceramic Materials, Cambridge CUP
219. Lappert MF, Power PP, Sanger AR and Srivastava RC (1980) 'Metal and Metalloid Amides', Ellis Horwood, Chichester
220. Maya L (1986) Inorg Chem 25: 4213
221. Maya L (1987) Inorg Chem 26: 1459
222. Brown GM and Maya L (1988) J Am Ceram Soc 71: 78
223. Sugiyama K, Pac S, Takahashi Y and Motojima S (1975) J Electrochem Soc 122: 1545
224. Dyagileva LM, Tsyganova EI, Gaidym TL and Aleksandrov YuA (1984) J Gen Chem USSR (Engl Transl) 54: 538
225. Kuramato N and Taniguchi H (1984) J Mater Sci Lett 3: 471
226. Interrante LV, Sigel GA, Garbauskas M, Hejna C and Slack GA (1989) Inorg Chem 28: 252
227. Janik JF, Deusler EN and Paine RT (1987) Inorg Chem 26: 4341
228. Paine RT, Janik JF and Narula C (1988) Mat Res Soc Symp Proc 121: 461
229. Bergstrom L and Pugh RJ (1989) J Am Ceram Soc 72: 103
230. Paine RT and Narula CK (1990) Chem Rev 90: 73
231. Narula CK, Paine RT and Schaeffer R (1988) ACS Symp Ser 360: 378
232. Narula CK, Schaeffer R, Paine RT, Datye AK and Hammetter WF (1987) J Am Chem Soc 109: 5556
233. Paine RT, Narula CK, Schaeffer R and Datye AK (1989) Chem Mater 1: 486
234. Narula CK, Schaeffer R, Paine RT and Datye AK (1989) Inorg Chem 28: 4053
235. Mehrotra RC and Bajaj P (1970) J Organometal Chem 24: 611 and 25: 359
236. Mehrotra RC and Mehrotra RK (1962) J Ind Chem Soc 39: 677
237. Mehrotra RC and Chandra G (1965) Ind J Chem 11: 497
238. Gaur DP, Srivastava G and Mehrotra RC (1974) J Organometal Chem 65: 195
239. Arai T and Kikuchi M (1984) Appl Optics 23: 3017
240. Saito M and Takizawa M (1986) J Appl Phys 59: 1450
241. Tanaka K, Imai Y and Odajina A (1985) J Appl Phys 57: 4897
242. Mehrotra RC, Gupta VD and Sukhani D (1968) Inorg Chim Acta Rev 2: 111
243. Nesmeyanov AN et al. (1961) Izv Akad Nauk SSR: 2146
244. Fitzer E and Schlichling (1980) High Temperature Science 13: 149
245. Czekaj CL, Rau MS, Geoffroy GL, Guiton TA and Pantano CG (1988) Inorg Chem 27: 3267
246. Guiton TA, Czekaj CL, Rau MS, Geoffroy GL and Pantano CG (1988) Mat Res Soc Symp Proc 121: 503
247. Guiton TA, Czekaj CL and Pantano CG (1990) J Non-Cryst Solids 121: 7
248. Guiton TA and Pantano CG (1988) Mat Res Soc Symp Proc 121: 509
249. Mehrotra RC (1989) Ref 18, pp 421–431
250. Mackenzie JD (1982) J Non-Cryst Solids 48: 1
251. Simonton TC, Roy R, Komarneni S and Breval E (1986) J Mater Res 1: 667
252. Carda J, Monros G, Escribano Pand Alarcon J (1989) J Am Ceram Soc 72: 160
253. Brinker CJ, Hurd AJ, Frye GC, Ward KJ and Ashby CS (1990) J Non-Cryst Solids 121: 294
254. Minehan WT and Messing GL (1990) J Non-Cryst Solids 121: 375
255. Gesser HD and Goswami PC (1989) Chem Rev 89: 765
256. Fujiu T, Messing GL and Huebner W (1990) J Am Ceram Soc 73: 85
257. Hench LL (1980) Science 208: 827
258. Hench LL and Ethridge EC (1982) 'Biomaterials—An Interfacial Approach', Academic, New York
259. Hulbert SF et al. (1973) Biomed Mater Symp 4: 137
260. Hench LL and Ethridge EC (1975) Adv Biomed Eng 5: 35
261. Weiner S (1986) CRC Critical Reviews in Biochemistry 20: 365
262. Tarasevich BJ, Liu J, Sarakaya M and Aksay M (1988) Mat Res Soc Symp 121: 225
263. Bose A and Bhandarkar S (1990) Better ceramics through chemistry IV Mater Res Soc (April 16–20) San Francisco
264. Voronkov MG, Zelchan GI and Lukevits (1978) 'Silicon and Life' Riga Zinatne

265. Birchall JD (1978) in 'New Trends in Bio-Inorganic Chemistry', Williams RJP, Dasilva JRRF (eds) Academic, New York
266. Williams RJP (1986) 'Silicon Biochemistry' CIBA Foundation Symposium 121; Wiley, New York
267. Hench LL (1989) J Biomed Mater Res 23: 685

Aerogels—Preparation, Properties, Applications

Jochen Fricke and Andreas Emmerling

Physikalisches Institut der Universität, Am Hubland, D-8700 Würzburg, FRG

About 1000 papers on aerogels have been published so far, with nearly 200 now being added per year. These numbers demonstrate the attraction which these low density materials exert on physicists, chemists and material scientists. With respect to basic science the (fractal) structure, the dynamics, the low temperature thermal properties and the interaction of guest molecules with the aerogel skeleton are of special interest. Applications of aerogels as low-n-material in Cerenkov detectors, as acoustic impedance matching layer, as catalytic substrate, as gas filter or as thermal insulant require the detailed knowledge of structure as well as of optical, thermal and acoustical properties; furthermore the technical utilization of aerogels requires the art of modifying the structural build-up, the porosity and composition in a controlled way.

1 Introduction

Aerogels [1, 2, 2a, 2b] are sol-gel derived solid materials with porosities from about 80% to 99.8%. The high porosity is achieved via supercritical drying of an alcogel or a hydrogel in an autoclave. Aerogels were first made by S. S. Kistler in the early thirties [3]. In his pioneer work Kistler demonstrated that aerogels in granular or monolithic form are excellent thermal insulants [4].

Progress in aerogel production was stimulated by prospects of using this material as a gelling agent for rocket propellants [5] or as catalyst [6]. Tiles of transparent monolithic silica aerogels for use in high energy Cerenkov particle detectors were produced in the late 1970s and the early 1980s [7–9]. The idea of employing aerogels as transparent cover in solar collectors and for windows was patented in 1982 [10].

Since that time interest in aerogels has grown progressively (Fig. 1). The total number of papers on this subject so far is about 1000. Interesting applications such as the production of pure glass from aerogels [11, 12], the use of granular aerogel [13] for passive solar usage [14], the substitution of CFC-blown polyurethanes by opacified aerogel powder [15], the production of fine metal powders [16] or of acoustic impedance matching layers [17] have attracted numerous physicists, chemists and engineers as well as R & D money. However, basic research on aerogels has also experienced an upsurge: scientists started to make and investigate aerogels with fractal structure [18, 19], to derive the density of states of aerogels [20, 21] or to measure their low temperature thermal conductivity or heat capacity [22]. Also spectroscopists became interested in aerogels, which they probed for Rayleigh scattering [23] or — if doped with fluorescent dyes — for decay lifetimes and spectral bandwidths [24, 25].

Kistler stated in his first papers [3], that aerogels can be fabricated not only from silica, but also from tungstic oxide, ferric and stannic oxide, nickel tartrate, cellulose, gelatine, agar, egg albumin and rubber. Kistler saw no reason why this

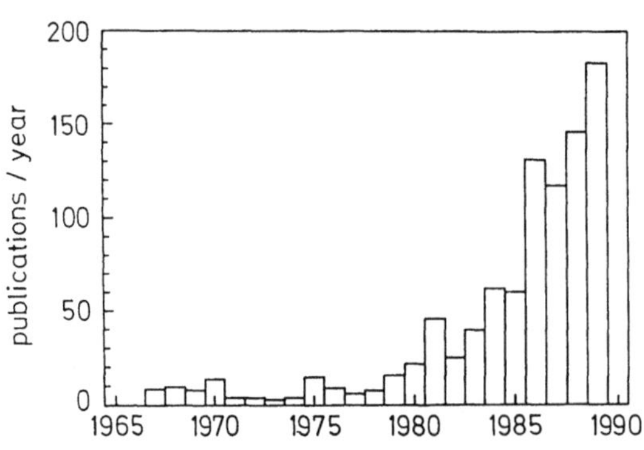

Fig. 1. The number of publications on aerogels produced annually since 1965

list should not be extendable indefinitely. Many research groups today are pursuing this idea.

2 Preparation

2.1 Sol-Gel Process

As numerous papers and books on sol-gel technology have been published [26 to 37], we want to restrict ourselves to a brief description of the sol-gel process, in order to understand the properties of the resulting aerogels.

The sol-gel process denotes the transition of a system of colloidal particles in a solution into a disordered, branched, continuous network, which is interpenetrated by a liquid. This gelation process can take place

- by the aggregation of colloidal particles via destabilizing the sol [38],
- by hydrolysis and polycondensation of organometallic precursors [27],
- by polycondensation of waterglass [27].

The second route is used most frequently for the preparation of aerogels, since it allows the most convenient variation of reaction parameters. Using this route, aerogels of main group oxides, transition and semimetal oxides, as well as organic compounds, can be produced:

Single oxide systems such as SiO_2 (see e.g. 1, 2), Al_2O_3 [39–41], TiO_2 [39], B_2O_3 [42], MoO_2 [43], MgO [43], ZrO_2 [44], SnO_2 [45], WO_3 [45], N_2O_5 [46], Cr_2O_3 [47], metal-on-aerogel materials (for catalytic purposes) such as Cu/Al_2O_3 [39], Ni/Al_2O_3 [39], Pd/Al_2O_3 [48], binary systems such as Al_2O_3/SiO_2 [49], B_2O_3/SiO_2 [50], P_2O_5/SiO_2 [50], Nb_2O_5/SiO_2 [45, 46], Dy_2O_3/SiO_2 [45], Er_2O_3/SiO_2 [45], Lu_2O_3 [45], CuO/Al_2O_3 [51], NiO/Al_2O_3 [51], PbO/Al_2O_3 [51], Cr_2O_3, Al_2O_3 [51], Fe_2O_3/Al_2O_3 [39], Fe_2O_3/SiO_2 [39], Li_2O/B_2O_3 [42], ternary systems such as $CuO/ZnO/ZrO_2$ [43], $CuO/ZnO/Al_2O_3$ [43], $B_2O_3/P_2O_5/SiO_2$ [50], $MgO/Al_2O_3/SiO_2$ [52], $Pb/Mg_{1/3}Nb_{2/3}O_3$ [53] and even the high-T_c superconductor $YBa_2Cu_3O_{7-x}$ have been prepared via aerogels [54]; pure metal powders can also be made from aerogels [16].

One of the most exciting recent achievements was the production of organic aerogels and of carbon aerogels [55–57]. Further organic or inorganic compounds or mixtures of both, which can be gelled [26–37] may possibly be transformed into aerogels via supercritical point drying.

In order to describe the principal features of the sol-gel process starting from organometallic precursors, we mainly consider silica aerogels, the most thoroughly studied system. The first reaction step is the hydrolysis of a liquid silicon alkoxide by addition of water and a catalyst:

$$Si(OR)_4 + 4H_2O \rightarrow Si(OH)_4 + 4ROH, \tag{1}$$

whereby R denotes an alkyl group (CH_3, C_2H_5, C_3H_7, . . .). The silicic acid condenses by expelling water:

$$2\,Si(OH)_4 \rightarrow (OH)_3Si\text{–}O\text{–}Si(OH)_3 + H_2O. \quad (2)$$

Further polycondensation reactions lead to additional (≡Si–O–Si≡) linkages. Finally a network of SiO_4-tetrahedra exists and this will be described later. The hydrolysis of the alkoxide need not be completed before condensation starts. On the other hand, hydrolysis can still take place at unhydrolyzed sites of partially condensed silica. The reaction rates of hydrolysis and condensation are most important to control the structural growth process. Though hydrolysis is of course influenced by the water-to-alkoxide ratio ($w = [H_2O]/[Si(OR)_4]$), both reactions are catalyzed by acids or bases (Fig. 2). The hydrolysis rate increases strongly both for low and high pH-values. The condensation rate shows a distinct minimum at the isoelectric point of silica ($pH \approx 2$)—where the silicate species are uncharged [38]; under neutral and basic conditions the condensation rate is large [59].

Nucleophilic substitution with inversion of one of the reaction species in base-catalysis and electrophilic substitution in acid-catalysis is supposed to be the dominant mechanism for hydrolysis and condensation [60]. This leads to the following conclusions concerning the primary structure of the gel:

- Under acidic conditions, hydrolysis is completed before condensation starts (except at very low pH). The slow condensation reaction of the silicic acid monomers leads to the formation of polymer-like silica chains, resulting in a weakly cross-linked, low density network [61].
- Under basic conditions, hydrolysis is the limiting step, i.e. whenever a monomer is hydrolyzed, it condenses rapidly. This behavior leads to the

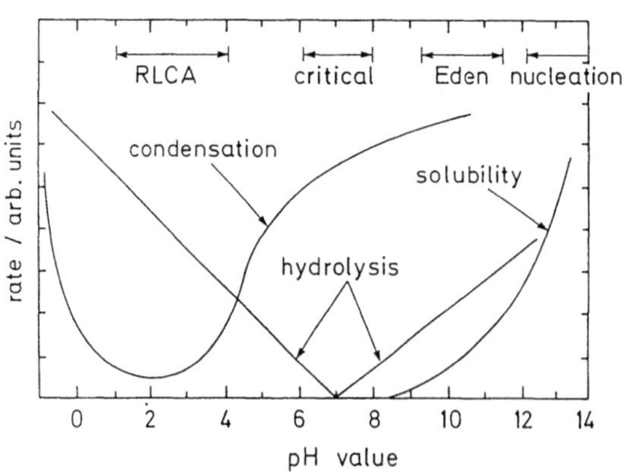

Fig. 2. pH-dependence of the reaction rates of hydrolysis and condensation; in addition the solubility of silica, assuming the latter to be the same for alcoholic and aqueous solutions, is shown. The different growth models and their range of applicability (upper part of the diagram) are explained in Sect. 3.2 [58]

growth of comparatively dense, colloidal particles [61]; their final size depends on the temperature and on the pH-value of the solution [38].

Generally it is assumed that once a siloxane bond (≡Si–O–Si≡) is formed, it will not dissociate. This is true for the pH-range below 8.5, where silica is practically insoluble. As shown in Fig. 2, however, the solubility strongly increases at higher pH values. The resulting dissolution and reprecipitation processes [27, 38] do not only influence the initial steps of structural growth but also later the aging and drying of the gel.

Irrespective of their substructure—low density polymerlike or high density colloidal—the sol particles aggregate and form clusters, which finally constitute the gel network (Fig. 3). The point of gelation is reached, when a continuous cluster extends throughout the vessel. The viscosity then increases sharply [62]. The gelation time is determined firstly by the reaction rates of hydrolysis and condensation, which depend both on the amount of catalyst as well as on temperature [63], and secondly by the degree of dilution, i.e. the amount of excess alcohol. The latter controls the aggregation rate and the final apparent density of the gel.

To prepare ultra-low-density aerogels, a highly diluted starting solution is required; however, a minimal alkoxide concentration is necessary (depending on pH) to enable the formation of a continuous network at all. This can in principle be achieved with acid catalysis; such gels, however, shrink dramatically upon aging and drying, due to their weakly crosslinked texture. Therefore, a two-step method has been introduced by Brinker et al. [64]: pre-hydrolysis with a sub-stoichiometric amount of water under acidic conditions, which leads to small, partly condensed silica particles; then addition of water and a base catalyst follows. The gels obtained consist of small, but highly cross-linked structural subunits. This allows one to produce aerogels with densities as low as 4 kg/m^3

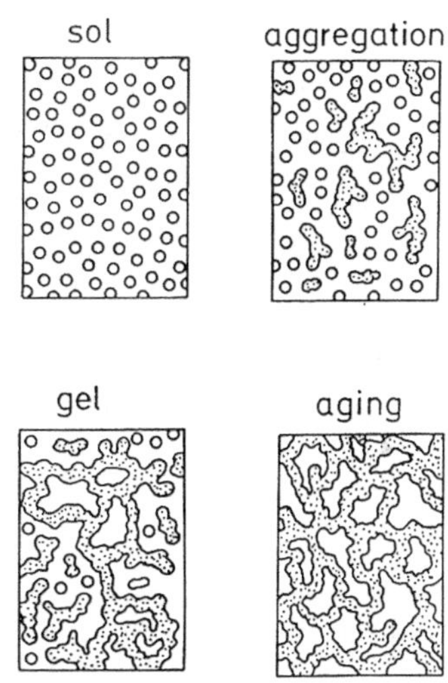

Fig. 3. Schematic representation of gel evolution [38]

[65, 66]. In addition, partially condensed alkoxysilanes may also be used as aerogel adhesives [67].

Reaching the gel point is not tantamount to the termination of the sol-gel process; after gelation the gel-body ages [68]: Sol particles and small clusters in the solution attach to the spanning cluster and extend the gel network (Fig. 3). Moreover, all the condensation reactions (Eq. 2) that caused gelation, continue; this holds especially for two neighbouring silanol groups (or one silanol and one ≡Si–OR group). If the condensing groups are located on neighbouring surfaces (Fig. 4), shrinkage and the expulsion of water (or alcohol) can take place. This process is called syneresis [68]. Shrinkage is particularly extensive for the flexible polymerlike network of acid-catalyzed gels, which can thus be prepared with bulk densities as high as 500 kg/m^3 [69]. Additional strengthening or stiffening of the network is produced by Ostwald ripening [38]. This irreversible process of dissolution and reprecipitation is caused by different solubilities of surfaces with different curvature: Molecules or small particles located at convex surfaces are more solvable compared to those at concave surfaces. As a result, larger particles grow at the expense of smaller ones and necks between the particles are filled up. Thus a narrowing of the particle size distribution with a larger mean particle diameter and a decrease in specific surface area, but no shrinkage is observed (Fig. 5). This mechanism, which is enhanced by increasing pH, temperature or pressure, proceeds continuously, leading to a coarsening of the gel structure. Finally a disintegration of the network may occur.

It is obvious that the sol-gel process allows us to tailor nano-structured materials by choosing the appropriate reaction parameters. In addition an ultrasonic treatment can be applied [70–72]. The sol-gel process also provides the possibility of doping gels with molecular compounds. This can be performed either chemically, by using suitable alkoxide derivates, or physically, by including the additive in the porous network. Examples are the integration of

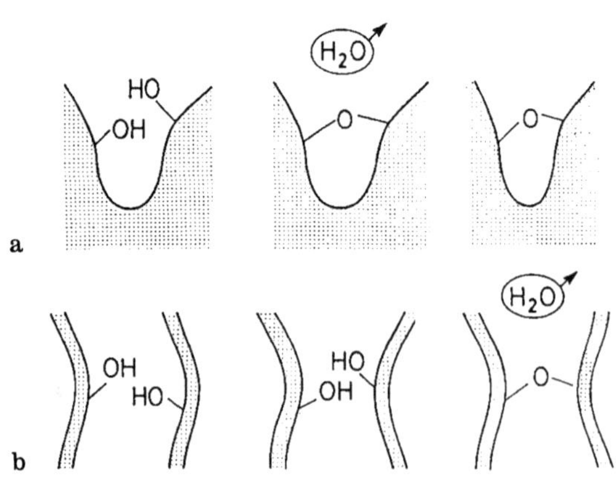

Fig. 4a. Syneresis is caused by condensation between silanol groups on neighbouring surfaces which results in shrinkage and expulsion of pore liquid. **b**. This is especially evident for a network of flexible chains, typical for acid-catalyzed gels [68]

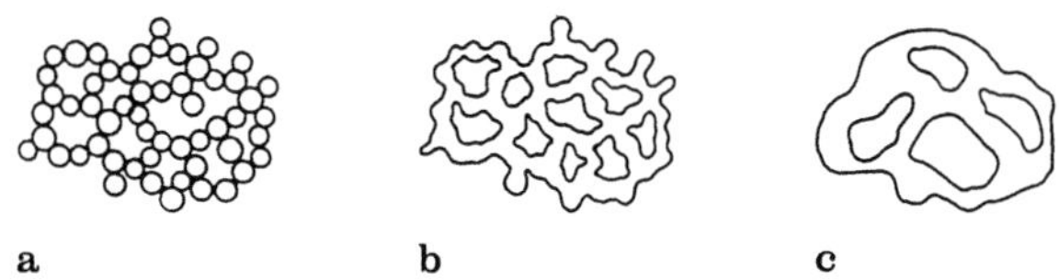

Fig. 5a–c. Continuous aging of the original gel body (**a**) leads to an increasing coalescence of particles (**b**), and a coarsening of the network structure (**c**) [26]

hydrophobic alkyl groups to improve the water resistance of aerogels [73] and the inclusion of soot to enlarge the infrared extinction [74]. Last but not least, aerogels with their tenuous disordered structure can serve as a host for fluorescent dyes [75].

2.2 Drying

The gel, still embedded in the solvent, which consists mainly of alcohol with some water and catalyst, is called an alcogel. To remove the liquid, several extraction methods can be employed [76]. Normal drying at ambient conditions leads to capillary tensions as the vapor/liquid interface retreats into the porous structure (Fig. 6). The resulting shrinkage (and cracking) of the gel network proceeds until the densified structure is able to withstand these forces. Depending on the aging, the structure of the dry "xerogel" consists of a more or less densely packed arrangement of gel particles or network chains; the remaining porosity amounts to about 50%. Shrinkage can be reduced by prolonged aging [77], by aging with additional catalyst [78], or by using chemical additives (DCCAs) [79].

To avoid shrinkage and cracking, i.e. to preserve the structure of the wet state, the build-up of capillary forces has to be prevented. As already discussed by Kistler [80], this can be achieved, if the liquid-vapor phase transition is

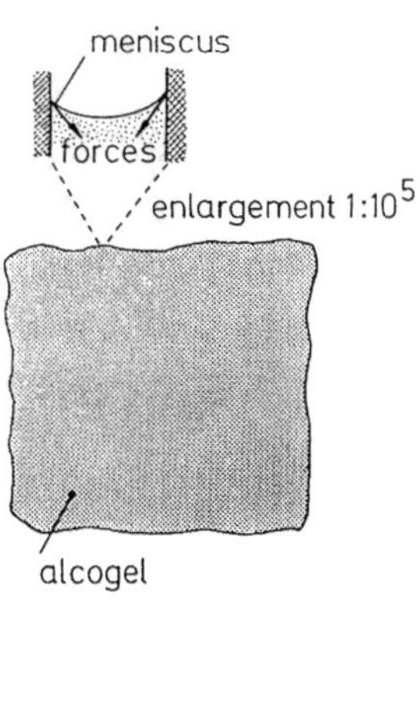

Fig. 6. Schematic representation of the menisci appearing on subcritical drying. The tensions originating from the vapor/liquid surfaces cause the fragile gel network to shrink and/or crack

avoided by removing the solvent supercritically [81, 82]. The typical path followed during supercritical drying (SCD), which has become an important method in ceramic processing [83], is shown in the phase diagram of Fig. 7. Hereby the alcogel is put in a pressure vessel (an autoclave). Then the pressure and the temperature are raised above the critical point of the solvent. As in a supercritical fluid, liquid and vapor phase become indistinguishable, no capillary forces occur. After releasing the fluid through the outlet valve and subsequent cooling, the aerogel can be taken from the autoclave. It is important that enough solvent is provided in order to guarantee supercritical conditions throughout the whole drying process; otherwise shrinking and cracking will occur [84]. Often the autoclave is pre-pressurized with nitrogen to prevent premature evaporation of the solvent [85].

Today large pieces of monolithic aerogels can be produced by SCD. They have mostly been employed in Cerenkov detectors [7–9, 86–88], but could also be used as transparent insulation in solar architecture [14, 89–91], or as precursors for the production of high purity glass [11, 12, 92].

The high temperature alcohol drying process (methanol: $T_c = 239.4\,°C$, $p_c = 80.9$ bar) is not free from risk. This can be learned from the explosion in Lund/Sweden in 1984. The main gasket of a 3000 l-autoclave failed and the escaping methanol vapor instantly caught fire [89].

Moreover, SCD, with respect to alcohol, may influence the gel structure as aging does. In the presence of water and (basic) catalyst, the heating phase of the autoclave process leads to Ostwald ripening and coarsening. Consequently a decrease of specific surface area [93], a narrowing of the particle size distribution [94] and a stiffening of the network [95] occur.

On the other hand, SCD can be performed safely near ambient temperatures [96, 97], if the pore liquid is exchanged for a liquid with a lower critical

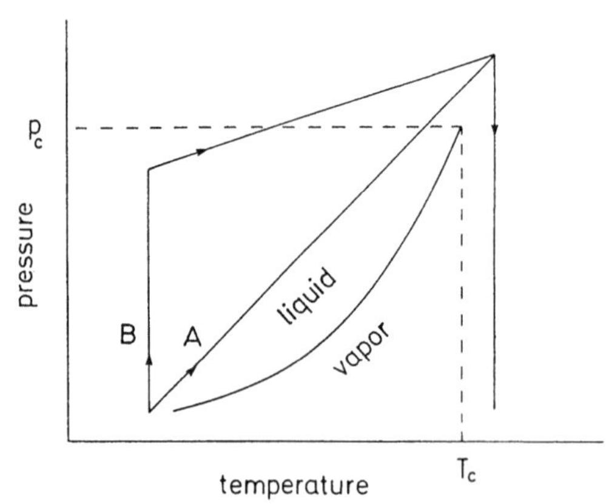

Fig. 7. Temperature-pressure paths during supercritical drying. The conventional path (**A**) makes use of a totally filled autoclave; (**B**) corresponds to the case of nitrogen prepressure. Above the critical point of the solvent (T_c, p_c) capillary tension ceases [76]

temperature, as e.g. carbon dioxide ($T_c = 31.0\,°C$, $P_c = 73.9$ bar). Depending on sample size, a complete solvent exchange, however, may require a long time. Here, a pre-exchange with an intermediate solvent can be helpful; this is particularly important, if components of the solvent and the liquid used for SCD are not miscible, as for instance water and CO_2 [42, 98]. The aerogels produced according to the low-temperature CO_2-route basically retain their wet state structure [94]; their hydrophilic surface, however, attracts ambient moisture, which can cause cracks and a milky appearance long after the SCD procedure [99]. In contrast, alcohol-dried aerogels are rather hydrophobic as re-esterification of silanol groups occurs at elevated temperatures [92, 99].

The CO_2 process is also used for the production of granular aerogels from water glass [13]. Such aerogels have a large potential for application in transparently insulated house walls [14, 100] and as substitutes for CFC-blown polyurethane foams [15].

Attempts have also been made to remove the solvent by sublimation after freezing it [101, 102]. These freeze-dried "cryogels" show a porosity of 80% at maximum [101] and only half the inner surface of a comparable aerogel [102]. Up to now monolithic samples could not be prepared. Special precautions for freeze-drying are long aging periods to stabilize the gel body, a solvent exchange to provide a low expansion coefficient and high sublimation pressure [101], and the addition of salts to achieve low freezing rates and freezing temperatures [103].

2.3 *Modifications*

As mentioned above, aerogels can be transformed into dense glass at relatively low temperatures [50, 52, 99, 104–109]. The advantage of this procedure compared to a conventional melt process is not only the lower temperature required, but also the higher degree of purity and homogeneity (in the case of multicomponent systems) that can be achieved [11, 12, 59, 106, 110]. Below temperatures at which densification takes place some weight loss occurs: physisorbed water molecules are rapidly evaporated at temperatures between 100 °C and 200 °C; above approximately 300 °C combustion of unhydrolyzed alkoxy groups starts. This process results in an active surface and the formation of new siloxane bonds. Further crosslinking occurs via condensation of neighbouring silanol groups, which is easy to observe by employing infrared spectroscopy [98, 99, 106, 109, 111–113]. Figure 8 shows the evolution of the spectral transmission of a base-catalyzed aerogel upon heating in air. The broad peak of adsorbed water around $\lambda \approx 2.9\,\mu m$ vanishes; the two absorption peaks at $\lambda \approx 3.4\,\mu m$ and $\lambda \approx 3.55\,\mu m$, which belong to methoxy groups in this case, receed. Simultaneously new silanol groups are formed, which is indicated by the intensification of the sharp hydroxyl absorption peak at $\lambda \approx 2.67\,\mu m$. After this oxidation heat treatment (baking), the network has become stiffer [95, 114] and the specific surface is slightly decreased [93] and smoothed out [115, 116].

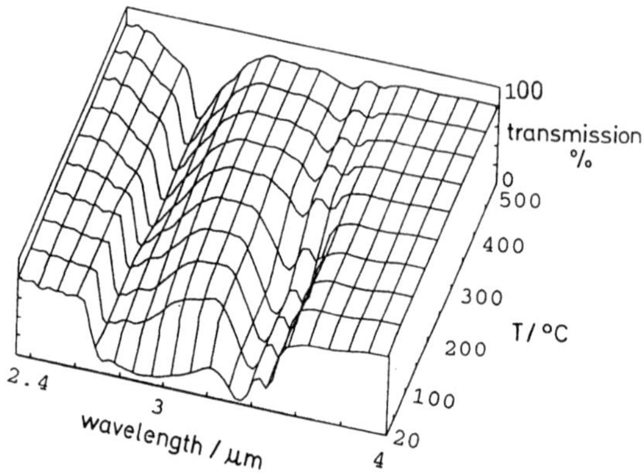

Fig. 8. Evolution of the IR-transmission spectrum upon baking for a base-catalyzed aerogel with initial density of 165 kg/m^3 in air. As organic methoxy groups are cracked (λ = 3.4 μm, λ = 3.53 μm) additional silanol groups (λ = 2.68 μm) appear. Adsorbed water ($\lambda \approx$ 2.8–3.0 μm) is already removed far below 300 °C [112]

The densification via sintering starts at higher temperatures—depending on the chemical composition of the aerogel. This proces is driven by the interfacial energy between solid and vapor phase. Material transport can in principle be realized both by diffusion or viscous flow, whereas the latter is regarded as the predominant mechanism for amorphous materials [117]. Hereby the inner surface area as well as the porosity is decreased. Sintering of aerogels has not only been described for silica systems [11, 12, 19, 71, 95, 98, 99, 104–109, 116, 118, 119], but also for non-silica or mixed-oxide systems [45, 46, 50, 52, 53, 120, 121]. Densification studies have been performed using either a constant heating rate [104, 105], or a constant sintering temperature [116, 119]. Generally a coalescence of particles and a reduction of pore sizes is observed [19, 107, 119]. This leads to a continuous stiffening of the gel network upon densification [95, 122–124]. Sintering of aerogels is controlled by a variety of parameters [99, 117]: the extent of hydrolysis and the amount and type of catalyst used in sol-gel processing, the initial density and the kind of heat treatment, as well as the content of alkalications and water. The textural evolution of a typical aerogel is represented in Fig. 9.

The kinetics of sintering [117] can be considered by means of simple geometric models. The first approach introduced by Scherer [125] idealizes the gel structure by a rectangular lattice of intersecting cylinders with uniform cell sizes (Fig. 10). Other geometries show qualitatively similar kinetics, indicating only a minor influence of the actual microstructure on sintering. With these models densification can be described on a dimensionless time scale. The time to achieve a certain density is controlled by the initial cell (pore) size, the specific surface energy and the viscosity, whereby the latter is strongly dependent on both sintering temperature and content of surface silanol groups [117, 126, 127]. According to an extension of these models, which also includes calculations on

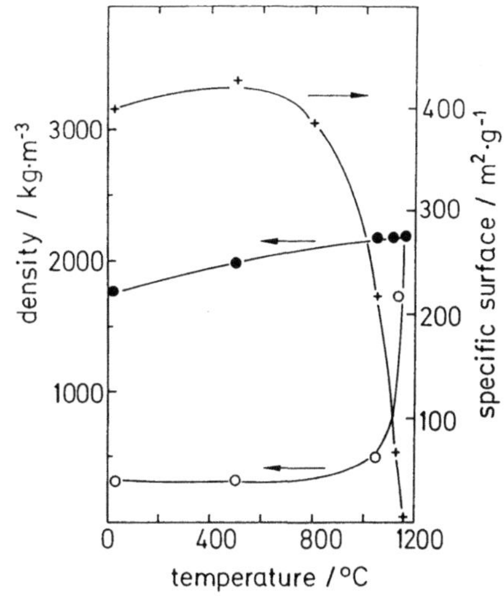

Fig. 9. Evolution of bulk density (○), skeletal density (●) and specific surface area (+) as a function of temperature for a typical aerogel upon densification via constant rate heating. The small increase in the BET-derived specific surface area and in the pycnometrically obtained skeletal density after oxidation heat treatment is due to a better accessibility of the nanoporosity for nitrogen molecules and helium atoms, respectively. Above 1000 °C rapid sintering with a drastic reduction of porosity and internal surface is observed; the final densification of the skeleton is attributed to structural relaxation [106]

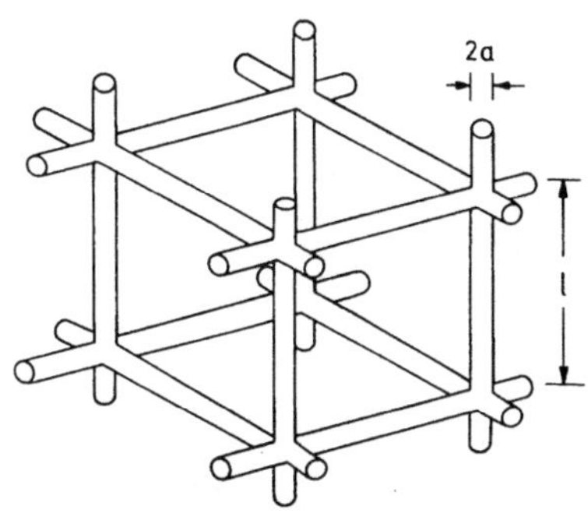

Fig. 10. Cylinder model consisting of a rectangular lattice of intersecting cylinders, drawn for a radius to length ratio x = 1/20

pore size distributions [128], sintering of micropores is expected to occur much faster than larger pores.

The predicted density variation could be confirmed in case of xerogels (see e.g. 129) and aerogels [118, 124]. Some measurements have shown indication for a preferential sintering of larger pores [130, 131]. A recent investigation of isothermal sintering of base-catalyzed aerogels [124] clearly demonstrates a fast collapse of the macroporous network and a slow continuous coalescence of the microporous particles. This becomes especially evident, if the specific surface as a function of the pore volume is considered (Fig. 11).

Partially densified aerogels seem to be becoming important for applications. As in isothermal sintering, where the structural changes upon densification depend only on the achieved density [124], the controlled preparation of aerogels with well defined properties is possible. Since the aerogel texture is strengthened and thus able to withstand greater tensions, it can be filled with liquids without destruction of the tenuous network [45, 132]. This is intriguing with respect to liquid phase catalysis. Densified aerogels can thus be doped and

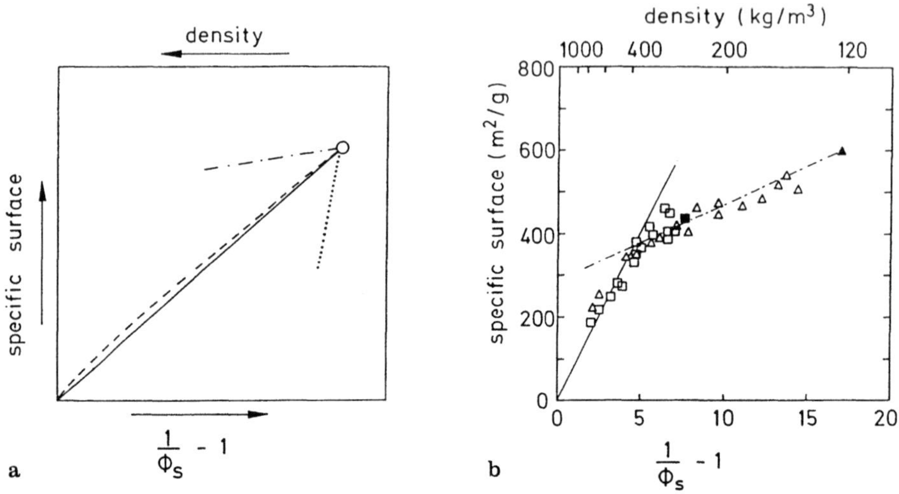

Fig. 11a. Different "sintering scenarios" can easily be distinguished by plotting the specific surface (s/m) versus the quantity $(1/\phi_s - 1)$. The latter is proportional to the pore volume; ϕ_s denotes the volume fraction of the solid phase. The specific surface area shows a characteristic decrease with increasing density, if sintering either depletes structural entities at all scale lengths (———), or leads to uniform shrinkage (-----); a predominant reduction of the micro- or macrostructure results in a significantly stronger (.....), or weaker (–·–·–) decrease of the specific inner surface, respectively. **b.** Sintering behaviour for two base-catalyzed aerogels with initial densities of 122 kg/m³ (△) and 256 kg/m³ (□); the *filled symbols* (▲, ■) mark the two original samples. Up to a density of about 500 kg/m³, densification essentially proceeds via reduction of the macroporous network structure. Subsequently, a uniform shrinkage of all remaining structural entities occurs [125]

serve as a porous host material [45, 106]. Many host–guest interactions have been investigated [24, 25, 132–138].

Doping can also be performed in the gas phase, which is even applicable to lower density, non-sintered aerogels. This allows one to improve hydrophobization by using vapors of dimethyldichlorsilane [139], or to stain aerogels with fluorescent dyes, as described by Haarer's group [25].

Pyrolysis of organic aerogels [140] allows the production of carbon aerogels [141]. As resorcinol–formaldehyde aerogels are highly crosslinked aromatic substances, the structure of the resulting carbon aerogel is similar to that of its precursor. Recently pyrolysis has also been used for the preparation of boron nitride aerogels from its polyborazinyl amine precursor [142].

3 Structure

3.1 Experimental Methods

Before discussing the aerogel structure itself, a brief survey is given of methods commonly used to reveal structural features [143]. Generally three basic techniques can be distinguished:

a) adsorption or penetration measurements (BET, pycnometry, porometry):

BET-adsorption/desorption [144], usually performed with liquid nitrogen, is a wide-spread technique for determining the specific surface area of porous materials. It is worthwhile mentioning the fact that only such surfaces that are accessible to the N_2 molecules are detected. Interesting in this respect would be the use of He, however, no such measurements have come to our knowledge so far.

The Kelvin equation correlates the curvature of the liquid surface with the applied partial pressure [145]. It thus allows us to derive the pore size distribution. While this method has been successfully used for xerogels, it fails for aerogels. The total pore volumes, determined for aerogels with $\rho \approx 100\,\mathrm{kg\,m^{-3}}$, are in the order of $2\,\mathrm{cm^3/g}$, instead of the expected value of about $10\,\mathrm{cm^3/g}$ [146]. Obviously the largest pores, which contribute most to the total pore volume, remain undetected by adsorption techniques.

Helium is preferably used for pycnometry. The sample is pressurized with helium and a subsequent expansion of this gas into a precisely defined volume results in a pressure drop. From the pressure measurements the skeletal volume of the sample, i.e. the volume occupied by the solid phase, and thus the skeletal density can be determined. Skeletal densities of silica aerogels obtained by He-pycnometry were found in the range of 1700 to $2100\,\mathrm{kg/m^3}$ [13, 147]. This indicates a porosity of the aerogel backbone, which is not accessible to the He atoms, in the order of 5%–20%.

Mercury is commonly used for porometry, which is often employed to determine the pore volume [11, 13, 39, 43, 102, 148]. Since Hg is a non-wetting liquid, it only penetrates the porous body, if pressure is applied. The higher the pressure, the smaller the pores, which can be filled up. For cylindrical pores the Washburn equation [149] allows one to determine the pore size distribution. Due to the low compressive modulus of aerogels, at low Hg pressures only compression of the body as whole occurs, instead of penetration [150]. Hence, pores much larger than those present in reality are being "derived". Non-existent pores as large as 10 μm [11, 13, 148] have been termed ultraporosity [151]. At elevated pressures the aerogel body is penetrated, possibly with a destruction of its delicate structure. The smallest pores that are detectable by Hg-porometry are about 2 nm, depending on the maximum pressure used.

b) direct methods (electron microscopy):

SEM/TEM techniques are instructive, since they yield a direct image of the structure. Morphological features such as particle shapes, particle arrangements, etc., can be recognized. Quantitative estimates for pore or particle sizes can be obtained, though the derivation of size distributions is rather laborious, especially as usually only a two-dimensional projection is available. A special problem in the case of aerogels is the preparation of thin samples for TEM pictures, or the coating with Au or Cu for SEM. The major problem, however, is the (destructive) interaction between the impinging electron beam and the tenuous aerogel skeleton. Even for small electron beam currents the deposited energy may cause sintering. Thus TEM/SEM results in general have to be interpreted with caution [152, 153]. Recently, a low-voltage, high-resolution SEM apparatus with a field-emission gun has been developed [154]. It can be operated without coatings and drastically reduces the thermal load.

c) indirect methods (scattering techniques, magnetic resonance):

The most powerful and non-destructive methods for probing structures are scattering techniques. The typical arrangement of a scattering experiment is shown in Fig. 12. The specimen is irradiated by monoenergetic photons (light, X rays) or particles (neutrons). Part of this incoming primary beam is scattered as a result of the inhomogeneity of the sample. The amount of scattered intensity depends on both the concentration of the scattering units as well as the contrast between the solid phase and the voids. The cross section for scattering of visible light shows a Λ^{-4}-dependence, due to Rayleigh scattering from inhomogeneities, which are smaller than the wavelength Λ; the atomic scattering cross sections for X rays and neutrons are in principle independent of wavelength. An important quantity for scattering experiments is the scattering wavevector

$$\vec{q} = \vec{k}' - \vec{k}. \tag{1}$$

k and k' are the wave vectors of the impinging and the scattered photon (neutron), respectively. In case of elastic scattering the absolute values of $\vec{k}$ and $\vec{k}'$ are unchanged ($|\vec{k}| = |\vec{k}'| = 2\pi/\Lambda$). If, in addition, the structure is statistically isotropic, the scattering intensity depends only on the absolute value of the scattering vector:

$$q = \frac{4\pi}{\Lambda} \sin\frac{\theta}{2}, \tag{2}$$

whereby θ denotes the scattering angle. The structure in real space and the scattered intensity observed in q-space are connected by a Fourier transformation; however, a simple back transformation cannot be performed, since

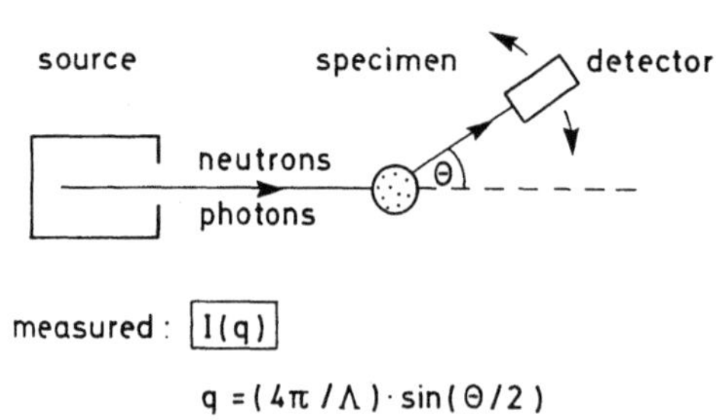

Fig. 12. Schematic representation of a scattering experiment. Due to inhomogeneities of the sample, part of the monochromatic, collimated primary beam of photons, or neutrons, is scattered. The scattered intensity is recorded either sequentially by scanning the angle with a point-like detector (as drawn), or parallelly, using a position-sensitive detector system

only the intensity, but not the phase shift of the scattered beam is measured. A general feature of scattering is the reciprocity behaviour: large structural entities show up at small q-values and vice versa.

In case of light scattering [155], using a customary He–Ne-laser ($\Lambda = 633$ nm), a q-range from 4×10^{-5} Å^{-1} to 2×10^{-3} Å^{-1} can be covered. Using Bragg's law

$$q\,L = 2\pi \tag{3}$$

to define a corresponding length scale L, structures from about 15 μm to 300 nm can be observed. Due to the fluctuation of dielectric constants in an inhomogeneous medium not only scattering, but also refraction occurs at the inner interfaces. Therefore, light scattering data are often hard to interpret and exact Mie-theory has to be applied [156]. Only on the assumption that the inhomogeneities are much smaller than the light wavelengths, refraction can be neglected and the simple Rayleigh-Gans-approximation applies [157]. The latter is always applicable for X-ray or neutron scattering, since the refractive indices are then close to 1.

Typical wavelengths are 1–2 Å in X-ray and 2–10 Å in neutron experiments; these wavelengths are smaller than the structural sizes of 1 to 100 nm to be probed. Thus, the scattered intensity has to be detected at small scattering angles (0.01°–10°) close to the primary beam. This technique is therefore called small angle scattering [158–161], abbreviated SAXS or SANS (depending on the kind of radiation being used). Commonly a q-range from 10^{-3} Å^{-1} to 1 Å^{-1} can be covered.

Figure 13 shows a typical scattering curve of a disordered material, consisting of a network of particles. The characteristics of the scattering pattern can be related to the different structural features on the respective scale length [162]. At smallest scattering vectors, corresponding to a scale length much larger than the inhomogeneities, a constant scattering intensity is observed. The absolute value is determined by number and size of the scattering units and by the contrast. Enlarging q, a scale length is reached, where the largest inhomogeneities are resolved. Since the phase shifts between scattered waves become larger, the intensity gradually drops. In a first order approach this decay, called Guinier scattering, can be described by a Gaussian profile [163]

$$I(q) \approx I(o)\exp\{-q^2 R^2/3\}. \tag{4}$$

The radius of gyration R can be used to estimate the size of the macropores or largest aggregates, if a special shape (sphere, cylinder, platelet) is assumed. Subsequently, a more or less extended linear decay is observed in the double-logarithmic plot. This behaviour can be explained in terms of fractals [164], considering the branched appearance of the aggregates. If the mass within an assumed sphere with radius R around the center of each aggregate increases with

$$M \propto R^D,\ D < 3, \tag{5}$$

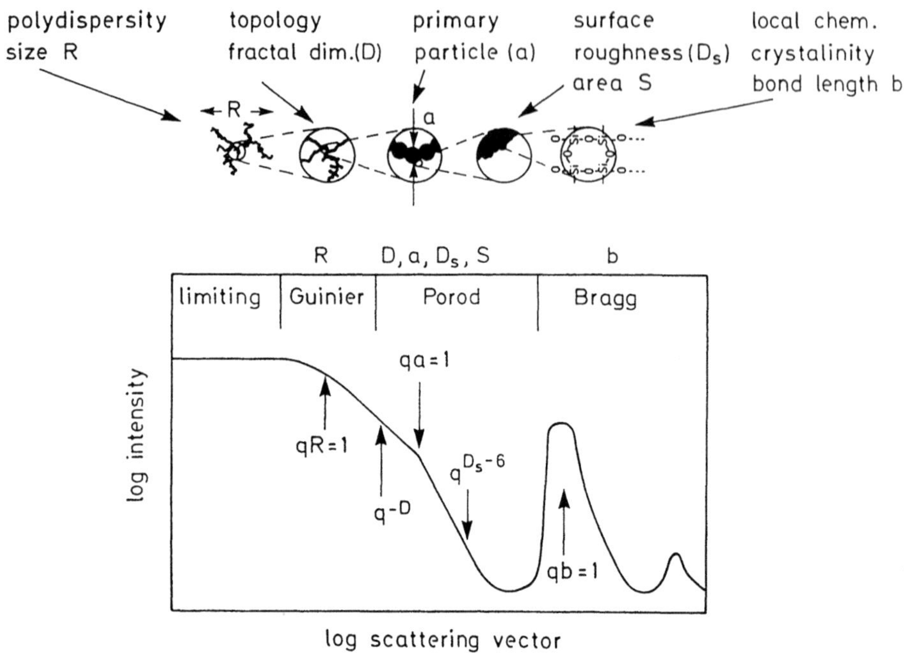

Fig. 13. Small angle scattering curve from a disordered material. All structural features at different scale lengths appear in the corresponding regions of scattering vector q. Several characteristic parameters can be obtained as described in text [162]

the intensity is also expected to obey a power law [165]

$$I \propto q^{-D}. \tag{6}$$

D is called the (mass) fractal dimension; below $D = 1$ no coherent structure can be formed. Within this fractal regime the structure looks the same at all lengths $R > L > a$; this is called self-similarity or dilation symmetry [164].

Upon increase of the scattering vector the scale length approaches the particle diameters. A second crossover is observed, from which the size a of these building blocks can be estimated by Guinier analysis. Hereafter, scattering becomes sensitive to the nature of the particle surfaces; again a power law decay of the intensity follows

$$I \propto q^{D_s-6}. \tag{7}$$

D_s is the surface fractal dimension, which equals 2, in the case of a smooth surface, where a decay proportional to q^{-4} is observed; this is known as Porod's law [166]. If the surface is fractally rough, it would scale with a non-integer power of the particle size a

$$S \propto a^{D_s}, \tag{8}$$

with $2 < D_s < 3$ [167]. Further increase in q resolves the structure on an atomic or molecular scale. Since these subunits show some short range order, broadened "Bragg" peaks are observed.

In addition to these values, which can be directly obtained from the scattering curve, further information becomes available, if a two-phase media model (TPM) is applied [168]. In this model the porous body is assumed to consist of two phases with constant mass (or electron) density: the solid phase with a mass density ρ_s, which occupies a volume fraction ϕ_s, and the voids ($\rho_p = 0$), which occupy a volume fraction $\phi_p = 1 - \phi_s$ (Fig. 14). ϕ_p and ϕ_s are identical with the porosity and relative density ρ/ρ_s, respectively. On the assumption of smooth interfaces, Porod scattering is valid for sufficiently large scattering vectors:

$$I = K\,q^{-4}. \tag{9}$$

Together with the q^2-weighted, integrated intensity

$$Q = \int_0^\infty I(q)\,q^2\,dq, \tag{10}$$

the specific surface area (of open and closed pores)

$$\frac{S}{m} = \pi \frac{\phi_p}{\rho_s} \frac{K}{Q} \tag{11}$$

can be derived. The latter represents the harmonic average over the mean chord length of each phase:

$$1/l_c = 1/l_s + 1/l_p \tag{12}$$

with

$$l_p = \langle l_1 \rangle = l_c/\phi_s \tag{13a}$$

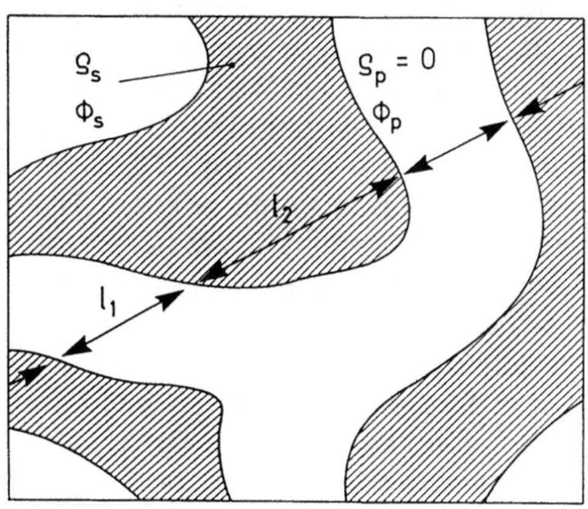

Fig. 14. Two phase media model: two phases of different, but within each phase constant density, are separated by smooth interfaces. If both volume fractions are known, the mean chord length $\langle l_1 \rangle$ and $\langle l_2 \rangle$ of each phase, as well as the specific inner surface, can be derived from small angle scattering experiments [168]

and

$$l_s = \langle l_2 \rangle = l_c/\phi_p. \tag{13b}$$

The brackets ⟨ ⟩ denote the average over all chords l_1 and l_2, cut out of the two regions by all lines crossing the system in arbitrary directions (see Fig. 14). For aerogels, l_p can be interpreted as a mean free path in the voids of the porous network, i.e. an effective pore size, and l_s as a mean diameter of the particles (or network chains).

Small angle scattering is considered one of the most suitable non-destructive techniques for investigating the nano-structure of porous materials in general, and of aerogels in particular.

Nuclear magnetic resonance spectroscopy (NMR) is often employed to monitor the structural evolution on the molecular level [169]. The extent and kinetics of hydrolysis and condensation can be studied and compared to models [170] from the early stages of polymerization until gelation and aging [171]. Generally, ^{1}H- and ^{29}Si-NMR are applied to silica systems. In addition, ^{17}O- and ^{13}C-NMR can be used to study the water and solvent components; ^{13}C-NMR has been applied to organic gels [172]. Most recently, Devreux et al. [173] have investigated silica aerogels, doped with paramagnetic impurities, by means of ^{29}Si-NMR, in combination with magic angle spinning. From the power-law decay of nuclear relaxation they derived the fractal dimension D, which agrees well with the one determined by SAXS.

3.2 Results

Scattering techniques allow us to follow the structural evolution during the sol-gel process on a nanoscopic scale; as with NMR there is no alteration of the growth process. Figure 15 shows a typical development of small angle X-ray scattering curves during gelation of two silica gels. Both samples (A and B) have been prepared by mixing tetramethoxysilane (TMOS), water and methanol at pH ≈ 9, in a molar ratio 1:2:15 (A) and 1:4:15 (B), respectively. The scattering intensity at zero angle $I(q \to 0)$, as well as the radius of gyration R reveal a continuous increase on gelation. The slope D in the Porod regime remains constant, disregarding the initial stages, where D cannot reasonably be calculated in the investigated measurement interval (Fig. 16). The time evolution of the Guinier radius R can be used to define the gelation time, as Craievich has recently shown for Zirconia gels [72]. From a double-logarithmic representation of R versus time, two distinct power-law regimes become visible: a steep slope before the sol-gel transition, followed by a slight increase after gelation.

A number of small angle scattering studies on sol-gel process of silica systems have been reported [175–179]. Schaefer and Keefer [175, 176] investigated the change of the scattering profiles of acid- and base-catalyzed gels prepared from tetraethoxysilane (TEOS) as a function of time. Concentrated as well as diluted solutions were studied. Since the radii of gyration derived from SAXS are

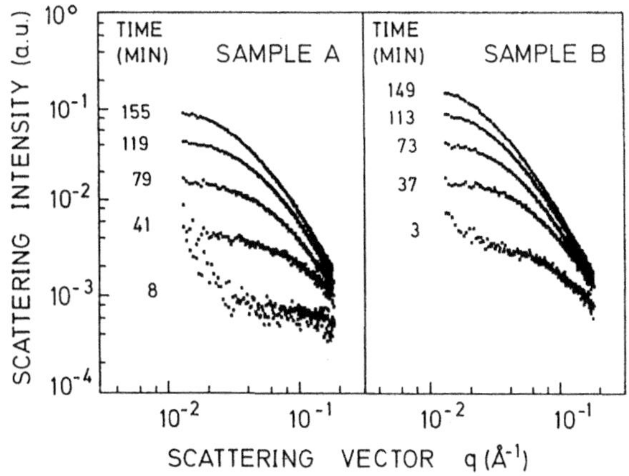

Fig. 15. Selected scattering profiles of two TMOS-solutions upon gelation. The increase in scattering intensity and the shift of the crossover towards smaller scattering vectors is indicative for the structural growth [174]

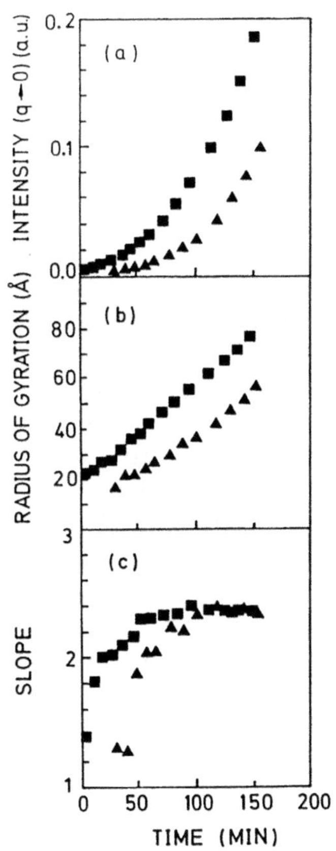

Fig. 16. Changes with time of zero angle intensity $I(q \to 0)$, radius of gyration R and or fractal dimension D upon gelation. The base-catalyzed samples A (▲) and B (■) were hydrolyzed by a stoichiometric and a double-stoichiometric amount of water, respectively [174]

independent of the degree of dilution, base-catalyzed gels are supposed to consist of compact, highly branched subunits. Acid-catalyzed gels, on the contrary, reveal increasing Guinier radii upon dilution; this suggests a less pronounced entanglement of a polymerlike network. Orcel et al. [177] followed the structural evolution of a TMOS-gel from a linear type polymer to a more highly branched polymer and then towards a particle. Hardmann-Rhyne et al. [178] studied the influence of H_2O or D_2O hydrolysis and HCl and DCl catalysis by SANS. A further SAXS study of the kinetics of aggregation was performed by Lours et al. [179]. Their particular interest was the evolution of the measured fractal dimension D′. Assuming the real fractal dimension D to be constant throughout gelation, the change in D′ was explained in terms of a changing polydispersity of the structural subunits. If the mass distribution of the particles obeys a power law [180]

$$N(M) \propto M^{-\tau}, \tag{14}$$

D′ and D are then connected via the relation

$$D' = D(3 - \tau). \tag{15}$$

Gelation of alumino-silicates was studied by Pouxviel et al. [181, 182], who found the hydrolysis of the Si–OR groups to be the rate-determining factor for the growth of this mixed-oxide system. Their scattering curves exhibit a maximum due to interparticle interference effects, whereby the particles are fractal aggregates with $D \approx 1.9$ and Guinier radii between 8 and 12 nm. At high water content, D increases even after gelation due to further condensation reactions. Mullite gels ($3\,Al_2O_3 \cdot 2\,SiO_2$) were also studied by Heinrich and Raether [183]. The authors found fractal dimensions $D \approx 2.2$, exponentially increasing Guinier radii in the initial stages and primary particle sizes, which inversely depend on the age of the precursor solution.

Light scattering studies of the TEOS gelation were performed by Tewari et al. [23], who observed the scattered intensity at an angle of 90°, with respect to the impinging laser beam. Using different water/TEOS ratios they generally found that the scattering intensity approaches a limiting value almost in the beginning of gelation for base catalysis, for acid catalysis the increase in intensity proceeds even after the sol-gel transition occurred [184]. Similar results have been obtained by Wood and Rabinovich [185] in the same type of experiment. They investigated TEOS-gelation depending on TEOS-concentration, type and amount of catalyst. Two-step preparation of TEOS-gels was studied by Boonstra et al. [186] with respect to different amounts of water in acid hydrolysis and basic condensation. Complete light scattering profiles have been recorded by Lourdin et al. [187], who studied diluted and thus slowly gelling base-catalyzed TMOS-gels. For all samples a fractal dimension of 1.66 and an inverse relationship between the final Guinier radii and original TMOS concentration is reported.

Disregarding questions of polydispersity, the fractal dimension might be used to estimate the kind of growth process on gelation [188]. Several computer

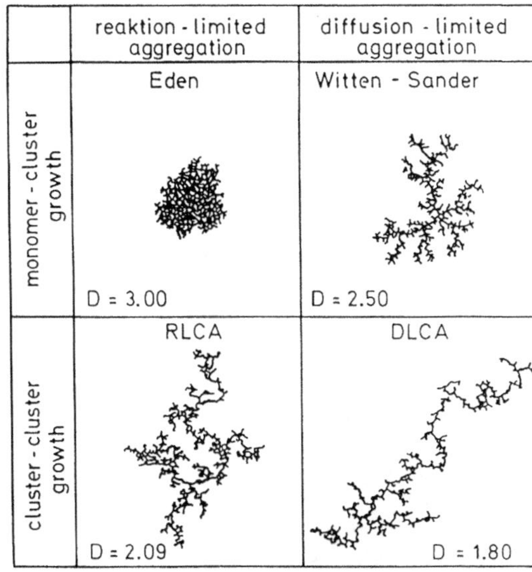

Fig. 17. Common kinetic growth models: Reaction and diffusion limited growth of clusters and cluster-aggregates. Though represented in a 2-d space, the fractal dimensions D are given for the 3-d analogs [191]

simulations have been carried out to simulate aggregation, which can be classified concerning mass transport and accretion (Fig. 17). Diffusion limited aggregation (DLA) occurs, if no repulsive potential is present and the diffusing particle adheres to the cluster on first trial. This phenomenon was described by Witten and Sander [189]. According to their model highly branched clusters with a fractal dimension $D = 2.5$ grow. The reaction-limited aggregation (RLA) operates with a sticking probability < 1. The diffusing particle is caught by an existing cluster only after numerous contacts. This behaviour, known as Eden growth [190], leads to compact units ($D = 3$), with rough surfaces. Both models describing monomer-cluster growth (MC) can be extended to the cluster–cluster growth (CC).

Hereby the clusters themselves undergo diffusional motion leading to cluster aggregates [191]. Thus large ramified structural entities are formed, which have no obvious center. Fractal dimensions around 1.7 . . . 1.8 and 2.0 . . . 2.1 are found for the diffusion- (DLCA) and reaction-limited (RLCA) cluster-cluster aggregation, respectively.

Alternatively, gels can be regarded as percolating systems [192], where not a kinetic, but an equilibrium growth is modeled. Space is considered as a lattice, which — in two dimensions — resembles a chessboard pattern. Each cell is occupied by a particle, with a probability p, whereby two neighbouring particles in a row or a column are thought to be bonded. Above a critical probability p_c at least one spanning cluster is formed, i.e. a continuous network exists (Fig. 18).

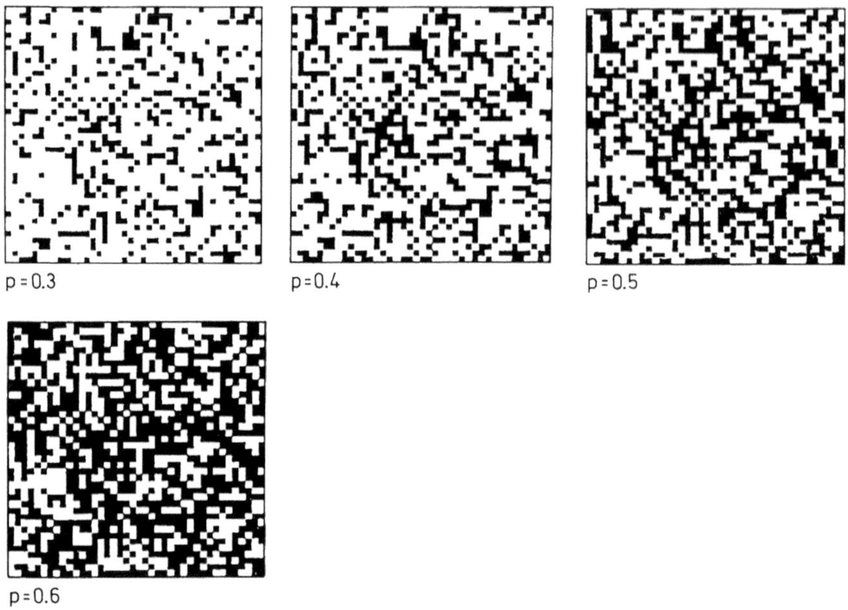

Fig. 18. Illustration of percolation on a two-dimensional square lattice. Each site is occupied with a probability p; patterns for $p = 0.3$, 0.4, 0.5 and 0.6 are given. Below the percolation threshold ($p_c = 0.59$) only separated clusters are present. For $p > p_c$ a coherent network appears [193]

The percolation threshold depends on the lattice geometry and on the dimensionality of space [194]. Experimentally p_c-values from 0.6 to 0.84 are found for silica sol-gel systems [195]. The fractal dimension D of the percolating cluster is expected to be 2.5; far above the percolation threshold more compact networks with higher Ds are formed, whereas well below percolation isolated clusters exist (lattice animals) with $D \approx 2.0$. With a modified dynamical algorithm [196], called kinetic percolation (KP), similar results are obtained. Starting from a seed site, a cluster is gradually generated by random occupation of peripheral sites.

Schaefer has reviewed [58, 110, 197] the structural evolution of silica gels with respect to the preparation conditions.

Under acidic conditions, when hydrolysis is fast and condensation slow, a burst of reactive monomers is produced. Random condensation occurs, which is best described by the RLCA model. Fractal dimensions close to 2 are observed for silica gels in solution [58]. This result is also found for the two-step processes, independent of the catalytic conditions of the second step [110, 188]. Since monomers are exhausted in the first step, the subsequent condensation proceeds via CC growth. As the corresponding aerogels show higher D values [58, 198], compaction as a result of syneresis upon aging and drying must have occurred.

Under neutral conditions the polymerization is limited by the extremely slow hydrolysis. Thus, whenever a hydrolyzed site either on a monomer or cluster appears, condensation takes place immediately. Consequently already existing clusters "hibernate", waiting for activation by hydrolysis. The fractal dimension measured for such gels is mostly 2.5, indicating either DLA or KP [58]. In the latter case, D would represent the fractal topology of a critically growing cluster. Here critical means that the percolating cluster has just enough active sites to sustain growth at all. This growth mechanism seems to be more likely compared to DLA, since D is nearly independent of the amount of water present in the system ($0.5 \lesssim w \lesssim 4$). Below a water/silicon-ratio w of 0.5, when not enough hydrolyzed sites are available, growth ceases ($p < p_c$) and the system approaches the fractal dimension of lattice animals ($D \approx 2.0$). Above the hydrolysis limit, i.e. in the presence of sufficient water, uniformly dense clusters, corresponding to $p > p_c$, are expected [110].

Under basic conditions hydrolysis becomes faster, but is still slower than condensation. A continuous reaction-limited MC growth is thus expected, since CC aggregation is reduced as an effect of the nucleophilic condensation mechanism, which requires the inversion of one of the two condensing species. Accordingly compact colloidal particles are formed via RLA, until all monomers are consumed. Subsequent growth has to proceed along the CC route, otherwise no gelation would occur. This is supported by the fractal dimension of the arising particle network of base-catalyzed gels. Mostly $D \approx 2.0$ is found in good agreement with RLCA [58, 110].

Only within the limit of fully hydrolyzed four functional monomers, i.e. under strongly basic conditions with large amounts of water, RLA can be described with the classical Eden model. At lower pH and water contents the probability of hydrolysis drops and reaction species with unhydrolyzed sites are formed. Therefore the Eden model must be slightly modified: a fraction of the potential growth sites of the monomers are blocked from forming bonds. Keefer [199] called this process poisonned Eden growth. Choosing appropriate distributions of blocked sites, dense or nanoporous non-fractal particles, fractally rough colloids and even mass fractals can be simulated [200, 201]. This has been confirmed experimentally for base-catalyzed gels, which were prepared with different water contents at constant pH [202, 203]. Figure 19 shows schematically the correlation between the water/silicon-ratio w and the mass (D) and surface (D_s) fractal dimension, as well as the Porod slope ($P = D_s - 2D$), observed in a scattering experiment.

The assumption that siloxane bonds (Si–O–Si) do not dissociate, once they are formed, is common to all kinetic models. Under exceedingly basic conditions, however, silica becomes increasingly soluble (Fig. 2). The late-stage evolution of the gel body is then often dictated by dissolution and reprecipitation phenomena; the resulting smoothening of the structure upon aging has been observed by Keefer [199].

The structure of base-catalyzed aerogels, produced under aggressive hydrolysis of TMOS for use in Cerenkov detectors (DESY/Hamburg, Germany; Airglass AB/Staffanstorp, Sweden), has been interpreted in terms of a branched network ($D = 2$), built-up of smooth colloidal particles ($D_s = 2$), as shown in

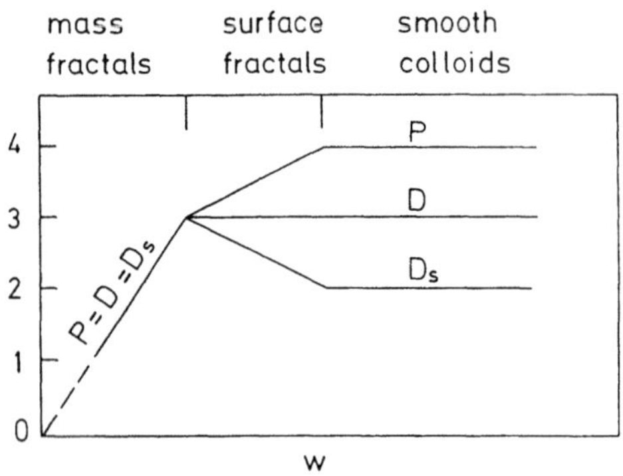

Fig. 19. Schematic representation of mass (D) and surface (D_s) fractal dimension together with the scattering exponent (P) as a function of the water/silicon-ratio (w). Three distinct regions of structural appearances are marked by the crossover points; the actual shape of the curves is still unknown [202]

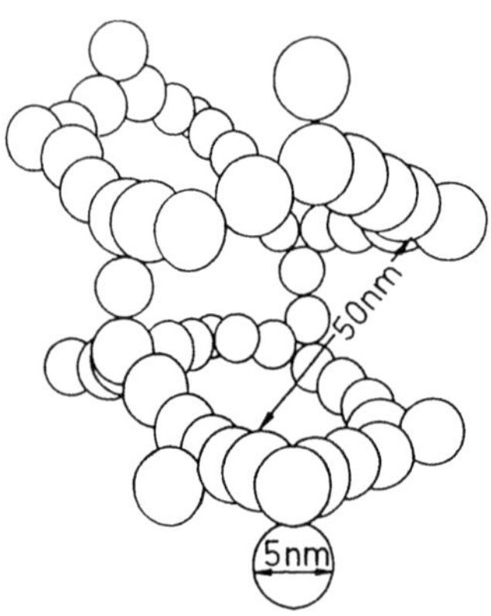

Fig. 20. Proposed structure of a base-catalyzed aerogel. Small colloidal particles build up a chain-like network structure

Fig. 20. Radii of gyration for particles and pores of 1 nm and 9 nm, respectively, are found for an aerogel density of $\rho = 88\,kg/m^3$ [204]. Such aerogels, with various densities ($\rho = 70$ to $290\,kg/m^3$), have also been investigated, applying either fractal analysis [205, 206], or TPM [206] to evaluate the SAXS data. No correlation exists between density and particle diameters ($l_s = 4.5\,..\,7\,nm$); the latter are mostly pH-dependent. The mean free path l_p, within the open aerogel network, shrinks ($l_p = 150\,..\,50\,nm$), with increasing density. Isothermal sintering of those aerogels induces a strong decrease in mean free path, but only a slight increase of chain diameter during the initial stages of densification ($\rho \lesssim 600\,kg/m^3$) [124].

Base-catalyzed aerogels have also been prepared and investigated with SANS, by Vacher and coworkers [19, 115, 116]. Their aerogels reveal a colloidal network structure; for densities as low as $15\,kg/m^3$ [19, 116] $D \approx 1.8$ has been found over more than one decade in q. Despite the still high porosity at "normal" densities ($\rho = 100$ to $200\ kg/m^3$), the fractal regime is confined to a rather small q-range. This is a result of the comparably large colloidal subunits. Neutrally prepared aerogels show a more extended fractal q-range at comparable densities [19], since the subunits are of molecular size. $D \approx 2.4$ is generally found independent of macroscopic density ($\rho = 95 \ldots 356\ kg/m^3$), which is in good agreement with critical growth [19, 115]. Their substructure—which determines the upper limit of the fractal q-range—is identical, but the location of the crossover into constant intensity scattering (the lower limit) scales with density (Fig. 21). Vacher et al. introduced the concept of mutual self-similarity [115]: At scale lengths within the fractal range, the structure of samples of different densities are indistinguishable. Applying an isothermal heat treatment, they observed a similar sintering behavior for both types of aerogels [19, 116]: a gradual growth in particle size and a reduction of the fractal domain.

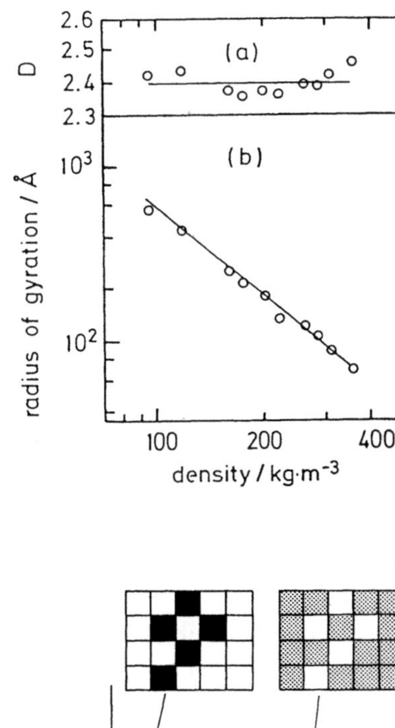

Fig. 21. Fractal dimension *D* and Guinier radii *R* as a function of density for neutrally reacting aerogels. Since *D* is constant for all samples *R* scales with density [115]

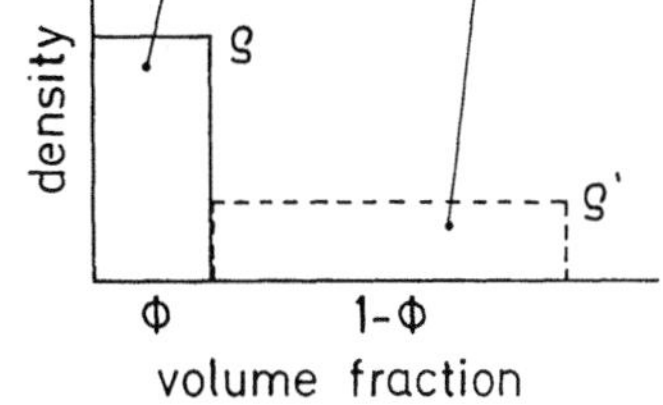

Fig. 22. Two complementary models with the same apparent density and interfacial area: *left*- network of dense particles with density ρ; *right*- voids in a low-density matrix (ρ') [207]

Craievich and coworkers have performed some combined SAXS- and BET-studies on differently catalyzed aerogels, in order to obtain detailed information on their hierarchical structure [207, 208], and the structural changes upon aging and densification [209]. They used the BET-derived specific surfaces to calculate the product of the volume fractions ϕ_s and ϕ_p, according to Eq. (11). Problems can be recognized from the generally occurring difference in SAXS- and BET-derived surfaces. In addition there is an inherent ambiguity if the volume fractions ϕ_s and ϕ_p are to be related to the solid matrix and the voids. Two extreme interpretations are possible (Fig. 22):

a) solid particles with a high skeletal density as the minor phase enclose large empty pores,
b) small pores are embedded in a low density solid matrix as the major phase.

Craievich et al. considered the aerogel to consist of a low-density solid matrix ($\rho \approx 100$ to $800\ kg/m^3$), including small air-filled voids. This is, however, in strong contradiction to the pycnometric results of Woignier and Phalippou [147].

Alumino-silicate aerogels have been studied by Chaput et al. [53]. Their SAXS measurements also reveal a two-level structure: smooth colloidal particles with an approximate diameter of 5 nm form a network with $D = 1.85$, independent of apparent density, i.e. the samples are mutually self-similar [210]. Similar results are obtained for cordierite aerogels, though both substructure and fractal topology seem to vary with density [52]. Structural changes upon sintering of these aerogels can once again be explained by particle growth and collapse of the fractal aggregates. Zirconia aerogels have been prepared and investigated by Vesthegem et al. [44]. The fractal dimension obtained $D = 2.8$ is interpreted as a result of MC growth. The structure of organic resorcinol/formaldehyde aerogels has recently been probed by Pahl et al. [211]. Guinier radii are shown to increase (3 nm . . 20 nm) with resorcinol/catalyst-ratio (50 . . 300), which agrees well with pore sizes observed by TEM. The fractal network ($D \approx 1.7$) is supposed to be built up of colloidal subunits.

Light scattering on base-catalyzed TMOS-aerogels and on acid-catalyzed TEOS-aerogels has been performed by Hunt and Berdahl [212, 213]. They observed a strong scattering anisotropy with a large amount of back-scattering for both polarization planes, which they tried to explain in terms of long-range correlations. However, due to the geometry of the scattering experiment, several important corrections to the measured angular scattering intensities must be applied. This was shown by Beck et al. [214, 215], who found correlation lengths l_c of 23, 17.5 and 13.5 nm, for TMOS-aerogels with densities $\rho = 71$, 105 and 143 kg/m^3. These TPM results are also explained by long-range density fluctuations of the aerogel network.

In spite of the many problems concerning experimental performance and image interpretation [152, 153], electron microscopy is often employed. This was done either to yield a nanostructural picture of aerogels [13, 42, 86, 184, 216, 217], or to follow structural changes, for instance, upon densification [12, 105].

A reliable TEM-study of low-density silica aerogels has been performed by Bourret [218]. In accordance with SANS [115] or SAXS [204] results, neutrally prepared aerogels are shown to reveal self-similarity at several levels of magnification, while the fractal regime of base-catalyzed aerogels is less extended and only a two-level hierarchical structure becomes visible. The latter is supposed to consist of a three-dimensional network of more or less bent chains, which are linked with a coordination number of three or four (Fig. 23).

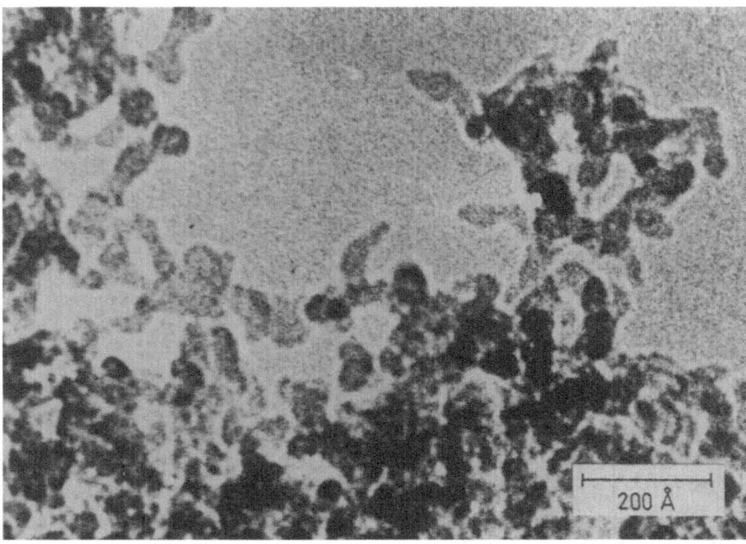

Fig. 23. Transmission electron micrograph of a base-catalyzed aerogel with density of 30 kg/m^3. (By courtesy of A. Bourret)

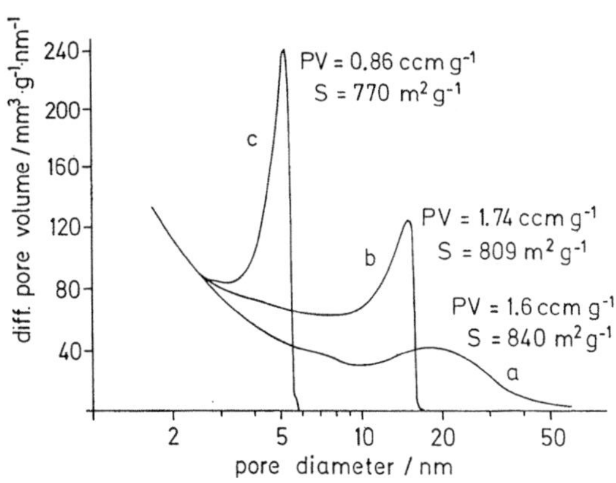

Fig. 24. Change in pore size distribution, pore volume (*PV*) and surface area (*S*) for a 145 kg/m^3-aerogel (*a*) induced by storage in water vapor (*b*) and penetration of liquid water (*c*) [146]

Moreover, BET measurements have often been used to obtain microstructural information. The goal of such measurements has been either to derive specific values for catalytic purposes [39, 40] or to investigate the influence of preparation conditions [13, 56, 59, 71], drying [93, 102, 103] or densification [11, 99]. A detailed BET and SAXS study of base-catalyzed aerogels of different density has been prepared by Schuck et al. [146]. Especially interesting is the dramatic loss of pore volume, if the aerogels are stored in water vapour or liquid water (Fig. 24).

4 Properties of Aerogels

With increasing porosity many solid state properties change drastically. As aerogels have porosities around 90% or above, it is to be expected that their acoustic, thermal and optical properties differ distinctly from those of vitreous (non-porous) silica.

4.1 *Elasto-Mechanical Properties*

It has been shown [17, 123] that the longitudinal sound velocities c_l in aerogels can be as low as $100\ \mathrm{m\,s^{-1}}$ (Fig. 25). The Young's modulus Y is of the order of $10^6\ \mathrm{N\,m^{-2}}$, i.e. about a factor of 10^4 down, compared to that of vitreous silica (Fig. 26). For both quantities in general a scaling behavior with density ρ can be derived [17, 123, 219–223]:

$$c_l \propto \rho^{\alpha}, \quad \alpha \approx 1.0 \ldots 1.4 \tag{16}$$

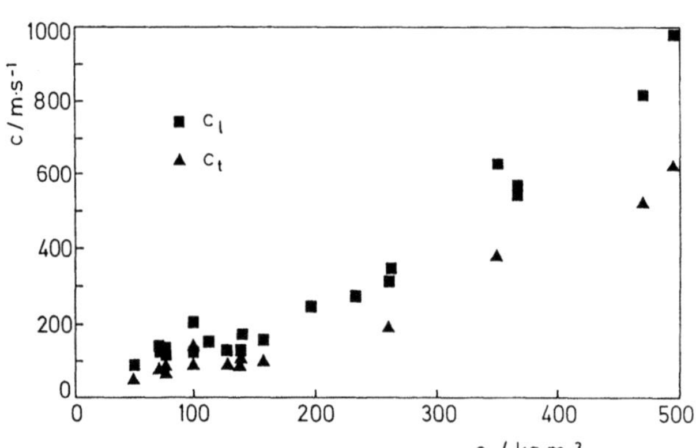

Fig. 25. Longitudinal and transversal sound velocities c_l and c_t versus density ρ for different silica aerogels made by S. Henning (Airglass, Staffanstorp/Sweden), G. Poelz (DESY, Hamburg/Germany) and L. W. Hrubesh (Lawrence Livermore Lab., Livermore/USA) [123]

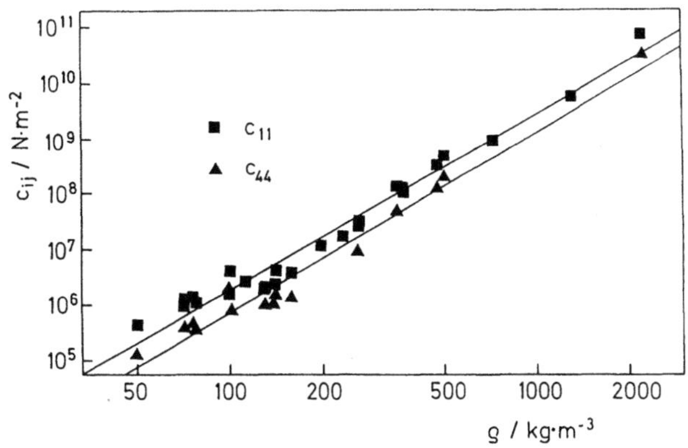

Fig. 26. Elastic constants of aerogels, silica gels (xerogels) and silica glass versus density ρ. The derived scaling exponents are $\beta_{11} = 3.17 \pm 0.11$ and $\beta_{44} = 3.25 \pm 0.15$; c_{11} is approximately equal to the Young's modulus [123]

and

$$Y \propto \rho^\beta, \quad \beta \approx 3.2 \ldots 3.8. \tag{17}$$

These scaling laws can either be explained in terms of percolation theory [224, 225], or as a result of a hierarchical structure [226]. In addition to sound velocities and Young's modulus, the shear [95, 227] and rupture [95, 228] moduli display scaling, while the Poisson ratio ($\mu \approx 0.22$) is independent of density.

For individual specimens considerable deviations from these scaling laws can be observed; most important in this respect is the connectivity on the nm scale length, resulting from the preparation conditions. At a given density aerogels, prepared under acidic or neutral conditions, are about twice as stiff as the base-catalyzed ones [227]. Aging [95] as well as baking [95, 123] approximately double the mechanical strength.

A striking feature of sound propagation in aerogels is the decrease of elastic constants by nearly 20% upon external uniaxial stressing with loads of 6×10^4 N m^{-2} (Fig. 27). This effect could be explained with "knee-bending" deformation of the microstructure. In such a configuration the restoring forces become weaker with decreasing angle between the legs [123]. Another explanation is given by the Montpellier group, which correlates this effect with the anomalous pressure dependence of the sound velocity of silica glass [229]. Stress–strain behavior has also been examined as a function of temperature [230, 231]: the rigidity as well as the range of linear elasticity are found to increase inversely with decreasing temperature.

An interesting question is the variation of the sound velocity with the pressure p_G of "trapped" air. As can be seen from Fig. 28 the experimental data

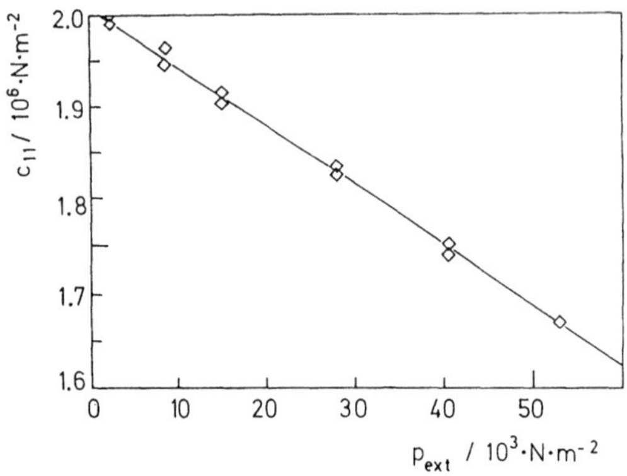

Fig. 27. Elastic modulus c_{11} versus external stress $p_{ext}(\rho = 100\ \mathrm{kg\,m^{-3}})$. The upper data points are derived from the loading, the lower from the unloading phase [123]

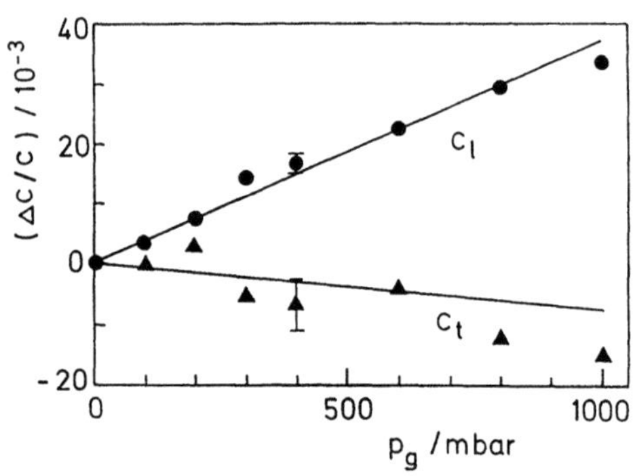

Fig. 28. Relative change of sound velocities c_l and c_t versus gas pressure p_G for an aerogel with $\rho = 105\ \mathrm{kg\,m^{-3}}$. The solid lines represent the theoretical curves for isothermal compression [232]

excellently fit the theoretical curves calculated assuming isothermal compression of air within the pore space [232]:

$$c_l(p_G) = \left(\frac{c_{11} + p_G}{\rho + \phi_p \rho_G}\right)^{1/2} \text{ and} \tag{18a}$$

$$c_t(p_G) = \left(\frac{c_{11} + c_{12}}{2(\rho + \phi_p \rho_G)}\right)^{1/2}. \tag{18b}$$

$c_{11} = \rho c_l^2$ and $c_{12} = c_{11} - 2\rho c_t^2$ are the elastic constants for evacuated aerogel, ρ its density and ρ_G the air density at pressure p_G and ϕ_p is the porosity of the specimens.

Via sintering the fragile aerogels can be transformed into high-strength solid bodies [95, 124, 219, 227]. The variation of elastic constants upon densification

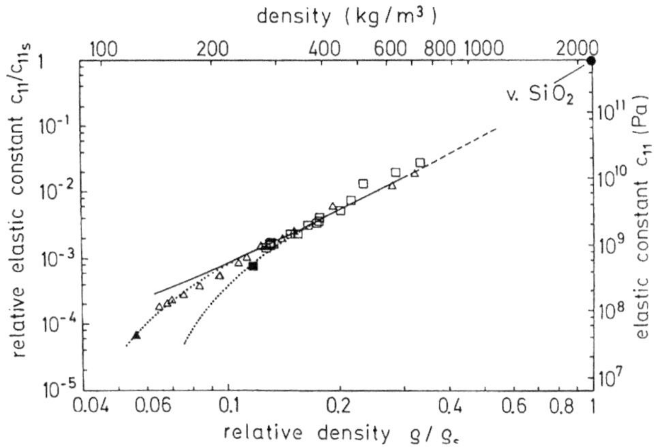

Fig. 29. Relative elastic constant c_{11}/c_{11s} as a function of relative density ρ/ρ_s (▲, ■ original samples with $\rho = 122\ kg/m^3$ and $256\ kg/m^3$ and △, □ sintered "daughter" samples). The *full dot* in the right upper corner marks the value for non-porous vitreous silica, c_{11s} refers to this material. Obviously the elasticity depends only on the achieved density ρ [124]

is shown in Fig. 29. It is remarkable that after an initial phase with an especially strong variation of c_{11} with ρ the resulting elastic constants can be unequivocally correlated with the achieved density on sintering—independent of the initial density [124]. The c_{11} versus ρ plot (Fig. 29) even shows scaling between $\rho \approx 250$ and $800\ kg\,m^{-3}$.

4.2 *Optical and Infrared-Optical Properties*

Despite the high porosity, silica aerogel—if properly made—is highly transparent in the visible spectral region [23]. At 630 nm the specific extinction coefficient e_{vis} is of the order of $0.1\ m^2\,kg^{-1}$ [214]. Thus in an aerogel with a density $\rho \approx 100\ kg\,m^{-3}$ the photon mean free path at that wavelength is

$$l_{photon} = 1/(e_{vis}\,\rho) \approx 0.1\ m. \tag{19}$$

Towards the blue and ultraviolet strong Rayleigh scattering is observed [23, 233], which is to be expected for a material with pore structures in the 1 to 100 nm size range. Accordingly the angular scattering distribution is nearly isotropic [214]. The variation of the extinction coefficient with wavelengths in this spectral region cannot be described satisfactorily by a $1/\Lambda^4$ dependence [234], valid for a medium containing isolated point scatterers. Reasons may be the occurrence of absorption or anisotropic scattering. Due to the strong scattering power at smaller wavelengths, extraction of data from transmission measurements is non-trivial as multiple scattering has to be taken into account. Particle diameters, which can be derived assuming Rayleigh scattering [213,

233], are in the order of 10 nm [23, 233, 235], whereas comparative electron micrographs show 5 nm-sized particles [233, 235].

In general a higher transmission is observed for low-temperature CO_2-dried aerogels, compared to their high-temperature alcohol-dried counterparts [23, 235]. Due to desorption of water and combustion of organic residues, the optical transmission of the latter is improved after baking [8, 23, 235, 236]. Further improvement might be achieved by choosing appropriate preparation conditions during sol-gel processing [8, 9, 237].

In the infrared, silica aerogels show the strong absorption bands typical for SiO_2 [112]; additional absorption is caused by water [238], chemi- or physisorbed at the huge internal aerogel surface (Fig. 30). Generally also organic groups can be recognized in the absorption spectrum; they result from compounds added to the network in the sol-gel or the autoclaving process.

Typical is a specific extinction of $\lesssim 10\,m^2/kg$ below 7 µm and above 30 µm and a specific extinction of $\gtrsim 100\,m^2/kg$ between 8 and 25 µm. If one compares these data with the extinction in the visible a ratio

$$e_{IR}/e_{vis} > 100 \tag{20}$$

is seen. Thus silica aerogels can be considered translucent insulation materials, which effectively transmit solar light, but strongly block thermal infrared radiation. Infrared spectroscopy has, for instance, also been applied to alumina [240], boron-phosphorus-silicon mixed oxide [50], zirconia [44], cordierite [241] and organic aerogels [55].

The refractive index of low density aerogels is very close to 1; experimentally one finds [86]

$$n - 1 \approx 2.1 \times 10^{-4}\ \rho/kg\,m^{-3}, \tag{21}$$

in agreement with a simple theoretical estimation (9). For $\rho \approx 100\,kg\,m^{-3}$ one thus gets $n \approx 1.02$. The factor in Eq. (21) depends on the number of OH-groups

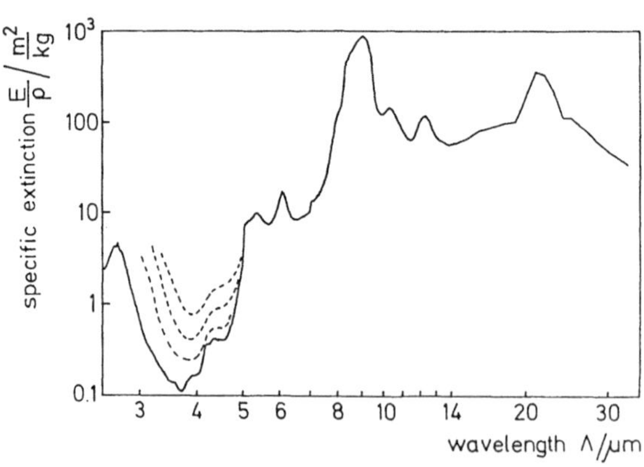

Fig. 30. Specific extinction coefficient of silica aerogels without (——) and with various water contents (– – –) in the infrared [239]

on the inner surface. For pure silica one would expect 1.9×10^{-4} [86]. Due to the small refractive index, light enters a piece of aerogel almost without reflective losses.

The dependence in Eq. (21) is often used to probe the local density variation in aerogel monoliths [242]; mostly employed is the minimal deflection method at 90° corners to determine n (9). However, for this procedure the aerogel surface has to be flat and the corners have to be squared. In addition, one has to be aware of the influence of adhering OH-groups to the silica skeleton, if absolute values are to be derived.

4.3 Thermal Properties of Aerogels

Thermal transport in monolithic and granular aerogels has been studied extensively by Fricke and coworkers [243–256]. In the following paragraph the main features of gas and solid conduction, as well as radiative heat transfer, are summarized.

The structural build-up in aerogel occurs in the 1 to 100 nm range, which is known from light scattering, SAXS, SANS, BET and gas diffusion measurements. Therefore gas conduction is not expected to be fully developed in aerogels, even at pressures of 1 bar. The thermal conduction under variation of gas pressure and its correlation with structure was thoroughly studied by S. S. Kistler [4]. Recent measurements [250] show, that for 80 kg/m³ aerogels the gas-induced rise of thermal losses only occurs for pressures above 50 mbar (see Fig. 31). The derived average size for the "pores" is about 100 nm.

In evacuated aerogels thermal transport occurs via solid conduction and radiative heat transfer. The former is expected to be drastically reduced with respect to vitreous silica. This is due to the large amount of "missing" mass in aerogels, which restricts the propagation of local excitations to the chains in the tenuous structure. Above 100 K one can describe the solid conductivity

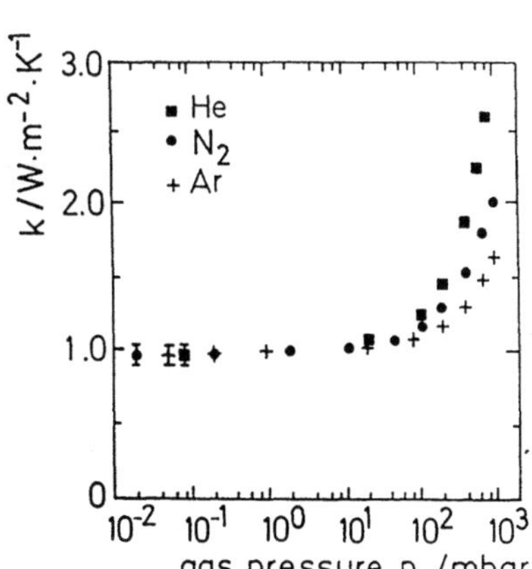

Fig. 31. Thermal transfer coefficient k for a 80 kg/m³ aerogel tile of 9 mm thickness versus gas pressure; the mean temperature was $T_r = 300$ K [250]

of aerogels λ_s by the solid conduction of vitreous silica λ_0 and a geometrical factor G

$$\lambda_0/\lambda_s = G. \tag{22}$$

G is strongly density dependent; for $\rho \approx 100\,\mathrm{kg\,m^{-3}}$ one gets $G \approx 500$.

The radiative heat transport in aerogels is governed by infrared absorption (see Fig. 30) [251–253]. While for low temperatures ($T \approx 200$ K) thermal radiation primarily penetrates this material in the wavelength region $\Lambda > 30\,\mu$m, at ambient and elevated temperatures thermal leakage occurs predominantly in the 3 to 5 μm range (Fig. 32) [254].

In thin layers of aerogel (e.g. 10 mm) with low emissivity boundaries a strong coupling occurs between conductive and radiative heat transfer modes. Thus the total conductivity is always larger than the sum of solid and apparent radiative conductivity [256]. Despite this effect the total conductivity of evacuated monolithic silica aerogel ($\rho \approx 100\,\mathrm{kg/m^3}$) at $T \approx 300$ K is only about $0.010\,\mathrm{W\,m^{-1}\,K^{-1}}$ [239]; introduction of air increases this value to about $0.020\,\mathrm{W\,m^{-1}\,K^{-1}}$. Upon increase of temperature the conductivity of silica aerogels rises strongly (Fig. 33).

The thermal conductivity can be reduced if an opacifier, e.g. carbon soot—a strong broad band absorber—is added to the silica sol and thus integrated into the gel. The specific extinction in this case is drastically increased, especially in the 2 to 8 μm range (Fig. 34). Under these circumstances powderous aerogels have total conductivities of about $0.020\,\mathrm{W\,m^{-1}\,K^{-1}}$ in air (Fig. 35), while for monoliths $\lambda \approx 0.015\,\mathrm{W\,m^{-1}\,K^{-1}}$ result (Fig. 36). These are by far the lowest conductivity values of all powders and solid bodies in air.

4.4 *Dynamics of Aerogels*

The dynamics of Debye solids is characterized by the vibrational density of states.

$$g(\nu) \sim \nu^2/v^3, \tag{23}$$

where ν is the frequency and v the sound velocity. Knowing that the latter is about 50 times smaller than in vitreous silica, one can estimate that g for aerogels for a given frequency in the Debye regime is about a factor of 10^5 larger than in non-porous SiO_2-glass (Fig. 37).

The Debye regime ends, when with rising frequency the wavelengths of the sound waves become comparable to the size R of the largest structural entities ($R \approx 100$ nm). This occurs for frequencies around

$$\nu_c = v/R = 100/10^{-7}\,\mathrm{s^{-1}} \approx 10^9\,\mathrm{s^{-1}}. \tag{24}$$

Above this frequency localized modes within the porous network are excited. At yet higher frequencies ($\nu \gtrsim 100$ GHz) modes of the network-building particles become dominant. If fractal aerogels are considered, localized excitations in the

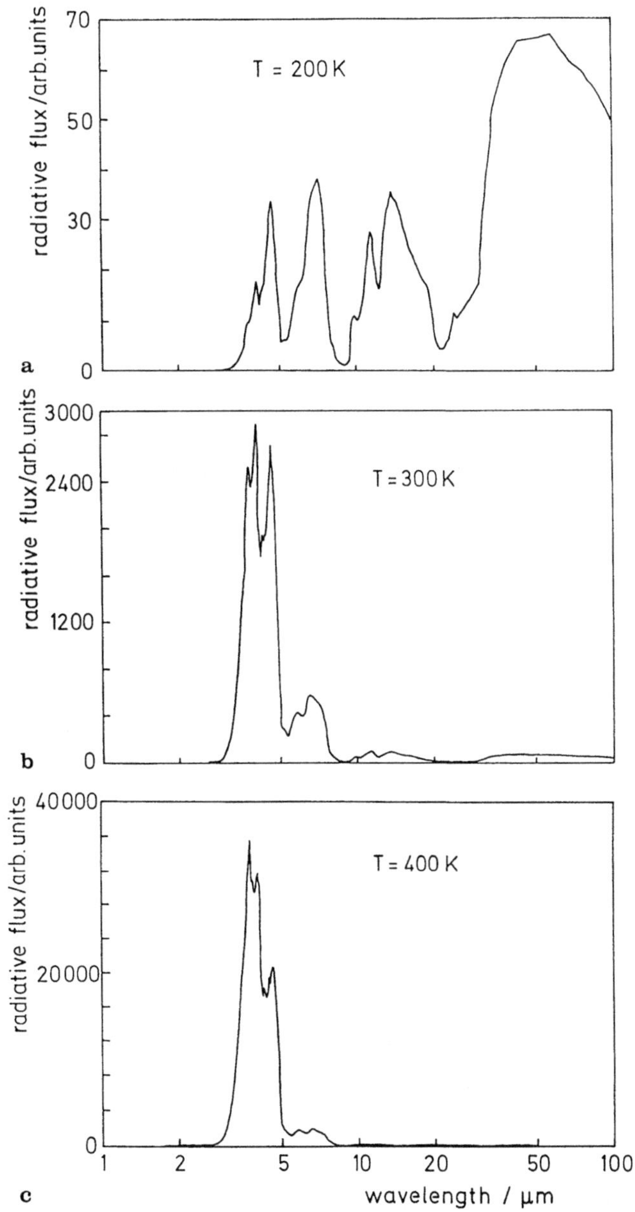

Fig. 32. Relative spectral flux through silica aerogel between black boundaries for T_r = 200, 300, and 400 K; note the different ordinate scales. For T_r = 200 K the major radiative contribution occurs above wavelengths of 30 µm; low emissivity boundaries would cut off the flux maxima [255]

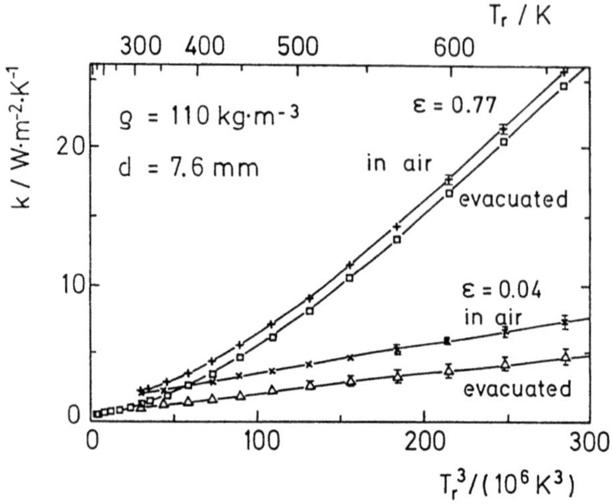

Fig. 33. Thermal transfer coefficient k versus mean temperature T_r cubed for an aerogel tile with $\rho = 110\,\mathrm{kg\,m^{-3}}$, $d = 7.6$ mm; parameter is the boundary emissivity ε; data for evacuated and non-evacuated specimens are shown [256]

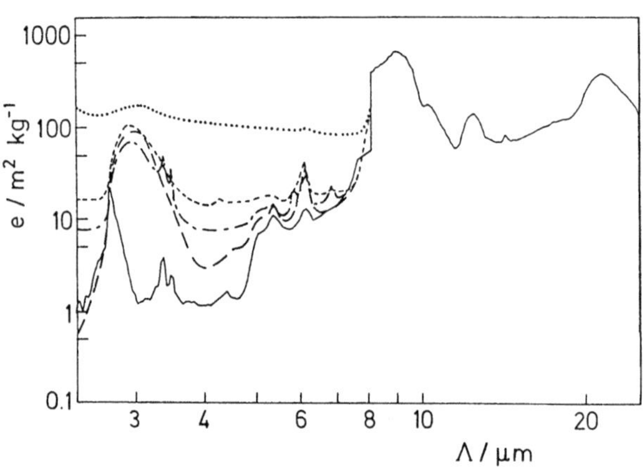

Fig. 34. Spectral specific extinction e of pure and opacified silica aerogels versus wavelength $\triangle$; —— pure SiO_2-aerogel, baked; — — — pure SiO_2-aerogel, —·—· aerogel plus 0.1% soot, - - - - - - 0.5% soot, • • • • • • 5% soot [74]

intermediate region around 10 GHz are called fractons [257–261]. Analogous to small angle scattering different structural features appear at corresponding frequency regimes.

The Debye regime and the crossover from phonons to fractons is mostly studied with Brillouin spectroscopy. This quasi-elastic light scattering technique measures the frequency shift of scattered photons, due to the excitation

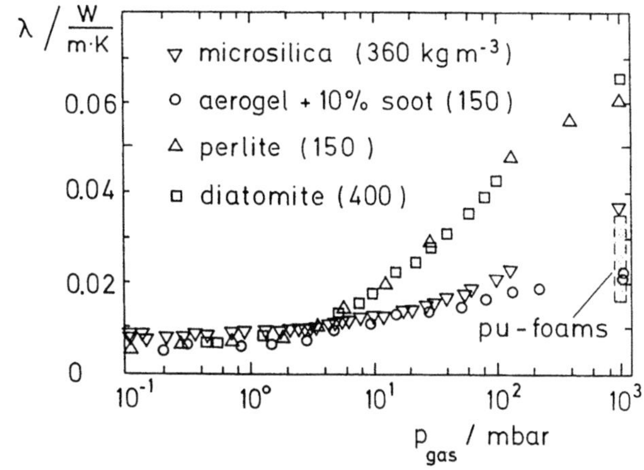

Fig. 35. Thermal conductivity λ of several powder insulations as a function of gas pressure p_g for $T \approx 300$ K; without evacuation only opaque aerogel powders have the potential to substitute F11-blown pu-foams [15]

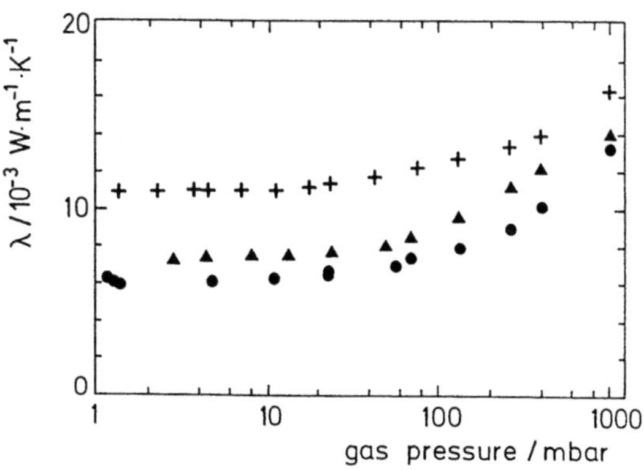

Fig. 36. Thermal conductivity λ of opacified monolithic silica aerogel as a function of gas pressure p_g; + 230 kg m^{-3}, 2.5% soot, ▲ 150 kg m^{-3}, 10% soot and ● 120 kg m^{-3}, 5% soot [74]

(or annihilation) of low-frequency acoustic phonons. The sound velocity c_l, the crossover frequency ν_c and the exponent of the fracton dispersion curve

$$\nu \propto q^{D/\bar{d}} \tag{25}$$

can be obtained. D is the fractal dimension of the network and $\bar{d}$ the spectral dimension of fractons. $\bar{d} \approx 4/3$ is expected for all kinds of fractals obeying scalar elasticity [257]. Experimentally $\bar{d}$ is found close to this predicted value for aerogels with fractal features [262–264], though influences of preparation might occur [265]. Moreover, both sound velocity and crossover frequency are observed to scale with density [263, 264].

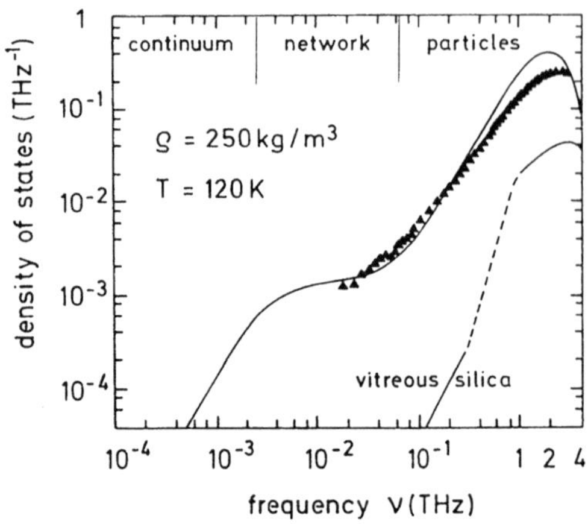

Fig. 37. Approximate behaviour of vibrational density of states $g(\nu)$ versus frequency for base-catalyzed aerogel ($\rho = 250$ kg/m^3) and vitreous silica [20]

Higher frequency excitations can be studied by Raman scattering, a related quasi-elastic technique. Power-law behavior of the fracton dispersion law (Eq. 25) is again confirmed [266–268]. Since the second crossover from network to particle modes is accessible, a mean particle size is obtained, if the transversal sound velocity c_t within this subunits is known [269]. Chain diameters of 5 to 7 nm, for instance, are derived for base-catalyzed silica aerogels in good agreement with SAXS data [270]. Thus influences of preparation and heat treatment on structure can be examined with this dynamical method [271–273]. Using SAXS-derived particle radii to estimate the transversal sound velocity, values of about 2200 m/s have recently been found for base-catalyzed aerogels by Pelous et al. [274]. Compared to $c_t = 3700$ m/s for vitreous silica, this indicates a microporosity within the building blocks of the network, which resists the initial stages of densification.

Fracton regime and particle modes can also be investigated by inelastic neutron scattering [275, 276]. Hereby the momentum and energy transfer of the scattered neutrons is determined, from which the density of states can be calculated in a straightforward manner [277]. Inelastic neutron scattering has been employed to neutrally reacted (fractal) aerogels by Vacher and coworkers [21, 278, 279] and base-catalyzed (colloidal) aerogels by Reichenauer et al. [20, 280, 281]. Recently, indications for two distinct fracton regimes have been observed and attributed to either bending or stretching of the network chains [282].

High-resolution neutron backscattering measurements allow us to probe the phonon-fracton crossover [283]. Applied to base-catalyzed aerogels spectral dimensions $\bar{d} \approx 2$ are obtained by Conrad et al. [283–285], though preliminary Brillouin spectroscopy [286] indicates rather small values ($\bar{d} \approx 0.5$). Neutron spin-echo spectroscopy is another high-resolution technique, which has recently been applied to investigate the dynamics of fractal aerogels [287] and aerogels with different connectivity as a result of various preparation conditions [288].

From the density of states, derived on an absolute scale, the heat capacity c can be calculated [20, 278–281]. This is extremely important, since low-temperature (T < 10 K) measurements of specific heat c and conductivity λ are hard to perform. With a large c and a small λ in the considered temperature range, long thermal time constants result, using classical steady state methods [22, 289, 290]. Lately, however, a dynamical technique for the simultaneous measurement of c and λ has been developed and successfully applied to aerogels [291]. A comparison between a calculation and experimentally derived heat capacity data is shown in Fig. 38. Note, that c for aerogels is one to two orders of magnitude larger than for vitreous glass at low temperatures.

4.5 Gasflow in Aerogels

A prerequisite for the application of aerogels as gas filters or catalysts is the knowledge of the size of the largest pores. This quantity can be estimated from SAXS measurements (see Sect. 3.1), from the variation of the thermal conductivity with gas pressure (see Sect. 4.3) and from gas flow measurements. The latter allows us to derive the flow permeability P of aerogel layers; recent measurements [292] have yielded permeabilities in the order of 10^{-8} m²/(Pa s) for nitrogen.

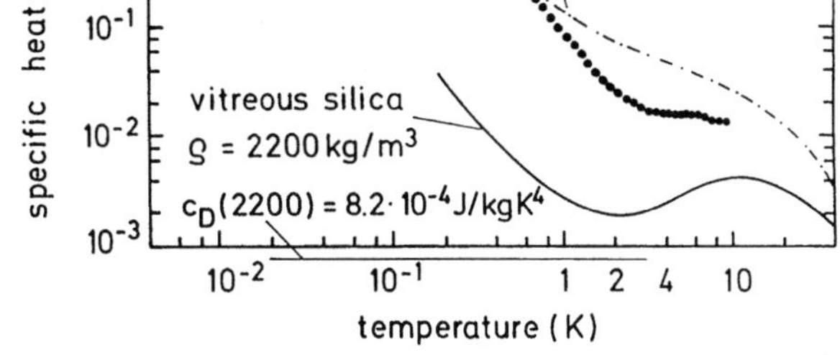

Fig. 38. Comparison between the specific heat / T^3 determined by a calorimetric measurement for an aerogel with macroscopic density $\rho = 270$ kg/m³ [289] (-- ● ●) and the calculation for a 250 kg/m³-sample using the density of states (-·-); c_D indicates the corresponding Debye-values [20]

5 Applications of Aerogels

All applications make use of the high porosity of the aerogels, which is responsible for the low index of refraction, the small Young's modulus, the low acoustic impedance, the low thermal conductivity and the excellent accessibility of the inner surface. In some applications in addition the high optical transparency is of importance.

5.1 Cerenkov Detectors

Most of the monolithic silica aerogels made until now have been used in Cerenkov detectors [7–9, 86–88]. The small index of refraction allows one to determine the momentum of relativistic particles within a momentum range, which is not covered by compressed gases or by liquids. The impinging particle produces an electromagnetic shock wave, the cone angle of which depends on the ratio of particle velocity and group velocity within the medium. Due to Rayleigh scattering the angular information is blurred. Thus aerogel tiles of different density generally serve as threshold detectors [293, 294]. A variation of the index of refraction n between $8 \times 10^{-3} < n - 1 < 0.4$ is possible, if sintered aerogels are included (Fig. 39). Today several high energy research groups are installing multi-cell aerogel detectors either for use in storage rings and accelerators [295, 296], or for the investigation of cosmic rays [87, 88].

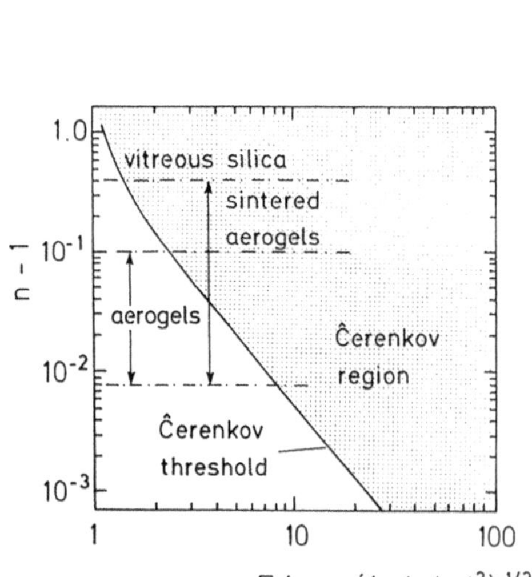

Fig. 39. The Cerenkov threshold can be effectively probed with aerogels, which can be made in the range $0.008 < n - 1 < 0.4$ for the refractive index n [8]

5.2 *Transparent Aerogels for Passive Solar Energy Usage*

One of the most promising applications of SiO_2-aerogels is their use in transparent thermal superinsulation (see e.g. 90). Extensive studies of transparent aerogels have shown, that the thermal resistance as well as the solar transmission are sufficiently high to make these materials most suitable for passive solar usage [14, 91]; the large thermal resistance is firstly caused by the high porosity of the gel-network, which consequently has a very small solid thermal conductivity [297]; secondly the gaseous thermal conduction is suppressed, as even at ambient pressure the mean free path of the gas molecules is comparable with the pore size [244]; thirdly, the thermal radiation at 300 K is effectively absorbed in aerogel layers 1 to 2 cm thick [245]. A convincing example is the use of granular silica aerogels as translucent insulation [298] in a two-family house in Ardon/Switzerland (Fig. 40). The energy consumption for heating purposes is exceptionally low (about 300 liters of oil per year); the total investment was smaller than for a conventionally insulated house [299].

Other possible applications for aerogels are in frosted windows or in superinsulating covers for solar panels [300–303]. Aerogel insulated solar storage collectors are also being investigated, since they can be operated without use of anti-freezing agents (see e.g. 304).

One ought to mention that a number of other materials are being installed and investigated for this purpose: capillary structures, foams, macro glass bubbles and low density fibers. However, these materials are either organic and thus combustible, or do not provide a high enough transparency. Moreover, aerogels are found to be the translucent thermal insulating material with the highest thermal resistivity for given thickness [305].

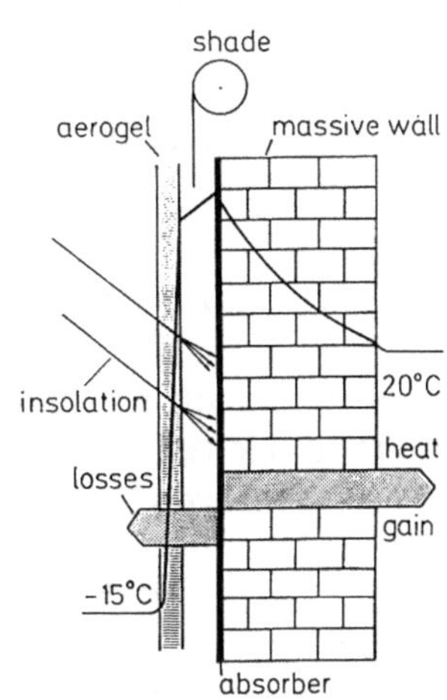

Fig. 40. Transparently insulated house wall using an aerogel-filled glass pane system [90]

5.3 Opaque Aerogels as Substitutes for CFC-blown Polyurethane Foams

A new idea is the application of monolithic, granular or powdery aerogels as opaque insulation in refrigerators or heat storage systems, where, up to now, CFC-blown polyurethane (pu) foams have been employed [15, 306]. The CFCs have to be abandoned for environmental reasons since they are responsible for the destruction of the ozone layer and make a considerable contribution to the greenhouse effect. After 1995 the production and use of CFC in Germany will be forbidden by law.

Typical thermal conductivities for pu insulations freshly blown with freon (F 11) are $\lambda \lesssim 0.018$ W/(m K) at 300 K; for comparison: the thermal conductivity of non-convecting air is $\lambda \approx 0.026$ W/(m K) and of F 11 $\lambda \approx 0.008$ W/(m K). Aging of pu-foams leads to diffusion of air into the foam, which causes a gradual increase of the thermal conductivity up to 0.025 W/(m K) and beyond [307]. It should be pointed out that pu-foams have a combined solid and radiative conductivity of around 0.008 W/(m K). Filled with air, the total thermal conductivity of pu-foams is thus above 0.030 W/(m K). The need for non-hazardous non-inflammable thermal insulations with conductivities in the 0.020 W/(m K) range or below is obvious. The substitution of pu-foams by SiO_2-aerogels is feasible, if the infrared (IR) radiative heat transfer through the highly porous material can be sufficiently reduced. This is possible via integration of materials like soot, Fe_3O_4, or TiO_2 into the gel to make it IR opaque. As was shown in Figs. 35 and 36, at 300 K non-evacuated powderous and monolithic opaque aerogels have thermal conductivities of about 0.020 and 0.014 $W m^{-1} K^{-1}$, respectively. Evacuation reduces these values to 0.004 and 0.007 $W m^{-1} K^{-1}$. Integration of an aerogel layer in a CFC-free refrigerator insulation is shown in Fig. 41. A non-evacuated all-aerogel insulation is also feasible, however, would require an additional stable casing.

Aerogels can also be used in heat storage systems. As a general rule, with rising temperature the content of the agent causing opacity has to be increased. The best values achieved so far are $\lambda \approx 0.020\ W m^{-1} K^{-1}$ for a monolithic aerogel with density of 150 $kg m^{-3}$ and 10% soot at 300 °C [74].

5.4 Acoustic Impedance Matching

Another interesting application of aerogels is for acoustic impedance matching in ultrasonic transducers operating in air [17, 224]. It is well known that low density aerogels have the smallest impedances of all solid materials ($Z \approx 10^4 \ldots 10^5\ kg m^{-2} s^{-1}$). This allows the use of quarter wave layers of aerogels to match the high impedance of piezo ceramics with the small impedance of air (Fig. 42). Theoretically this can lead to a sound level increase of about 40 dB. Experimentally a gain of 25 dB has been realized [308].

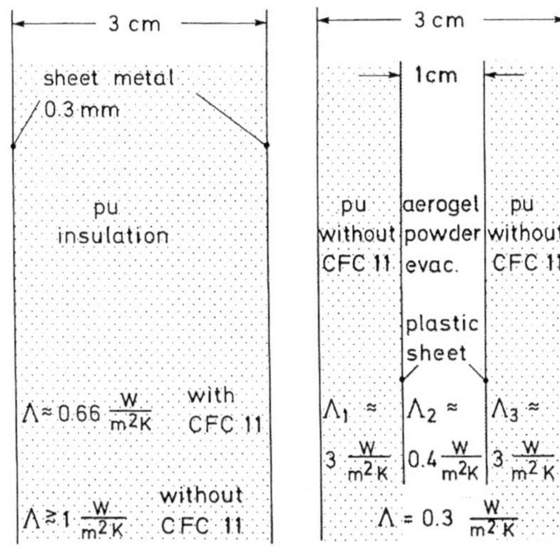

Fig. 41. Heat transfer coefficient Λ of a standard pu-insulation with and without F11 for a refrigerator (*left*) and proposed CFC-free insulation with evacuated aerogel powder sandwich (*right*); the latter has about half the thermal losses compared to the CFC-containing insulation on the left [15]

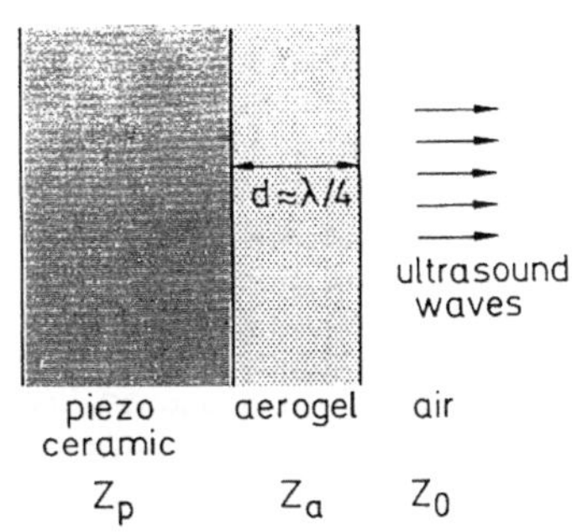

Fig. 42. Acoustic impedance matching with aerogel layers [17]. The acoustic impedances have to fulfil the condition $Z_a = (Z_p \cdot Z_o)$. This corresponds to the relation $n = (n_1 \cdot n_2)^{1/2}$ for the refractive indices in optics

5.5 *Aerogels as Gas Filters*

As the pore structures of aerogels can be modified in a controlled way by sintering [124], these materials are very suitable as gas filters in the 20 to 100 nm region. Such filters are becoming important as the lateral dimensions of electronic components in microchips are being reduced to about 0.5 μm or below. The critical size of defects accordingly has to be kept below 0.1 μm. From theoretical considerations on the effectivity of the aerogel structure as a filter medium high quality factors are expected [309].

If aerogels are to be used as filters for liquids, the skeleton has to withstand the surface tension of the liquid used. This requirement can be met with partially densified aerogels [45], a correlation between sustainable surface tension and aerogel density has recently been published for neutrally prepared aerogels [106].

5.6 *Aerogels as Catalytic Substrates*

Aerogels have large internal surfaces, which are readily accessible, due to the open porosity of these materials. In addition the production of mixed aerogels, containing catalyzing agents, is possible. As was pointed out by Pajonk and Teichner [51], Fe_2O_3-on-silica or -alumina aerogels, for instance, show a catalytic activity in the Fischer-Tropsch synthesis of hydrocarbons, which is two to three orders of magnitude higher than that of conventional iron catalysts. The use of aerogels in liquid catalysis requires that the skeleton withstands the surface tension. As stated in Sect. 5.5 this can be achieved by sintering; however, at the same time the large specific surface has to be maintained. This problem has already been addressed in Sect. 2.3.

5.7 *Miscellaneous*

Small high velocity particles in space can be captured by low density aerogels [310]. As these particles easily penetrate the porous material and are gradually decelerated, a "soft landing" can be achieved. If transparent aerogels are used, the stopped particles can even be inspected by eye, or with a microscope.

Aerogels from low-Z elements may be used in laser fusion experiments [311]. Li_2O/B_2O_3 aerogels have been prepared for this purpose [42]; organic aerogels have also been tested with promising results [312]. Recently, TEOS-aerogels were used to produce hollow spheres as containers for fusion fuels [313, 314].

Aerogels can be used to produce extremely pure glass, as already mentioned in Sect. 2.3. Philips researchers sintered 160 $kg\,m^{-3}$ SiO_2 aerogels at temperatures below 1250 °C and achieved full density [12].

Fine metallic powders have also been produced using supercritical drying [16]. These powders show submicron sized spheres, which seem to have undergone extensive sintering during the autoclaving at 275 °C. Cu, Pd and Au powders have been made.

Aerogels may serve as host for atoms and molecules. From fluorescent decay times and spectral line widths valuable information about the guest–host interaction can be derived [24, 25]. Also diffusion of polymers [137] and the behavior of liquid helium [138, 315, 316] in the porous aerogel body, and even the consequences of shock compression treatment [317] have been studied.

Last but not least, aerogels have been used as non-hazardous, effective insecticides [318–321]. Due to their hydrophilic character rapid loss of body fluids from the organic tissues occurs, when insects come into contact with aerogel powder.

6 References

1. Fricke J (1986) Aerogels, Springer Proc. Phys. 6, Springer, Berlin Heidelberg New York
2. Vacher R, Phallippou J, Pelous J, Woignier T (1989) Proceedings of the 2nd Int. Symposium on Aerogels. (Revue de Physique Appliquée, Colloque C4), Les Editions de Physique, Les Ulis Cedex

2a. Fricke J (1988) Sci. Am. 256: 92

2b. Gesser HD, Goswani PC (1989) Chem. Rev. 89: 765

3. Kistler SS (1931) Nature 127: 741
4. Kistler SS, Caldwell AG (1934) Ind. Eng. Chem. 26: 658
5. Nicolaon GA, Teichner SJ (1968) Bull. Soc. Chem. Fr. 1906 and US Patent 3672833
6. Teichner SJ, Nicholaon GA, Vicarini MA, Gardes GEE (1976) Adv. Colloid and Interface Sci. 5: 245
7. Cantin M, Casse M, Koch L, Jouan R, Mestran P, Roussel D, Bonnin F, Moutel J, Teichner SJ (1974) Nucl. Instr. Meth. 118: 177
8. Poelz G, Riethmüller R (1982) Preparation of Silica Aerogel for Cerenkov Counters, DESY 81-055 Hamburg/Germany (1981) and Nucl. Instr. Meth. 195: 491
9. Henning S, Svensson L (1981) Phys. Scr. 23: 697
10. v. Dardel G, Henning S, Svensson L (1982) European Patent Spec. 0018 955 13 1
11. Zarzycki J, Woignier T, in Ref [1], p 42
12. Mulder CAM, van Lierop JG.in Ref [1], p 68
13. Broecker FJ, Heckmann W, Fischer F, Mielke M, Schroeder J, Stange A, in Ref [1], p 160–166
14. Goetzberger A, Wittwer V, in Ref [1], p 84–93
15. Fricke J, Arduini-Schuster MC, Büttner D, Ebert H-P, Heinemann U, Hetfleisch J, Hümmer E, Kuhn J, Lu X. Opaque Silica Aerogel Insulations as Substitutes for Polyurethanes Foams. 21st Int. Thermal Cond. Conf., Oct. 15–18, 1989, Lexington/Kentucky
16. Armor JN, Carlson EJ, Carrasquillo G (1986) Mater. Lett. 4: 373
17. Gronauer M, Fricke J (1986) Acustica 59: 177
18. Schaefer DW, Martin JE, Hurd AJ, Keefer KD (1985) Springer Proc. Phys. 5: 31
19. Vacher R, Woignier T, Phalippou J, Pelous J, Courtens E, in Ref [2], p 127–131
20. Reichenauer G, Buchenau U, Fricke J, in Ref [2], p 145–150
21. Coddens G, Vacher R, Woignier T, Pelous J, Courtens E, in Ref [2], p 151–156
22. Maynard R, Calemczuk R, De Goer AM, Salce B, Bon J, Bonjour E, Bourret A, in Ref [2], p 107–112
23. Tewari PH, Hunt AJ, Lofftus KD, in Ref [1], p 31–37
24. Hof M, Schleicher J, Schneider FW (1989) Ber. Bunsen-Ges. Phys. Chem. 93: 1377
25. Reul S, Richter W, Haarer D (1990) Lochbrennspektroskopie in porösen anorganischen Matrizen. DPG-Frühjahrstagung, March 26–30, 1990, Regensburg. Verhandl. DPG (VI) 25: 1070
26. Hench LL, West JK (1990) The Sol-Gel Process, Chem. Rev. 90: 33
27. Brinker CJ, Scherer GW (1990) Sol-Gel Science. Academic, Boston
28. Aegerter MA, Jafelicci Jr M, Souza DF, Zanotto ED (1989) Sol-Gel, Science and Technology. World Scientific, Singapore
29. Klein LC (1988) Sol-Gel Technology for Thin Films, Fibers, Preforms, Electronics and Specialty Shapes. Noyes Publication, Park Ridge/New Jersey
30. Hench LL, Ulrich DR (1984) Ultrastructure Processing of Ceramics, Glasses and Composites. John Wiley, New York
31. Hench LL, Ulrich DR (1986) Science of Ceramic Chemical Processing. John Wiley, New York
32. Mackenzie JD, Ulrich DR (1988) Ultrastructure Processing of Advanced Materials. John Wiley, New York
33. Uhlmann DR, Weinsberg M, Ulrich DR (1989) Ultrastructure Processing IV. John Wiley, New York
34. Brinker CJ, Clark DE, Ulrich DR (1984) Better Ceramics Through Chemistry, Mat. Res. Soc. Symp. Proc. 32: North-Holland, New York
35. Brinker CJ, Clark DE, Ulrich DR (1986) Better Ceramics Through Chemistry II, Mat. Res. Soc. Symp. Proc. 73: North-Holland, New York
36. Brinker CJ, Clark DE, Ulrich DR (1988) Better Ceramics Through Chemistry III, Mat. Res. Soc. Symp. Proc. 121: North-Holland, New York

37. Brinker CJ, Clark DE, Ulrich DR (1990) Better Ceramics Through Chemistry IV, Mat. Res. Soc. Symp. Proc. XXX: North-Holland, New York
38. Iler RK (1979) The Chemistry of Silica. John Wiley, New York
39. Teichner SJ, in Ref [1], p 22–30
40. Armor JN, Carlson EJ (1987) J. Mater. Sci. 22: 2549
41. Cutting PA, Parkyns ND, Sing KSW (1972) J. Catal. 27: 222
42. Brinker CJ, Ward KJ, Keefer KD, Holupka E, Bray PJ, Pearson RK, in Ref [1], p 57 67
43. Teichner SJ, in Ref [2], p 1–6
44. Vesthegem H, Jaccon T, Lecomte A, in Ref [2], p 59–64
45. Bourret D, Sempere R, Bouaziz J, Sivade A, in Ref [2], p 71–78
46. Sivade A, Bourret D, Bouaziz J, Sempere R, in Ref [2], p 79–84
47. Armor JN, Carlson EJ, Conner WC (1987) React. Solids 3: 155 and US patent 87-124372
48. Armor JN, Carlson EJ, Zambri PM (1985) Appl. Catal. 19: 339 and US patent 82-449720
49. Chaput F, Lecomte A, Dauger A, Boilot JP, in Ref [2], p 137–142
50. Woignier T, Phalippou J, Zarzycki J (1984) J. Non-Cryst. Solids 63: 117
51. Pajonk GM, Teichner SJ, in Ref [1], p 193–199
52. Vesthegem H, Fargeot D, Dauger A, in Ref [2], p 65–70
53. Chaput F, Boilot JP, Lejeune M (1989) J. Am. Ceram. Soc. 72: 1355
54. Pommier B, Teichner SJ, Lejay P, Sulpice A, Tournier R, in Ref [2], p 41–44
55. Pekala RW (1989) J. Mater. Sci. 24: 3221 and US patent 88-199404
56. Pekala RW, Kong FM. in Ref [2], p 33–40
57. Pekala RW, Alviso CT, A New Synthetic Route to Organic Aerogels, to be published in ref. [37]
58. Schaefer DW, in Ref [2], p 121–126
59. Schmidt H, Scholze H, in Ref [1], p 49–56
60. Keefer KD, in Ref [34], p 15–24
61. Brinker CJ, Drotning WD, Scherer GW, in Ref [34], p 25–32
62. Sakka S, in Ref [28], p 76–102
63. Colby MW, Osaka A, Mackenzie JD (1988) J. Non-Cryst. Solids 99: 129
64. Brinker CJ, Keefer KD, Schaefer DW, Ashley CS (1982) J. Non-Cryst. Solids 48: 47
65. Hrubesh LW, Tillotsen TM, Poco JF. Characterization of Ultralow-Density Silica Aerogels made from Condensed Silica Precursor, to be published in ref. [37]
66. Pool R (1990) Science 247: 807
67. Tillotsen TM, Hrubesh LW, Thomas JM, in Ref [36], p 685–689
68. Scherer GW, in Ref [27], p 357–405 and in Ref [28], p 153–180
69. Woignier T, Phalippou J, Vacher R (1989) J. Mater. Res. 4: 688
70. Esquivias L, Zarzycki J, in Ref [32], p 255–270
71. de la Rosa-Fox N, Esquivias L, Zarzycki J, in Ref [2], p 233–238
72. Craievich A, Aggregation, Kinetics and Structure of Precursor Gels of Zirconia under Different Preparation Conditions. VIII Int. Meet. on Small Angle Scattering, Aug 6–9, 1990, Leuven/Belgium
73. Mielke M, private communication
74. Lu X, Wang P, Büttner D, Heinemann U, Nilsson O, Kuhn J, Fricke J, Thermal Transport in Opacified Monolithic Silica Aerogels. XII Europ. Conf. Thermophys. Prop., Sept. 24–28, 1990, Vienna
75. Avnir D, Kaufman VR, Reisfeld RJ (1985) J. Non-Cryst. Solids 74: 395
76. Scherer GW, in Ref [27], p 453–513 and in Ref [28], p 181–220
77. Zarzycki J, Prassas M, Phalippou J (1982) J. Mater. Sci. 17: 3371
78. Mizuno T, Nagata H, Manabe S (1988) J. Non-Cryst. Solids 100: 236
79. Hench LL, in Ref [31], p 52–64
80. Kistler SS (1932) J. Phys. Chem. 36: 52
81. Phalippou J, Woignier T, Prassas M (1990) J. Mater. Sci. 25: 3111
82. Mukherjee SP, in Ref [32], p 747–759
83. Matson DW, Smith RD (1989) J. Am. Ceram. Soc. 72: 871
84. Laudise RA, Johnson DW, Jr (1986) J. Non-Cryst. Solids 79: 155
85. van Lierop JG, Huizing A, Meermann WCPM, Mulder CAM (1986) J. Non-Cryst. Solids 82: 265
86. Poelz G, in Ref [1], p 176–187
87. Koch-Miramond L, in Ref [1], p 188–192

88. Rasmussen JL, in Ref [2], p 221–226
89. Henning S, in Ref [1], p 38–41
90. Fricke J, in Ref [1], p 94–103
91. Schreiber E, Boy E, Bertsch K, in Ref [1], p 133–139
92. Prassas M, Phalippou J, Zarzycki J (1984) J. Mater. Sci. 19: 1656
93. Boonstra AH, Baken JME (1989) J. Non-Cryst. Solids 109: 1
94. Wang P, Emmerling A, Tappert W, Spormann O, Fricke J, Haubold H-G, High-Temperature and Low-Temperature Supercritical Drying of Aerogels—Structural Investigation with SAXS. VIII Int. Meet. on Small-Angle Scattering, Aug 6–9, 1990, Leuven/Belgium
95. Woignier T, Phalippou J, Hdach H, Scherer GW. Mechanical Properties of Silica Alcogels and Aerogels, to be published in ref. [37]
96. Woignier T (1984) thesis, Université des Sciences et Techniques du Languedoc, Montpellier/France
97. Tewari PH, Hunt AJ, Lofftus KD (1985) Mater. Lett. 3: 363
98. Pauthe M, Phalippou J, in Ref [2], p 215–220
99. Phalippou J, Woignier T, Prassas M, in Ref [2], p 47–52
100. Boy E, Mundig M, Wittwer V, in Ref [2], p 99–105
101. Degn Egeberg E, Engell J, in Ref [2], p 23–28
102. Klvana D, Chaouki J, Repellin-Lacroix M, Pajonk GM, in Ref [2], p 29–32
103. Pajonk GM, in Ref [2], p 13–22
104. Phalippou J, Woignier T, Zarzycki J, in Ref [30], p 70–87
105. Prassas M, Phalippou J, Zarzycki J, in Ref [31], p 156–167
106. Woignier T, Phalippou J, Prassas M (1990) J. Mater. Sci. 25: 3118
107. Mulder CAM, van Lierop JG, Frens G (1986) J. Non-Cryst. Solids 82: 92
108. Papanikolau E, Meerman WCPM, Aerts R, van Roy TL, van Lierop JG, Meeuwsen TPM (1988) J. Non-Cryst. Solids 100: 247
109. Iura J, Kawaguchi T, in Ref [2], p 53–58
110. Schaefer DW (1989) Science 243: 1023
111. Su SR, Onorato PJK, in [10], p 237–244
112. Ebert H-P (1990) diploma thesis, Report E21-0490-1 Universität Würzburg
113. Baraton M-J, Merle-Mejean T, Quintard P, Lorenzelli V, in ref [2], p 239–243
114. Gross J (1987) diploma thesis, Report E12-1087-4 Universität Würzburg
115. Vacher R, Woignier T, Pelous J, Courtens E (1988) Phys. Rev. B 37: 6500
116. Vacher R, Woignier T, Phalippou J, Pelous J, Courtens E (1988) J. Non-Cryst. Solids 106: 161
117. Scherer GW, in Ref [27], p 675–742 and in Ref [28], p 221–256
118. Onorato PJK. unpublished results, see also Ref [129]
119. Lours T, Zarzycki, Craievich AF, Aegerter MA, dos Santos DJ (1988) J. Non-Cryst. Solids 106: 157
120. Rahaman MN, de Jonghe LC, Shinde SL, Tewari PH (1988) J. Am. Ceram. Soc. 71: C338
121. Vesthegem H, di Giampaolo AR, Dauger A (1987) J. Mater. Sci. Lett. 6: 1187 and Sci. Ceram. 14: 321 (1988)
122. Woignier T, Phalippou J, Vacher R. in Ref [36], p 697–702
123. Gross J, Goswin R, Gerlach R, Fricke J, in Ref [2], p 185–190
124. Emmerling A, Gross J, Gerlach R, Goswin R, Reichenauer G, Fricke J, Haubold H-G. (1990) J. Non-Cryst. Solids 125: 230
125. Scherer GW (1977) J. Am. Ceram. Soc. 60: 236
126. Gallo TA, Brinker CJ, Klein LC, Scherer GW, in Ref [34], p 85–90
127. Gallo TA, Klein LC, in Ref [35], p 245–250
128. Scherer GW (1988) J. Am. Ceram. Soc. 71: 447
129. Scherer GW (1987) J. Ceram. Soc. Jpn. Inter. Ed. 95: 20
130. von Nordstrand RA, Kreger WE, Ries HE (1951) J. Phys. Coll. Chem. 55: 612
131. Iura J, Hishikura H, Kamikatano M, Kawaguchi T (1988) J. Non-Cryst. Solids 100: 241
132. Bouaziz J, Woignier T, Bourret D, Sempere R (1986) J. Non-Cryst. Solids 82: 225
133. Bouaziz J, Bourret D, Sempere R, Regnier J (1986) J. Non-Cryst. Solids 82: 183
134. dos Santos DJ, Aegerter MA, Cruz CH, Scarparo M, Zarzycki J (1986) J. Non-Cryst. Solids 82: 165
135. Sempere R, Bourret D, Bouaziz J, Sivade A, in Ref [2], p 227–232
136. Gokturk HS (1989) thesis, Stevens Inst. Techn., Hoboken/New Jersey
137. Kollmar A, Legrand JF. Experimental Report 9-11-200, ILL Grenoble (1989)

138. Maynard R, Deutscher G (1989) Europhys. Lett. 10: 257
139. Mielke M, private communication
140. Pekala RW, Aerogels and Xerogels from Organic Precursors, to be published in Ref [33]
141. Pekala RW, Kong FM (1989) Polym. Prepr. 30: 221
142. Lindquist DA, Borek TT, Kramer SJ, Narula CK, Johnston G, Schaeffer R, Smith DN, Paine RT (1990) J. Am. Ceram. Soc. 73: 757
143. Fricke J, Reichenauer G (1987) J. Non-Cryst. Solids 95 & 96: 1135
144. Brunauer S, Emmett PH, Teller E (1938) J. Am. Chem. Soc. 60: 309
145. Gregg SJ, Sing SW (1982) Adsorption, Surface Area and Porosity. Academic Press, New York
146. Schuck G, Dietrich W, Fricke J, in Ref [1], p 148–153
147. Woignier T, Phalippou J (1987) J. Non-Cryst. Solids 93: 17
148. Jura J, Hishikura H, Taneda N, Kawaguchi T (1988) Rep. Res. Lab. Asahi Glass Co. Ltd., 38: 15
149. Washburn EW (1921) Proc. Nat. Acad. Sci. 7: 115
150. Smith DM, Glaves CL, Gallegos DB (1988) Am. Chem. Soc. Div. Petr. Chem. Preprints 33: 618
151. Mukherjee S, Cordaro JF, Debsikdar JC (1988) Adv. Ceram. Mater. 3: 463
152. Tewari PH, Hunt AJ, Lieber JG, Lofftus K, in Ref [1], p 142–147
153. Lampert CM, Mazur JH, in Ref [1], p 154–159
154. Price LW, McCarthy PL (1988) Scanning 10: 29
155. Kerker M (1969) The Scattering of Light. Academic, New York
156. Mie G (1908) Ann. Phys. 25: 377
157. van de Hulst HC (1981) Light Scattering by Small Particles. Dover, New York
158. Guinier A, Fournet G (1955) Small Angle Scattering of X rays. John Wiley, New York
159. Glatter O, Kratky G (1982) Small Angle X-ray Scattering. Academic, London
160. Dachs H (1978) Neutron Diffraction. Springer, Berlin Heidelberg New York
161. Kostorz G (1978) Neutron Scattering in Material Science. Academic, New York
162. Schaefer DW (1987) Mat. Res. Soc. Symp. Proc. 79: 47
163. Guinier A (1939) Ann. de Phys. 12: 161
164. Mandelbrot BB (1982) The Fractal Geometry of Nature. Freeman, San Francisco
165. Sinha SK, Freltoft T, Kjems J (1984) in: Family F, Landau DP (eds.) Kinetics of Aggregation and Gelation, Elsevier, Amsterdam, p 87–90
166. Porod G (1951) Kolloidz. 124: 83
167. Bale HD, Schmitt PW (1984) Phys. Rev. Lett. 53: 596
168. Porod G, in Ref [159], p 17–51
169. Brinker CJ, in Ref [2], p 97–233
170. Assink RA, Kay BD (1988) J. Non-Cryst. Solids 99: 359 and 104: 112 and 107: 35 (1988)
171. Devreu F, Boilot JP, Chaput F, Lecomte A (1990) Phys. Rev. A 41: 6901
172. Ward RL, Pekala RW (1990) Polym. Prepr. 31: 167
173. Devreux F, Boilot JP, Chaput F, Savopal B (1990) Phys. Rev. Lett. 65: 614
174. Emmerling A, Fricke J, Haubold H-G, Reichenauer G, Wang P. SAXS-Investigations of Sol-Gel Process. HASYLAB-Jahresbericht 1988, p 319–320
175. Schaefer DW, Keefer KD. in Ref [34], p 1–14
176. Schaefer DW, Keefer KD (1984) Phys. Rev. Lett. 53: 1383
177. Orcel G, Gould RW, Hench LL, in Ref [35], p 289–294
178. Hardmann-Rhyne KA, Coyle TD, Lewis EP, in Ref [35], p 173–178
179. Lours T, et al. (1988) J. Non-Cryst. Solids 100: 207
180. Martin JE, Ackerson BJ (1985) Phys. Rev. A31: 1180
181. Pouxviel JC, Boilot JP, in Ref [32], p 197–209
182. Pouxviel JC, Boilot JP, Dauger A, Wright A (1988) J. Non-Cryst. Solids 103: 331
183. Heinrich T, Raether F. SAXS-Measurements of the Condensation in Mullite-Precursors. VIII Int. Meet. on Small Angle Scattering, Aug. 6–9, 1990, Leuven/Belgium
184. Tewari PH, Hunt AJ, Lofftus KD, Lieber JG, in Ref [35], p 195–205
185. Wood DL, Rabinovich EM (1989) J. Non-Cryst. Solids 107: 199
186. Boonstra AH, Meeuwsen TPM, Baken JME, Aben GVA (1989) J. Non-Cryst. Solids 109: 153
187. Lourdin P, Appel J, Pelous J, Woignier T, in Ref [2], p 197–201
188. Schaefer DW, Keefer KD, in Ref [35], p 277–288
189. Witten TA, Sander LA (1981) Phys. Rev. Lett. 47: 1400
190. Eden M (1961) in Neyman F (ed.) Proc. 4th Berkeley Symposium on Math., Stat. and Prob., University of California Press, Berkeley, p 223

191. Meakin P (1986) in Stanley E, Ostrowsky N (eds.) On Growth and Form, Martinus-Nijhoff, Dordrecht, p 111
192. Stauffer D (1985) Introduction to Percolation Theory. Taylor and Francis, London
193. Fricke J (1986) Physik in unserer Zeit 17: 101 and 151
194. Brinker CJ, in Ref [27], p 319–330
195. Zarzycki J, in Ref [31], p 21–36
196. Leach PL (1976) Phys. Rev. B 14: 5046
197. Schaefer DW (1988) Bull. Mat. Res. Soc. 13–2: 22
198. Schaefer DW, Brinker CJ, Wilcoxon JP, Wu D-Q, Phillips JC, Chu B, in Ref [36], p 691–696
199. Keefer KD, in Ref [34], p 15–21
200. Keefer KD, in Ref [31], p 131–139
201. Keefer KD, in Ref [35], p 295–304
202. Keefer KD, Schaefer DW (1986) Phys. Rev. Lett. 56: 2376
203. Craievich A, dos Santos DJ, Aegerter MA, Lours T, Zarzycki J (1988) J. Non-Cryst. Solids 100: 424
204. Schaefer DW, Keefer KD (1986) Phys. Rev. Lett. 56: 2199
205. Boukenter A, Champagnon D, Dumas J, Duval E, Quinson JF, Rousset JL, Serughetti J, Etienne S, Mai C, in Ref [II], p 133–136
206. Emmerling A (1989) diploma thesis, Report E 21-0589-1 Universität Würzburg
207. Craievich AF, Aegerter MA, dos Santos DJ, Woignier T, Zarzycki J (1986) J. Non-Cryst. Solids 86: 394
208. dos Santos DJ, Aegerter MA, Craievich AF, Lours T, Zarzycki T (1987) J. Non-Cryst. Solids 95 & 96: 1143
209. Lours T, Zarzycki J, Craievich A, dos Santos DJ, Aegerter MA (1987) J. Non-Cryst. Solids 95 & 96: 1151
210. Chaput F, Boilot JP, Dauger A, Devreux F, de Geyer A (1990) J. Non-Cryst. Solids 116: 133
211. Pahl R, Bonse U, Pekala R, Kinney JH. SAXS-Investigation on Organic Aerogels. VIII Int. Meet. on Small Angle Scattering, Aug. 6–9, 1990, Leuven/Belgium
212. Hunt AJ, in Ref [30], p 549–557
213. Hunt AJ, Berdahl P, in Ref [34], p 275–280
214. Beck A, Caps R, Fricke J (1989) J. Phys. D 22: 730
215. Beck A, Gelsen O, Wang P, Fricke J, in Ref [2], p 203–208
216. Hunt AJ, Lofftus KD, in Ref [36], p 679–684
217. Mukherjee SP, in Ref [32], p 747–759
218. Bourret A (1988) Europhys. Lett. 6: 731
219. Fricke J, in Ref [28], p 482–505
220. Woignier T, Pelous J, Phalippou J, Vacher R, Courtens E (1987) J. Non-Cryst. Solids 95 & 96: 1197
221. Woignier T, Phalippou J (1988) J. Non-Cryst. Solids 100: 404
222. Michard F, Zarembowitch A, in Ref [2], p 173–178
223. Woignier T, Phalippou J, in Ref [2], p 179–184
224. Gronauer M, Kadur A, Fricke J, in Ref [1], p 167–173
225. Woignier T, Phalippou J, Sempere R, Pelous J (1988) J. Phys. Fr. 49: 289
226. Maynard R (1989) Physica A 157: 601
227. Woignier T, Phalippou J, Vacher R (1989) in Ref [36], p 697–702 and J. Mater. Res. 4: 688
228. Phalippou J, Woignier T, Rogier R, in Ref [2], p 191–196
229. Xhonneux P, Courtens E, Pelous J, Vacher R (1989) Europhys. Lett. 10: 733
230. Arvidson JM, Scull LL (1986) Adv. Cryog. Eng. 32: 342
231. Scull LL, Arvidson JM (1988) Adv. Cryog. Eng. 34: 413
232. Gross J, Reichenauer G, Fricke J (1988) J. Phys. D 21: 1447
233. Rubin M, Lampert CM (1983) Sol. Energy Mater. 7: 393
234. Wang P. private communication
235. Tewari PH, Hunt AJ, Lofftus KD, Lieber JG, in Ref [35], p 195–205
236. Platzer W, Wittwer V, Mielke M, in Ref [1], p 127–132
237. Russo RE, Hunt AJ (1986) J. Non-Cryst. Solids 86: 219
238. van Roosmaleu AJ, Mol JC (1978) J. Phys. Chem. 82: 2748 and 83: 2485 (1979)
239. Scheuerpflug P, Caps R, Büttner D, Fricke J (1985) Int. J. Heat Mass Transfer 28: 2299
240. Parkyns ND (1972) J. Catal. 27: 34
241. Merle-Mejean T, Vesthegem H, Quintard P, Lorenzelli V (1988) J. Mol. Struct. 175: 401

242. Hrubesh LW, Alviso CT, in Ref [36], p 703–709
243. Caps R, Banmeister K, Fricke J (1986) High Temp.—High Pressures 18: 589
244. Büttner D, Caps R, Heinemann U, Hümmer E, Kadur A, Scheuerpflug P, Fricke J, in Ref [1], p 104–109
245. Caps R, Fricke J, in Ref [1], p 110–115
246. Büttner D, Hümmer E, Fricke J, in Ref [1], p 116–120
247. Heinemann U, Hümmer E, Büttner D, Caps R, Fricke J (1986) High Temp.-High Pressures 18: 517
248. Fricke J (1986) Umschau 7: 374
249. Fricke J, Caps R, Büttner D, Heinemann U, Hümmer E (1987) J. Non-Cryst. Solids 95 & 96: 1167
250. Büttner D, Caps R, Heinemann U, Hümmer E, Kadur A, Fricke J (1988) Solar Energy 40: 13
251. Caps R, Fricke J (1984) Int. J. Sol. Energy 3: 13
252. Büttner D, Fricke J (1985) Int. J. Sol. Energy 3: 89
253. Büttner D, Caps R, Fricke J (1985) High Temp.-High Pressures 17: 375
254. Caps R, Fricke J (1986) Sol. Energy 36: 361
255. Fricke J, Hümmer E, Morper H-J, Scheuerpflug P, in Ref [2], p 87–97
256. Caps R, Döll G, Fricke J, Heinemann U, Hetfleisch J, in Ref [2], p 113–118
257. Alexander S, Orbach R (1982) J. Phys. Lett. 43: L 625
258. Entin-Wohlmann O, Orbach R, Polatsek G (1989) Springer Proc. Phys. 37: 288
259. Orbach R (1989) Physica D 38: 266
260. Alexander S (1989) Phys. Rev. B 40: 7953
261. Courtens E, in Ref [2], p 143–144
262. Courtens E, Pelous J, Phalippou J, Vacher R, Woignier T (1987) J. Non-Cryst. Solids 95 & 96: 1175
263. Courtens E, Pelous J, Phalippou J, Vacher R, Woignier T (1987) Phys. Rev. Lett. 58: 128
264. Pelous J, Vacher R, Woignier T, Sauvajol JL, Courtens E (1989) Philos. Mag. B 59: 65
265. Vacher R, Courtens E, Coddens G, Pelous J, Woignier T (1989) Phys. Rev. Lett. 39: 7384
266. Boukenter A, Champagnon B, Duval E, Dumas J, Serughetti J, Quinson JF (1987) J. Non-Cryst. Solids 95 & 96: 1175
267. Tsujimi Y, Courtens E, Pelous J, Vacher R (1988) Phys. Rev. Lett. 60: 2757
268. Hassan HA, Smith SRP, Page JH (1988) Solid State Comm. 68: 733
269. Rousset JL, Boukenter A, Champagnon B, Duval E, Quinson JF, Chatelut M, Dumas J, Serughetti J, in Ref [2], p 163–166
270. Boukenter A, Champagnon B, Duval E, Rousset JL, Dumas J, Serughetti J. J. Phys. C 21: 1 1097, Ref [88]
271. Rousset JL, Duval E, Boukenter A, Champagnon B, Monteil A (1988) J. Non-Cryst. Solids 107: 27
272. Sauvajol JL, Pelous J, Woignier T, Vacher R, in Ref [2], p 167–169
273. Pelous J, Woignier T, Vacher R, Sauvajol JL (1989) Physica A 157: 625
274. Pelous J, Woignier T, Vacher R, Sauvajol JL. J. Phys. Fr. 51: 433 Ref [90]
275. Courtens E, Vacher R (1987) Z. Phys. B 68: 355
276. Vacher R, Courtens E (1989) Phys. Scr. T 29: 239
277. Buchenau U (1985) Z. Phys. B 58: 181
278. Vacher R, Woignier T, Pelous J, Coddens G, Courtens E (1989) Europhys. Lett. 8: 161
279. Courtens E, Stoll E, Vacher R (1989) Physica D 38: 41
280. Reichenauer G, Buchenau U, Fricke J (1989) Springer Proc. Phys. 37: 310
281. Reichenauer G, Fricke J, Buchenau U (1989) Europhys. Lett. 8: 415
282. Vacher R, Courtens E, Coddens G, Heidemann A, Tsujimi Y, Pelous J, Foret M (1990) Phys. Rev. Lett. 65: 1008
283. Conrad H, Fricke J, Reichenauer G, in Ref [2], p 157–162
284. Conrad H, Reichenauer G, Fricke J (1989) Springer Proc. Phys. 37: 304
285. Conrad H, Buchenau U, Schaetzler R, Reichenauer G, Fricke J (1990) Phys. Rev. B 41: 2573
286. Pelous J, Courtens E, Vacher R, Woignier T, in Ref [2], p 209–212
287. Courtens E, Lartique C, Mezei F, Vacher R, Coddens G, Foret M, Pelous J, Woignier T (1990) Z. Phys. B 79: 1
288. Schaefer DW, Brinker CJ, Richter D, Farago B, Frick B (1990) Phys. Rev. Lett. 64: 2316
289. Calemczuk R, de Goer AM, Salce B, Maynard R, Zarembowitch A (1987) Europhys. Lett. 3: 1205

290. de Goer AM, Calemczuk R, Salce B, Bon J, Bonjour E, Maynard R (1989) Phys. Rev. B 40: 8327
291. Bernasconi A, Sleator T, Posselt T, Ott HR (1990) Rev. Sci. Instr. 61: 2420
292. Reichenauer G, Wolter E, Stumpf C, private communication
293. Carlson P (1986) Nucl. Instr. Meth. A 248: 110
294. Poelz G (1986) Nucl. Instr. Meth. A 248: 118
295. Vincent P, Debbe R, Pfoh A, Abreu M (1988) Nucl. Instr. Meth. A 272: 660
296. Holl I, Lorenz E, Mageras G (1990) A Concept for Particle Identification for a B-Factory Detector, MPI-PA-E/Exp. E1-224, MPI f. Physik und Astrophysik, Garching/Germany
297. Nilsson O, Fransson Å, Sandberg O, in Ref [1], p 121–126
298. Wolff B, Seybold G, Krueckau FE (1989) European Patent Appl. 340707
299. Novakov: Report on the Cité Solaire, Ardon/Switzerland. 2nd Internat. Workshop on Transparent Insulation, March 1988, Freiburg/Germany
300. Jesch LF (1987) Transparent Insulation Materials for Passive Solar Energy Utilization. Franklin Comp. Cons. Birmingham
301. Fricke J, Caps R, Büttner D, Heinemann U, Hümmer E, Kadur A (1987) Sol. Energy Mater. 16: 267
302. Rommel M, Wittwer V, Goetzberger A (1988) Adv. Sol. Energy Technol. 4: 1553
303. Hunt AJ, Lofftus KD (1988) Adv. Sol. Energy Technol. 4: 3146
304. Schmidt Ch, Goetzberger A, Schmid J. Test Results and Optimization of Integrated Collector Storage with a Transparent Cover. 2nd Internat. Workshop on Transparent Insulation, March 1988, Freiburg/Germany
305. Pflüger A (1987) Sol. Energy Mat. 16: 255
306. Fricke J (1989) Phys. in unserer Zeit 20: 189
307. Kuhn J (1989) diploma thesis, Report E21-0889-4 Universität Würzburg
308. Gerlach R. private communication
309. Cooper DW (1989) Part. Sci. Tecnol. 7: 371
310. Hrubesh LW (1989) Development of Low-Density Silica Aerogel as a Capture Medium for Hyper-Velocity Particles. Report UCRL-21234, LLNL, Livermore/USA
311. Fearon EM, Coronado PR, Garza RG, Darling DH (1987) J. Nucl. Mater. 149: 105
312. Hair LM, Pekala RW, Stone RE, Chen L, Buckley SR. J. Vac. Sci. Technol. A 6: 255
313. Kim NK, Kim K, Payne DA (1989) J. Vac. Sci. Technol. A 7: 1181
314. Jang KY, Kim K, Uphadye RS (1990) J. Vac. Sci. Technol. A 8: 1732
315. Chan MHW, Blum KJ, Murphy SQ, Wong GKS, Reppy JD (1988) Phys. Rev. Lett. 61: 1950
316. Snow WM, Sokol PE (1990) J. Low-Temp. Phys. 80: 197
317. Holmes NC, Radousky HB, Moss MJ (1984) Appl. Phys. Lett. 45: 626
318. Ebeling W, Reierson DA, Pence RJ, Viray MS (1975) Pestic. Biochem. Physiol. 5: 81
319. Sen-Sarma PK, Mishra SC (1976) Indian J. Entomol. 38: 77
320. Vrba CH, Arai HP, Nosal M (1983) Can. J. Zool. 61: 1481
321. Loschiavo SR (1987) J. Econ. Entomol. 81: 1231 and 81: 1237 (1988)

Sol–Gel-Derived Coating Films and Applications

Sumio Sakka and Toshinobu Yoko

Institute for Chemical Research, Kyoto University, Gokasho, Uji, Kyoto-Fu 611, Japan

Processing of sol–gel-derived coating films and their recent applications to preparation of high technology inorganic materials have been reviewed. Usually, dip-coating technique is suitable for preparing thin films on substrates of large areas. The thickness of the film can be controlled by the composition and the viscosity of the coating solution, and the withdrawing speed. Preferred crystal orientation can be given depending on the conditions of preparation. Thus, the technique can be applied to accomplishing a large variety of advanced, functional films having optical, electronic, chemical, and mechanical functions. Many examples of such films have been cited and discussed.

Structure and Bonding 77

1 Introduction

In sol–gel methods [1–4], inorganic materials are prepared from solutions containing metal compounds, such as metal alkoxides, metal acetylacetonates, and inorganic metal compounds as sources of cations in the aimed oxides, and water as hydrolysis agent and alcohols as solvents. In the solution, metal compounds undergo hydrolysis and polycondensation, forming polymers or particles, and the solution becomes a sol. Further reaction connects the particles, solidifying the sol into a gel. The dried gels may be final products in some cases. In this case, the maximum processing temperature may be lower than 100 °C or 160 °C and so one does not have to worry about the decomposition of the organic compounds embedded in the gel which have been added as indispensable components. On the other hand, heating of pertinent dried gels to several hundred degrees celsius produces glasses and ceramics which have prescribed shapes, such as plate and cylinder, fiber and coating film. Thus, in general, inorganic materials can be obtained without powder processing.

Dip-coating or spin-coating of glass, ceramic, metal, and plastic substrates by sol–gel method is very useful for modifying properties of the substrates with a large or small surface area and provides the substrates with new active functions [5, 6]. In the years around 1970, dip-coating was applied by Schroeder [7] to glasses, in order to modify their optical properties. In recent years, the dip-coating technique has again attracted much attention as the demands for developing various advanced materials.

The sol–gel method of fabricating thin films offers potential advantages over traditional techniques as shown in Table 1. Considering that a substrate is used for coating, the low processing temperature is particularly important in the application to electronic, optoelectronic, and photonic devices, because the substrate and other active elements on the substrate are not necessarily highly heat-resistant. Easy coating of large surfaces makes it possible to apply the sol–gel coating to display panel and windows as substrates. The small thickness may be advantageous for coating films for some optical and electronic devices. For other purposes, however, thicker films may be needed. Also high optical quality films can be provided.

Table 1. Advantages of sol–gel coating

Term	Remarks
Low-temperature processing	Coating of glasses, semiconductors, and integrated electronic and optical devices is possible
Easy coating of large surfaces	Application can be made to display and windows
Small thickness	Very thin coating films are possible
High optical quality	Transparent films are possible

2 Formation of Sol–Gel Coating Films

2.1 Coating Processes

The dip-coating consisting of three steps of dipping, withdrawing, and heating may be a representative procedure in making coating films by sol–gel method. A drawing apparatus is shown in Fig. 1. The advantages and disadvantages of dip-coating discussed below may be shared, however, with other liquid-phase coating procedures such as spin coating and dripping of a solution on the substrate [8].

In dip-coating, the substrate is dipped in a homogeneous solution containing various metal compounds which would finally form desired simple or complex oxides, as shown in Fig. 2. A liquid film attached to the substrate becomes a gel

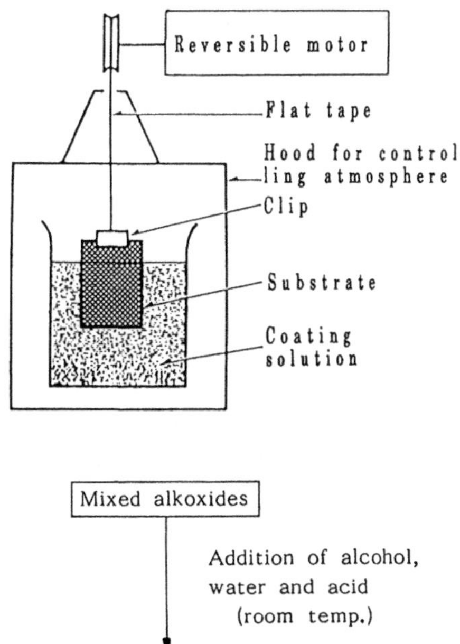

Fig. 1. Apparatus for dip coating

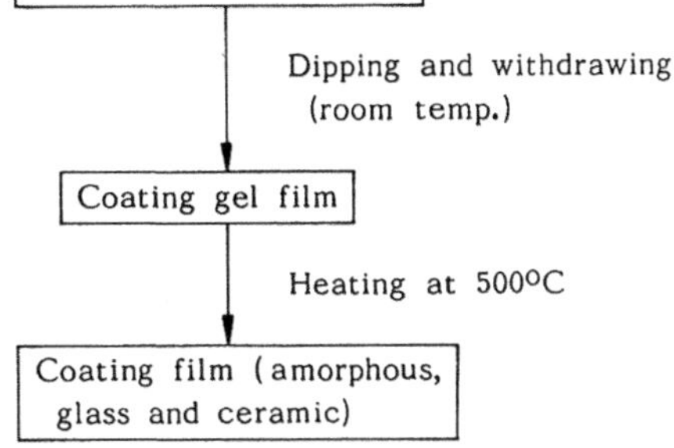

Fig. 2. Procedure for dip coating using metal alkoxides

film as a result of the gelling reaction in the ambient atmosphere upon withdrawal from the solution. The substrate with a gel film is heated to a certain high temperature, say to 500 °C, in order to produce chemical bondings between the film and the substrate. The chemical bondings –M–O–M′– may be formed as a result of dehydration. Subsequent heating at higher temperatures will change the gel film to a glass or ceramic film with desired crystalline species.

Brinker et al. [9, 10] analyzed the process of formation of a gel film from the solution on withdrawal of a substrate on the basis of a model shown in Fig. 3. It was assumed that during sol–gel thin film formation sols are deposited on the substrate surface by a complex process combining gravitational draining, solvent evaporation, and continued condensation reactions.

2.2 Bonding between Film and Substrate and Film Thickness

In typical sol–gel coatings, where glasses and ceramics such as SiO_2, Al_2O_3, and $BaTiO_3$ are made from metal alkoxides dissolved in mixed solutions of water and alcohol, as withdrawn semi-dried coating films consisting of particles with metalloxane bondings are further dried and bond to the substrate at higher temperatures during heating. Formation of the films which firmly bond to the substrate can be assured only when metal–oxygen–metal bonds as shown in Fig. 4 are formed. It is noted that the film remains brittle throughout the drying and bonding processes.

The thickness of the film achieved in one coating procedure should be small at 0.1–0.3 μm at the maximum in such typical sol–gel coatings. In a condition in

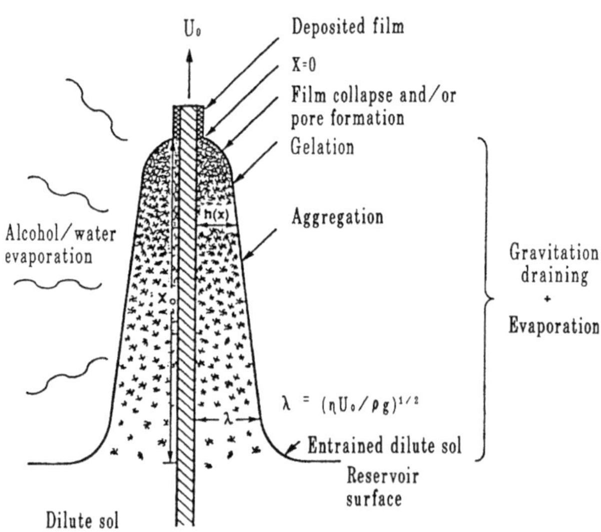

Fig. 3. Schematic of the sequential stages in sol–gel dip-coating process; after Brinker and Scherer [10]

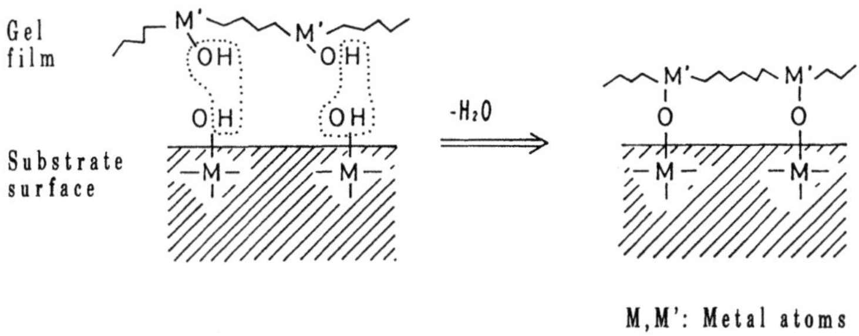

Fig. 4. Formation of chemical bonding (metalloxane bonding) between the coating film and the substrate

which the film might become thicker than the above limit, the film would come off the substrate entirely or partially. This peeling off may be caused by the shrinkage of the coating film during heating. When a semi-wet gel film as applied to the substrate happens to be too thick, the coherent force within the film would cause the shrinkage due to drying of the film in the direction parallel to the substrate surface. If this coherent force exceeds the force of bonding of the film to the substrate surface, the film may come off. In thinner films, the bonding force between the film and the substrate prevents the shrinkage in the direction parallel to the surface, and the film is supposed to shrink in the direction perpendicular to the substrate surface, resulting in the formation of the film that is firmly attached to the substrate.

Figure 5 shows the thickness of SiO_2 films prepared from $Si(OC_2H_5)_4$–H_2O–C_2H_5OH–HCl solution as a function of the viscosity of the coating solution [11]. The substrate is a soda-lime-silica glass slide for optical microscopy. The viscosity increases as the hydrolysis-polycondensation reaction progresses. It is seen that the thickness increases with increasing viscosity. When the viscosity is larger than 6 centipoise or 20 centipoise in the solutions without

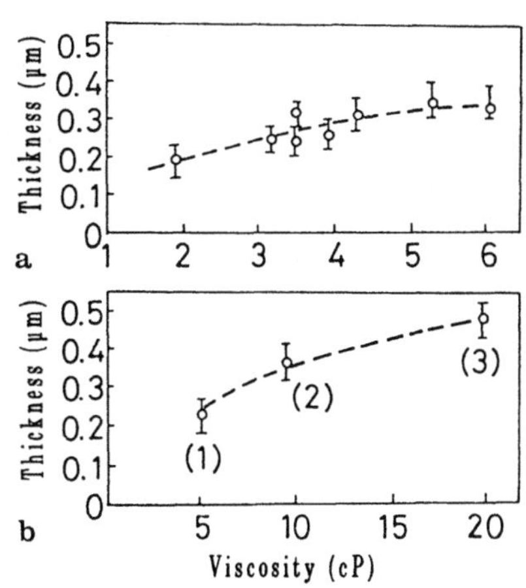

Fig. 5a,b. Change of the thickness of SiO_2 films with the viscosity of tetraethoxysilane solutions in the range where uniform coating films are obtained [11]. (*a*) without and (*b*) with hydroxypropylcellulose which increases viscosity of the solution

and with hydroxypropylcellulose, a viscosity-increasing agent, respectively, however, the film cannot be obtained. This is because the film is too thick to remain adherent to the substrate. It is noted that a difference in thermal expansion coefficient between silica film ($\alpha = 5 \times 10^{-7}/°C$) and the soda-lime-silica glass substrate ($\alpha = 100 \times 10^{-7}/°C$) does not cause a serious problem for the formation of a uniform film which remains firmly attached to the substrate.

In order to give an idea on the thicknesses achieved in one stroke of coating operation in various coatings, a classification of coating films in terms of the thickness is shown in Table 2.

In the cases of hard coating on plastics (point 3 in Table 2), relatively thick films can be achieved in spite of the presence of a shrinkage, probably due to the stress relaxation as a result of plastic flow of the drying film. In paints (point 4 in Table 2), relaxation relieves the stress, and so a very thick film as thick as 0.5 mm is easily obtained without peeling off or cracks. In the vapor-phase deposited film, almost no volume change (shrinkage) is seen during deposition and adherence, which permits the formation of films about 10 μm thick. The larger thickness might cause peeling-off due to the difference in thermal shrinkage during cooling. Very thick films can be achieved in porcelain enamels (point 6 in Table 2), due to no shrinkage taking place at the temperature where the molten glass solidifies.

In typical sol–gel coating (point 1 in Table 2), the volume shrinkage during the adhesion is large and no stress relaxation takes place, and so the maximum possible thickness is small at 0.1–0.3 μm. If the sol–gel derived gel is melted during heating and shrinks at the same time, a significant volume change would not occur at the time of bonding of the film to the substrate and thicker films reaching 1–2 μm would be possible (point 2 in Table 2), as shown for high T_c superconducting coating films [12].

It can be stated that a typical sol–gel coatings are suitable for preparing very thin films. An increase of thickness can be attained, however, by repeating the coating procedure consisting of dipping, withdrawing of the substrate and heating to several hundred degrees. The relation between the film thickness and

Table 2. Classification of coating films on the basis of the order of the magnitude of film thickness

	Thickness*	Shrinkage**	Remarks
(1) Sol–gel film (Thin)***	0.1–0.3 μm	Large	Brittle throughout the coating processes
(2) Sol–gel film (Thick)	1–2 μm	Almost none	Melting
(3) Hard coat on plastics	0.1 mm	Small	Plastic flow
(4) Paint	0.5 mm	Small	Plastic flow
(5) Vapor-phase deposited film (CVD, PVD)	10 μm	Almost none	No shrinkage
(6) Porcelain enamel	0.5 mm	Almost none	Melting

* Maximum possible thickness achieved in one coating procedure
** Shrinkage during adhesion
*** Typical sol–gel coating discussed in this article

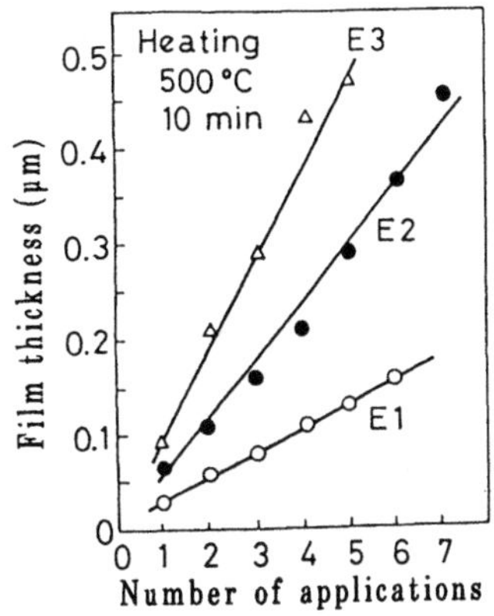

Fig. 6. Film thickness vs. number of applications for $BaTiO_3$ films by dip-coating technique [48]. Solutions E1, E2, and E3 contain 1.72 wt %, 4.28 wt %, and 6.34 wt % $BaTiO_3$, respectively

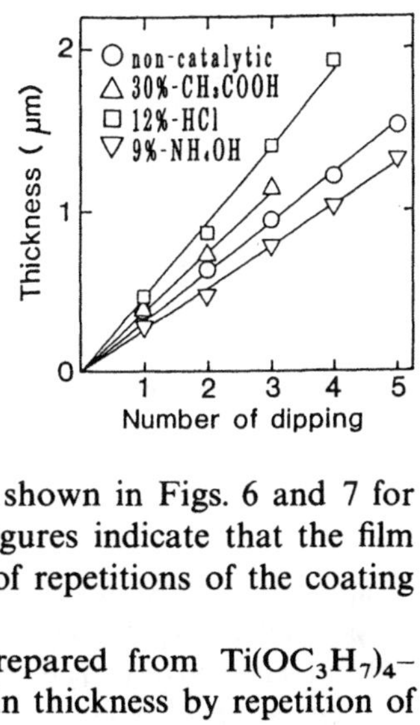

Fig. 7. Relation between the number of applications and the thickness of the TiO_2–CeO_2 coating films; after Makishima et al. [14]

the number of repetition of coating procedure is shown in Figs. 6 and 7 for $BaTiO_3$ [13] and CeO_2–TiO_2 [14] films. Both figures indicate that the film thickness increases in proportion to the number of repetitions of the coating procedure.

The CeO_2–TiO_2 coating films have been prepared from $Ti(OC_3H_7)_4$–$CeCl_3$–CH_3COOH–H_2O solutions. The increase in thickness by repetition of coating procedures has also been shown for TiO_2 films [15] prepared from $Ti(OC_3H_7)_4$–C_2H_5OH–H_2O–HCl solutions. In these cases, it is essential to repeat the whole procedure consisting of dipping, withdrawing and heating to, say 500 °C. Omitting of the last process of heating would lead to peeling-off of the film from the substrate. It is assumed that the film becomes a part of the original substrate after heating, meaning that one coating procedure is repeated every time in each subsequent coating operation.

2.3 *Factors Affecting the Film Thickness*

The following factors affect the thickness of glass and crystalline dip-coating films within the thickness range where uniform films attached to the substrate are obtained [11, 13, 16–21].

(1) The viscosity of the coating solution: The thickness of the film increases with increasing viscosity.

(2) The speed of withdrawing: The thickness increases with increasing speed of withdrawing.

(3) The oxide concentration of the solution: The thickness increases with increasing oxide concentration in the coating solution.

(4) Heating temperature and time: Gel coating films are porous and sintered when they are heated. Accordingly, the film thickness decreases with increasing heating temperature and time.

Very complicated structural changes of the solution take place during the process of withdrawing of the substrate from the solution, as indicated by Brinker et al. [9, 10], and so it is suspected that the film thickness should change with the drawing condition in a very complicated way. Bearing this in mind, the following relationship should be considered.

It is reported that the thickness of the film t is related to the drawing speed v in the formula [16]:

$$t = \mathrm{K}\left(\frac{\eta v}{\rho g}\right)^{1/2} \tag{1}$$

where η is the viscosity of the solution, ρ the density of the solution, g the gravitational acceleration, and K a constant. Yang et al. [17] applied the following relationship to the analysis of the t–v relationship for toluene solutions of styrene–hexylmethacrylate copolymers.

$$t = \mathrm{J}\xi\left(\frac{\eta - \eta_s}{\eta_o}\right)^{0.84}\left(\frac{\eta v}{\rho_s g}\right)^{1/2} \tag{2}$$

where J is a constant, ρ_s and ρ_p are the densities of a solvent and polymers, respectively; $\xi = \rho_s/\rho_p$ and, η and η_s are the viscosities of the solution and the solvent, respectively; η_o is the coefficient appearing in $\eta = \eta_s + \eta_o \cdot C_p d$ (C_p is the concentration of polymers), and g is the same as cited above.

Equations (1) and (2) indicate that the slope of the $\log t - \log v$ plot is 1/2. Dislich and Hussmann [6] obtained 2/3 for sols prepared from metal alkoxides, instead of 1/2 predicted from Eqs. (1) and (2). Mukherjee [21] and Strawbridge and James [18, 19] showed that the $v^{1/2}$ rule of the thickness is obeyed for the SiO_2 films made from tetraethoxysilane sols. Figure 8 shows the thickness-withdrawing speed relations for $BaTiO_3$ films [13], indicating that the slope is 0.52–0.62 which is close to theoretical values in Eqs. (1) and (2). According to Guglielmi and Zenezine [20], different investigators report n values in the $t - v^n$ relationship which range from 0.1 to 1, although n is close to 1/2 in most cases. This scattering of n values has to be checked bearing the complexity of structure and property change during withdrawing in mind.

The relations [11] between the film thickness t and the solution viscosity η for SiO_2 films shown in Fig. 5 are replotted in logarithmic scales in Fig. 9. It is seen that the slope is 0.5 and 0.56, respectively, for the solution without and with

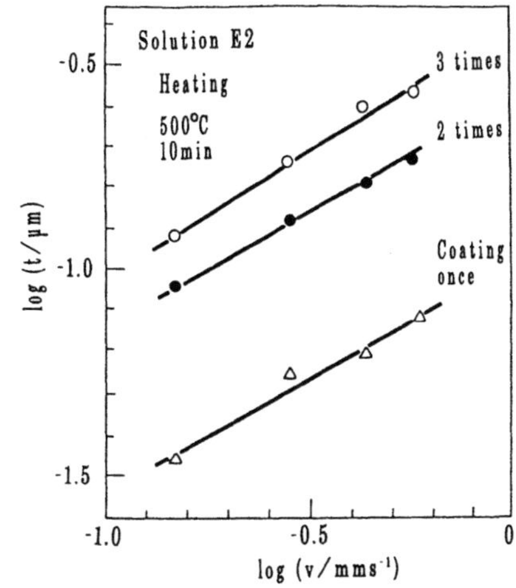

Fig. 8. Relation between the thickness t and the withdrawing speed v in logarithmic scale for $BaTiO_3$ films [13]

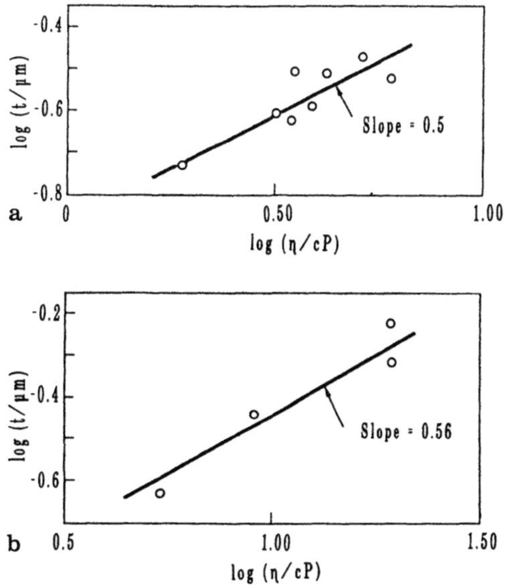

Fig. 9. Relation between the film thickness t and the solution viscosity η for SiO_2 films in logarithmic scale in the range where uniform coatings are produced

addition of hydroxypropylcellulose. These values of the slope are close to 1/2 as found in Eq. (1).

2.4 Ceramic Films with Preferred Crystal Orientation

Crystalline coating films are very useful as electronic, optoelectronic, and magnetic materials. If there is crystalline orientation in the coating films, highly efficient ferroelectric and ferromagnetic materials would be obtained. Hirano and Kato [22] have shown the possibility of producing thin coating films with preferred orientation by sol–gel method. They have shown that homogeneous and stoichiometric $LiNbO_3$ films with preferred orientation can be prepared by crystallizing the film at 400 °C on α-Al_2O_3 single crystal (sapphire) substrate. Double alkoxide solutions have been used. Okuwada et al. [23] prepared perovskite-type ferroelectric $Pb(Mg_{1/3}Nb_{2/3})O_3$ (PMN) crystalline film with preferred crystal orientation on platinum and single-crystalline MgO substrates by sol–gel method. Preferred orientation was achieved in the sol–gel coating film of $Li_2B_4O_7$ crystalline phase applied on Si and Al_2O_3 single crystal substrates [24].

Zhuang et al. [25] observed a certain crystalline orientation in Bi–Sr–Ca–Cu–O superconducting films prepared from metal acetate solutions on polycrystalline yttria-stabilized-zirconia substrates.

Such crystal orientations are assumed to be achieved when crystals are epitaxially precipitated from amorphous or glassy films formed on the substrates. Although the detailed mechanism is not known yet, the possibility of occurrence of crystal orientation may open a great opportunity for the sol–gel preparation of coating films.

3 Applications of Coating Films

A great variety of coating films have been prepared so far by sol–gel methods. Some of the sol–gel coating films are in practical use. Most of such films are used for optical purposes. Many optical and electronic films are promising for the use in electronic, optoelectronic, and photonic devices and many efforts are being made to develop them. Those for chemical and mechanical purposes are also important.

In this section, the coating films are classified into four categories, that is, the coating films of optical, electronic, chemical, and mechanical applications, according to their functions.

3.1 *Optical, Optoelectronic, and Photonic Coating Films*

Coating films with optical functions are shown in Table 3. As shown in this table, there are many transparent coating films which provide the substrate with new optical and optoelectronic functions.

3.1.1 *Optical Waveguide*

Sol–gel technique gives coating films of high optical quality, that is, the films which are transparent, homogeneous, and firmly sticking to the substrate. In

Table 3. Sol–gel coating films with optical functions

Optical effect	Application	Examples of composition
Light transmission	Optical waveguide	$PbO–TiO_2$ [26], $SiO_2–TiO_2$ [27, 28]
Optical absorption	U.V. shielding	$TiO_2–SiO_2$ [6]
	Optical absorption	Fe_2O_3, CoO, NiO [29]
	Coloration of glass	$SiO_2–R_mO_n$ (R = Cr, Mn, Fe, Co, Ni, Cu) [30, 31], $CeO_2–TiO_2$ [32–34], $GeO_2–V_2O_5$ [35]
Interference	Coloration	TiO_2 [36]
Reflection	Radiation shield	$In_2O_3–SnO_2$ [37], $VO_2–SiO_2$ [38]
	Solar energy reflecting	$PbO–TiO_2$ [26, 39], $Bi_2O_3–TiO_2$ [26]
	Combiner of HUD (head up display)	$TiO_2–SiO_2$ [40]
Anti-reflection	Increase of damage threshold of fusion laser glass, Clear vision	$Na_2O–B_2O_3–SiO_2$ [41], SiO_2 [42, 43] $TiO_2–SiO_2$ [44]
Electro-optic and nonlinear optic	Optical shutter, switch, optical IC	CdS, $CdS_{1-x}Se_x–Al_2O_3$ [46], SiO_2 [47] $Sr_{1-x}Ba_xNb_2O_6$ (SBN) [48], $LiNbO_3$, PZT [56–59]
Fluorescence	Luminescent solar concentrator	SiO_2 doped with organic dyes [45, 46, 49, 50]
	Lasers	Al_2O_3 doped with organic dyes [46]
Electrochromism	Display	WO_3 [51, 52]
	Transparent electrode for ion passing	$CeO_2–TiO_2$ [53]
Patterning	Optical ROM disk	B_2O_3, TiO_2, $ZrO_2–SiO_2$ [54]

utilizing this high optical quality, La Serra et al. [26] made $PbTi_4O_9$ planar waveguide on a glass substrate by sol–gel coating from solutions containing lead acetate $Pb(C_2H_3O_2)_2$, titanium isopropoxide $Ti(OC_3H_7)_4$, methoxymethanol $CH_3OCH_2CH_2OH$ and water. The refractive index of $PbTi_4O_9$ film is pretty large at 2.15 and the film is useful as planar optical waveguide. Using a rutile prism coupler, a He–Ne laser was coupled to a thin planar film 0.82 μm thick deposited on a glass substrate as shown in Fig. 10.

Attempts to fabricate $SiO_2–TiO_2$ planar waveguide [27] and lens-like waveguides [28] have been made. In order to make the latter, the dipping of substrate and withdrawing from the solution have been controlled by a microprocessor.

3.1.2 Optical Absorption and Coloring

$TiO_2–SiO_2$ coating films are used for U.V. shielding [6], since they have a strong optical absorption in the U.V. region of the spectrum. Sheet glasses are colored with thin coating films containing coloring agents such as transition metal elements [29–31]. Yamamoto et al. [30] made $SiO_2–R_mO_n$ (R = Cr, Mn, Fe, Co,

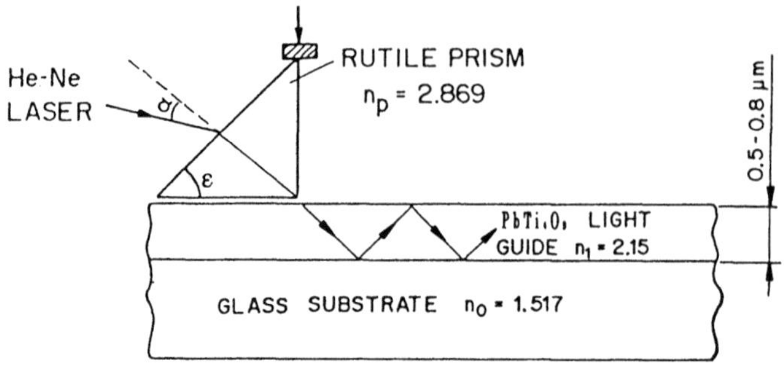

Fig. 10. $PbTi_4O_9$ planar waveguide prepared by sol–gel coating. He–Ne laser light beam is coupled to the waveguide by prism coupler method; after Le Serra et al. [26]

Ni, and Cu) coating films from tetraethoxysilane-metal nitrate-water-acid solutions. Most metal nitrates are soluble in water. In this method very high concentrations up to 30–50 mol % of metal oxides could be dissolved in the starting solution and coating films. Some of the transition metal elements for example Co and Ni give very thin films about 0.3 μm thick an unexpectedly strong color [30]. The optical absorption is much higher than that calculated from the extinction coefficient of the coloring element assuming that the element is dissolved in gel or oxide matrix as ion. It is then suggested that the strong color is caused by colloidal particles of the oxide of the coloring element.

Transparent, yellow-colored coating film of the TiO_2–CeO_2 system [32–34] can be obtained by sol–gel coating using titanium isopropoxide and cerium chloride as starting materials. This film has a bright yellow color and is also highly UV-absorbent [34], and so is applied to shielding of U.V. light. Blue-colored GeO_2–V_2O_5 coating films have been made from germanium and vanadium alkoxides [35]. The coating films have been heated at 500 °C in reducing atmosphere to reduce the V^{5+} ions to V^{4+} ions. The optical absorption of the film measured as a function of temperature shows a hysteresis in the temperature range of 40–80 °C. This might be related to semiconductor-metal transition of crystalline VO_2, although no such change in electrical conductivity is seen.

The color caused by interference of light is favored by Japanese people, because it is soft-tuned like that of rainbow. This interference color is achieved with the sol–gel coating films consisting of high refractive index oxide. The change of optical absorption spectrum [36] of the TiO_2 film as a function of the film thickness is shown in Fig. 11. It is seen that the number of maximums and minimums in light intensity increases with increasing thickness of the film, as the theory of interference shows.

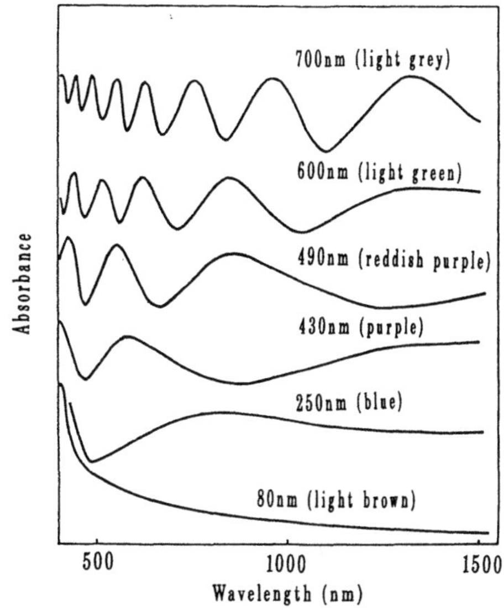

Fig. 11. Optical absorption spectra of sol–gel-derived TiO_2 coating films with different thicknesses [36]. The major part of the variation in the light transmission must be due to interference effects like those known from soap bubbles of differing wall thicknesses

3.1.3 Reflecting Films

Coating of sheet glass with a reflecting film serves in cutting the thermal radiation of sunlight coming into a room of building, lowering the power required for cooling the room during summertime. In_2O_3–SnO_2 [37] and VO_2–SiO_2 [38] films are made for this purpose by sol–gel coating.

La Serra et al. [26] have proposed PbO–TiO_2 and Bi_2O_3–TiO_2 coating films for this purpose. Coating films of these systems have practically no optical absorption in the visible region. Therefore, control of *n-t* values, that is, the refractive index times the film thickness will provide almost flat transmission and reflection characteristics. Accordingly, the films are suitable for solar-energy-reflecting windows. The cutting of solar energy by reflection is more advantageous than the sun-shielding glass window which absorbs sunlight. The calculation of *n-t* value for reflection of the light is made by taking the interference of the lights reflected from both surfaces of the film.

$PbTiO_3$ ceramics can be prepared from a few different combinations of the starting compounds. Gurkovich and Blum [39] used the reaction of lead acetate $Pb(CH_3COO)_2$ dissolved in methoxyethanol $CH_3OCH_2CH_2OH$ with titanium isopropoxide $Ti(OC_3H_7)_4$ in order to make a $PbTiO_3$ precursor. La Serra et al. [26] used Ti(iso-$OC_3H_7)_2$(acac)$_2$ prepared by modifying Ti(iso-$OC_3H_7)_4$ by acetylacetone Hacac and lead acetate $Pb(CH_3COO)_2$. For compositions of the Bi_2O_3–TiO_2 system, a bismuth sol prepared from $Bi(NO_3)_3 \cdot 5H_2O$ in acetic acid and lead acetate were mixed.

Another interesting application of reflecting films can be found in the display of the speed of a running car at the windshield [40]. TiO_2–SiO_2 films with thickness of about 2000 Å were applied to the windshield by sol–gel coating using titanium and silicon alkoxides as starting materials. The coating film of about 10 cm × 10 cm in size selectively reflects the light signal showing the speed which is generated at a high intensity fluorescence display tube. It is noted that the wavelength of the maximum reflectivity in the spectrum for the reflecting film is controlled so that it agrees with that of the maximum intensity of the fluorescence light for display. This type of coating film for head-up display (HUD) is now practically used in some Japanese automobiles.

3.1.4 Antireflecting Films

Antireflecting films applied on glasses prevent loss of lights and occurrence of stray lights in optical systems and provide clear vision of pictures through a glass sheet or window. So far, multilayer films have been prepared by vapor-phase deposition, such as sputtering, vacuum deposition and chemical vapor deposition. The sol–gel coating technique gives a chance for easy application of single layer antireflecting films and multilayer films on large surfaces.

Two types of single-layer antireflecting coating films have been designed for the purpose of protecting optical systems from damage by laser light and increasing the threshold of a glass laser element in high intensity optical lasers for nuclear fusion. Mukherjee and Lowdermilk [41] developed a technique in which a Na_2O–B_2O_3–SiO_2 film is applied on glass by sol–gel processing, the glass film is phase-separated and etched with hydrofluoric acid (HF) or ammonium fluoride $(NH_4)F$. This etching dissolves the B_2O_3–Na_2O phase and SiO_2 skeletons to a less extent, producing silica film of high porosity. The porous silica film, thus prepared has a refractive index gradient in such a way that the refractive index becomes small towards the free surface of the film. This antireflecting film raises the damage threshold of optical laser glass lenses for the laser nuclear fusion systems from 4–5 $J\,cm^{-2}$ for multilayer antireflecting coating to 20 $J\,cm^{-2}$, increasing the efficiency of the laser.

The hydrofluoric acid etching of silica film prepared by coating the glass with a tetraethoxysilane solution makes a film having the similar refractive index gradient [42] as that mentioned above. The etching solution enters the pores from the free surface, starting to etch the silica skeleton near the entrances of pores and making the porosity in the position near the free surface larger than that for the deeper positions. This results in the antireflecting film in which the refractive index becomes lower towards the free surface. This coating provides antireflectivity over the entire spectral range with better than 99% transmission and increases the laser damage resistance to 9 $J\,cm^{-2}$.

A multiple-layer antireflecting film consisting of SiO_2 and TiO_2 layers has been applied to sheet glass [44], which makes it possible to see the picture covered by the glass sheets very clearly.

3.1.5 *Electrooptic and Nonlinear Optic Films*

There will be a growing number of electrooptic and nonlinear optic applications of thin films in the field of integrated optics. The most important requirement for the application is that films must be transparent. Fortunately, the sol–gel-derived films can meet this requirement in most cases. Glass films are of course transparent. Crystalline films also remain transparent, since single crystal films are possibly obtained or crystallite size is usually much smaller than the wavelength of lights. Accordingly, sol–gel method has a great potential for preparation of good quality electrooptic and nonlinear optic thin films.

So far, only a few researches have been made in this field. Avnir et al. [45] reported the preparation and optical properties of silica and silica–titania films doped with various fluorescent and laser organic dyes using sol–gel method. It was clearly demonstrated that the sol–gel-derived host materials are capable of dispersing homogeneously organic dye molecules without forming agglomerates.

Kurokawa et al. [46] prepared amorphous Al_2O_3 films doped with various organic dyes and ultrafine semiconductor particles and observed lasing (Rhodamine B), photochemical hole burning (Naphthoquinone), second harmonic generation (SHG) (2-methyl-4-nitroaniline), and third harmonic generation (THG) and phase-conjugated wave (CdS, CdSe, PbS). Metal chalcogenides were formed by reacting Al_2O_3 gel films containing Cd or Pb with H_2S or H_2Se gas.

Yoko et al. [47] reported the preparation of CdS-doped SiO_2 glass films by gelation of the $Si(OC_2H_5)_4$ sols containing $(NH_2)_2CS$ as source for S and $Cd(NO_3)_2$ as source for Cd. CdS ultrafine particles precipitate by heating the dip-coated SiO_2 film at 300 to 500 °C. The CdS crystals thus obtained are cubic, whereas the reaction of Cd containing gels with H_2S gas forms hexagonal CdS crystals. A difference in nonlinear optical properties between two different CdS crystals would be of great interest. The absorption edge of the CdS-doped films shifts to shorter wavelengths as the CdS content in the films decrease, which may be ascribed to blue shift probably due to the quantum size effect.

Xu and Mackenzie [48] reported that the sol–gel derived, transparent $Sr_{1-x}Ba_xNb_2O_6$ (SBN) ferroelectric film with preferential crystal orientation on silica glass substrate shows two wave mixing effect due to a Kerr effect. These indicate that sol–gel-derived ferroelectric thin films will play an important role in electrooptic and photonic applications in near future.

3.1.6 *Fluorescent Films*

The film flat plate LSC (luminescent solar concentrator) [49] has a coating film containing fluorescent dye molecules, as shown in Fig. 12. A LSC absorbs solar radiation in the flat coating film containing one or more fluorescent species. The fluorescent emission is trapped in the coating film by total reflection and

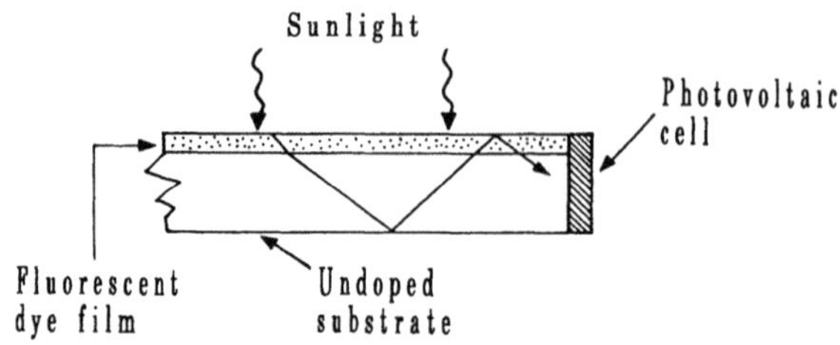

Fig. 12. Diagram of a flat plate luminescent solar concentrator with thin film; after Reisfeld et al. [45]

concentrated at the edges of the collector which constitute photovoltaic cells.

A LSC has the following advantages [49] over a conventional solar concentrator in that both direct and diffuse lights are collected, tracking the sun is not necessary and the luminescent species can be chosen to allow matching of the concentrated light to the maximum sensitivity of the photovoltaic cell. The advantage of the configuration of the doped thin film with the plate over the luminescent plates in which the dye is incorporated in the entire bulk is that the luminescent light emitted from the thin film is trapped in the plate with reduced self-absorption and scattering.

Reisfeld et al. [50] showed on the basis of the experimental data of absorption and emission of rhodamine 6G in sol–gel derived gel that the quantum efficiency of Rhodamine 6G is 0.75 and its molar extinction coefficient is 82,000. The calculation based on the overlap of the absorption of rhodamine 6G with the solar spectrum using a 50 μm thick film deposited on a plate having refractive index of 1.5 and area of 1 m^2 gives 1.5% in the optical efficiency. A larger efficiency can be obtained by using the photostable dye and combination of dyes which increases the overlap of absorption with the solar spectrum.

3.1.7 Electrochromic Films

An electrochromic film is colored when the electric field is applied and the color fades on reversing the field. Therefore, electrochromic devices can be applied to optical displays and windows controlling the light intensity.

Amorphous WO_3 films are an important candidate for electrochromic displays. In an electrochromic cell, both sides of a WO_3 film contacted with an electrolyte have electrodes. The electric field applied between the two plane electrodes causes the following coloring reaction accompanied by incorporation of electrons and ions into the film:

$$WO_3 + xe^- + xM^+ \rightleftarrows M_xWO_3 \qquad (3)$$

where M^+ represents H^+ or Li^+ and M_xWO_3 is a colored tungsten bronze. The color is blue for H_xWO_3. The use of H^+ ions gives fast-responding element and the use of Li^+ gives a chemically inactive, and stable element.

So far, WO_3 films have been made by vapor deposition methods. In order to make large area electrochromic windows, the application of sol–gel coating to glass sheets coated with electroconducting transparent SnO_2–Sb_2O_5 (Nesa film) or In_2O_3–SnO_2 (ITO film) films is useful. Livage [51] has applied the sol–gel method using inorganic tungstates to the preparation of WO_3 films. The resultant films show coloring and fading in relatively short times in the electrochromic cell using H_2SO_4 as electrolyte. Unuma [52] prepared colorless and transparent amorphous WO_3 thin films by sol–gel method using tungsten hexaethoxide as starting substance. Oxygen plasma treatment made it possible to decompose and eliminate organic matters at 200 °C, which keeps the amorphous nature required for preventing the deterioration of electrochromic property of the film. The system consisting of an amorphous WO_3 film deposited by dip-coating on a Nesa glass, an electrolyte of propylene carbonate containing lithium perchlorate and a Pt counter electrode showed electrochromic property, that is, blue coloring-bleaching property due to Eq. (3).

Baudry et al. [53] constructed an all solid state electrochromic cell as shown in Fig. 13. The cell consists of ITO coated glass sheets, vapor-deposited amorphous WO_3 film as electrochromic material, complex polyethylene oxide (PEO)–$LiN(SO_2CF_3)_2$ as polymer electrolyte, and TiO_2–CeO_2 film as counter-electrode deposited onto ITO coated glass. The TiO_2–CeO_2 coating film has been made by the sol–gel dip-coating process from the starting solution containing $Ce(NH_4)_2(NO_3)_6$ and titanium isopropoxide $Ti(O\text{-iso-}C_3H_7)_4$. The CeO_2–TiO_2 film works for the lithium insertion as a transparent counter electrode. The color is blue when the WO_3 film is polarized cathodically and transparent and colorless when it is polarized anodically.

3.1.8 Films for Patterning

Tohge et al. [54] have developed a novel process based on the sol–gel method to form submicron fine patterns on soda-lime-silica glass substrate which are applicable to diffraction gratings and optical ROM disks with extremely high reliability. The whole process is schematically shown in Fig. 14, which consists of (*1*) coating of gel film on a glass substrate, (*2*) patterning of the gel film by pressing

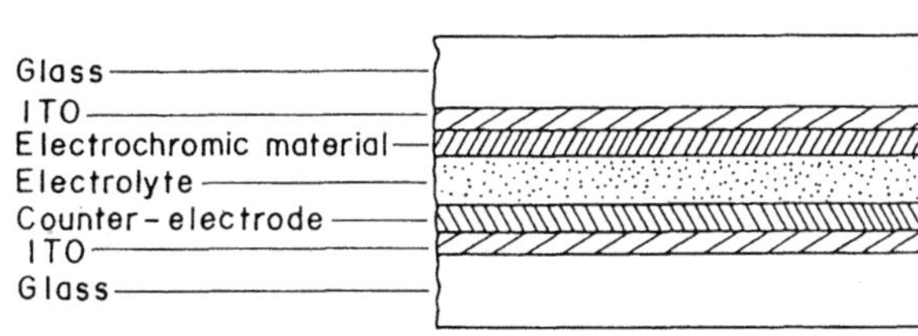

Fig. 13. Cross-section of a transmissive electrochromic plate; after Baudry et al. [53]

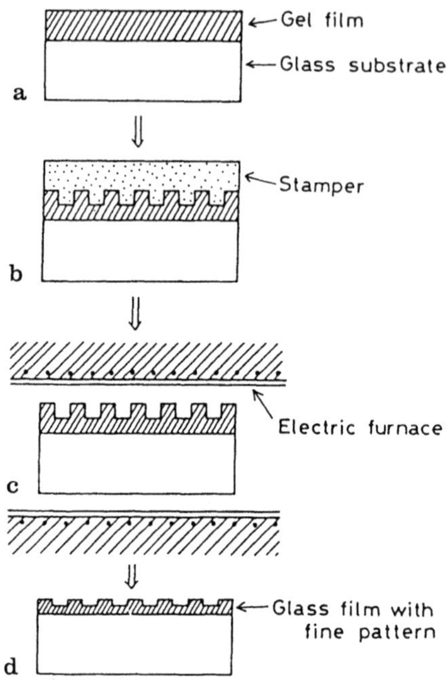

Fig. 14a–d. Fine patterning process by the sol–gel method; (*a*) formation of gel film on a substrate, (*b*) patterning on the gel film, (*c*) heat treatment and (*d*) glass film with fine patterning; after Tohge et al. [54]

a stamper and (*3*) drying and heating. The most important point in the process is to control the hardness of gel film just prior to stamping. This is achieved by adding a relevant amount of organic substances such as polyethylene glycol (PEG) to a sol. The chemical durability especially to moisture is pronouncedly improved by choosing the glass composition, for example, TiO_2–SiO_2 and ZrO_2–SiO_2 systems.

3.2 Sol–Gel Coating Films with Electronic and Magnetic Functions

Many attempts have been made to apply the sol–gel method to preparation of coating films with various electronic and magnetic functions, as shown in Table 4. Ferroelectric and piezoelectric coating films are becoming of great importance in electronics. Coating films of electronic conductor-oxides can be used as electrodes and for other purposes. Preparation of ion conducting coating films have been attempted by several investigators. The application of sol–gel coating film to the preparation of high T_c superconducting oxide films is very important and significant from the standpoint of superconducting wire formation. Magnetic film formation by sol–gel coating will be useful in near future.

Table 4. Sol–gel coating films with electronic and magnetic functions

Effect	Application	Examples of composition
Ferroelectric	Capacitor, IC memory	$BaTiO_3$ [13, 55], PZT [56–59], Ta_2O_5 [60, 61], PMN [23], PFN [62]
Piezoelectric	SAW device	$LiNbO_3$ [22, 63], $Li_2B_4O_7$ [64]
Electronic conduction	Transparent electrode and prevention of charge-up	Sb_2O_3–SnO_2 [65, 66], CdO–SnO_2 [67] SnO_2–TiO_2 [68], V_2O_5 [69]
Ion conduction	Solid-state battery	β-Al_2O_3 [70], Li_2O–SiO_2 [71], NASICON [72], $LiAlSi_mO_n$ [73]
Superconduction	Superconductor superconducting wire	$YBa_2Cu_3O_{7-x}$ [75–78], Bi–Pb–Sr–Ca–Cu–O [25, 79–82]
Magnetic	Recording tape, disk	Fe_2O_{3-x} [83], $NiFe_2O_4$ [84], Fe_3O_4 [85]

PZT: $PbZr_{1-x}Ti_xO_3$
PMN: $Pb(Mg_{1/3}Nb_{2/3})O_3$
PFN: $Pb(Fe_{1/2}Nb_{1/2})O_3$
NASICON: $Na_3Zr_2Si_2PO_{12}$

3.2.1 Ferroelectric and Piezoelectric Films

Much attention has been paid to ferroelectric thin films not only for optoelectronic applications stated in Sect. 3.1.5 but also for capacitor and nonvolatile IC-memories in microelectronics. The sol–gel coating process has also been applied to the preparation of thin films of $BaTiO_3$ [13, 55], $PbZr_{1-x}Ti_xO_3$ (PZT) [56–59], Ta_2O_5 [60, 61], $Pb(Mg_{1/3}Nb_{2/3})O_3$ (PMN) [23], and $Pb(Fe_{0.5}Nb_{0.5})O_3$ (PFN) [62]. It is experimentally confirmed that those films except Ta_2O_5 show a hysteresis loop characteristic of ferroelectrics. The dielectric constants are listed in Table 5. As seen from this table, $Pb(Fe_{0.5}Nb_{0.5})O_3$ film shows a dielectric constant as large as about 2000 at 60 Hz [62].

There have been several attempts to apply the sol–gel-derived piezoelectric films to surface acoustic wave (SAW) device [22, 63, 64]. Hirano and Kato [22] prepared $LiNbO_3$ films on various substrate materials with different crystal planes by the sol–gel method. They found that the $LiNbO_3$ films with preferential crystal growth can be obtained on sapphire single crystal substrates but not

Table 5. Dielectric constants for sol–gel-derived ferro-electric films

Substance	Dielectric constant
PZT	260 at 100 kHz [59]
	1200 up to 10 MHz [58]
Ta_2O_5	20 at 1 MHz [60]
$Pb(Fe_{0.5}Nb_{0.5})O_3$	2000 at 60 Hz [62]

on silicon single crystal. This can be easily understood if one takes into account that $LiNbO_3$ has the same crystal structure as α-Al_2O_3 used as the substrates.

Yamashita et al. [24, 64] have prepared preferentially oriented $Li_2B_4O_7$ films for SAW device from corresponding metal alkoxide solution. This material is characterized by a lower temperature coefficient of delay time than $LiNbO_3$ and higher electro-mechanical coupling constant than α-quartz. It was found that a sufficient amount of water, a proper amount of acid and use of Si(111) single crystal substrate lead to transparent single crystal $Li_2B_4O_7$ films preferentially oriented along (122) plane as shown by the X-ray diffraction pattern in Fig. 15.

3.2.2 Coating Films Showing Electronic Conduction

The electronic conducting films of In_2O_3–SnO_2 system (ITO film) are very useful as optically transparent electrode. Ogiwara and Kinugawa [65] and Gonzalez-Oliver and Kato [66] prepared ITO films by sol–gel method using, for example, tin and indium acetylacetonate. As electronic-conductor, cadmium stannate films have been made by Dislich [67] from tin acetylacetonate, cadmium nitrate and acetylacetone.

It is generally told that the In_2O_3–SnO_2 films prepared from solutions by dip-coating or spin coating show conductivities lower by an order of magnitude or so than those prepared by vapor deposition technique. This can be explained if it is assumed that the solution-derived films may be more or less porous which results in insufficient connection of constituent conducting particles. This

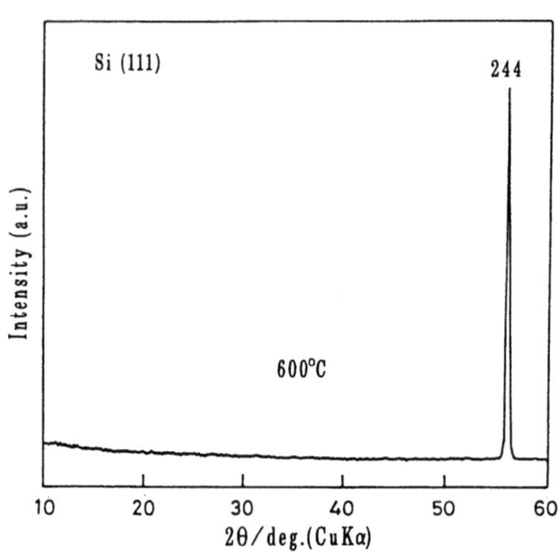

Fig. 15. X-ray diffraction pattern of $Li_2B_4O_7$ film on Si(1 1 1) heat-treated at 600°C; after Yamashita et al. [64]

problem of lower conductivity has to be solved, in order to make sol–gel technique more useful, although a uniform film can be applied to a substrate of large surface area by sol–gel coating.

Maddalena et al. [68] have prepared tin oxide coating films by sol–gel dip coating with absolute ethanol solution of $SnCl_2 \cdot 2H_2O$ [68]. The ordinary soda-lime glass sheets were dipped into the solution and withdrawn at a constant speed. Tin oxide gel films were dried at a reduced pressure at 60 °C for 2 h and heated at 500 °C for 10 min in air. Doping of SnO_2 films with ZrO_2 and TiO_2 can be made by adding calculated amounts of $Zr(OC_3H_7)_4$ and $Ti(OC_4H_9)_4$ to solutions of tin chloride. Figure 16 shows that the doping increases the electrical conductivity of the undoped film.

V_2O_5 coating films made by sol–gel method [69] are highly conductive, showing the similar conductivity as the crystalline vanadium oxide. The coating film may be used as a film for preventing occurrence of static electricity on plastic materials.

3.2.3 Coating Films Showing Ionic Conduction

Ion conducting films are applicable to solid electrolyte for solid state batteries and electrochromic display. Films of β-Al_2O_3, an Na^+-ion conductor, have been made by Dislich and Partlow [70]. The gel film can be transformed to β-Al_2O_3 dense ceramic film by heat treatment at 1200 °C which is much lower compared with conventional sintering temperature at 1700 °C. Low temperature heating can avoid extraordinary grain growth as well as Na loss.

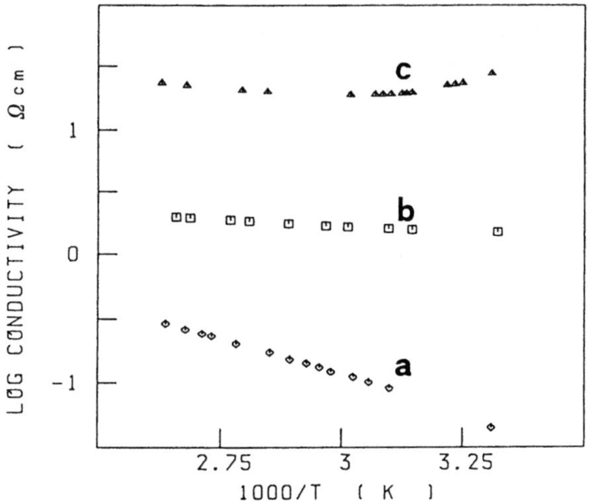

Fig. 16a–c. Temperature dependence of electrical conductivity of film: (*a*) undoped, (*b*) zirconium oxide doped; (*c*) titanium oxide doped; after Maddalena et al. [68]

Tatsumisago et al. [71] reported the sol–gel formation of Li^+-ion-conducting glass films which have the same conductivity as that of the corresponding rapidly quenched Li_2O–SiO_2 glasses. Too much addition of lithium to the film, however, resulted in poorer conductivity due to Li_2CO_3 formation. This problem may be avoided by adding a proper amount of water to a sol for hydrolysis.

Perthius et al. [72, 73] have made an attempt to prepare superionic conducting thick films 50 μm such as $Na_3Zr_2Si_2PO_{12}$ (NASICON), $Na_{3.1}Zr_{1.55}Si_{2.3}P_{0.7}O_{11}$, $Li_{0.8}Zr_{1.8}Ta_{0.2}(PO_4)_3$, β-$Li_2Al_2Si_4O_{12}$ (β-spodumene), and β-$Li_2Al_2Si_2O_8$ (β-eucryptite). They adopted silk-screen printing technique using corresponding superionic conducting oxide fine powders which were synthesized by the sol–gel method. This technique permits relatively low temperature sintering at 950–1050 °C, which, for example, prevents the precipitation of monoclinic zirconia in the case of NASICON-type conductors. The low-temperature processing also makes the application to the preparation of elaborated microelectronic devices possible.

Ravaine et al. [74] prepared organically modified silicates (ORMOSILS) by heating gels at 150 °C prepared from mixtures of silicon alkoxide and polyethylene glycol (PEG). Subsequent impregnation of gels with various lithium salts resulted in gels with high Li^+ ion conductivity.

3.2.4 *Superconducting Oxide Coating Films*

Coating films of high T_c superconducting oxides are important as substitutes for superconducting wires. At present, coating films of two kinds of high T_c superconducting oxide systems $YBa_2Cu_3O_x$ [75–78] with T_c at about 94 K and Bi–Ca–Sr–Cu–O [25, 79–82] with T_c at about 104 K have been made. In the sol–gel method, organic salts, metal alkoxides and metal nitrates are used as starting compounds for the superconducting oxides. As substrates, yttria-stabilized zirconia (YSG), $SrTiO_3$, and MgO are used.

As an example, Monde et al. [75] have prepared superconducting oxide films of the $YBa_2Cu_3O_x$ system on yttrium-stabilized zirconia substrates by sol–gel technique using metal alkoxide solution. Yttrium butoxide, barium methoxide, and copper methoxide in a given atomic ratio are dissolved in a triethanolamine-methanol solution. Coating with the resultant solution, heating at 900 °C in air and annealing in an oxygen atmosphere result in superconducting oxide thin films. The onset of superconductivity is seen at 98 K and the resistance rapidly decreases with decreasing temperature, reaching substantially zero at 56 K. It is noted that the formation of 5–7 μm-thick films by repeating coating procedure several times was required for the occurrence of superconductivity.

Bi–Ca–Sr–Cu–O superconducting oxide films of T_c(end) at 84 K and 79 K have been made from organic salts, respectively by Agostinelli et al. [79] and Nasu et al. [80]. Zhuang et al. [25] have prepared coating films of the Bi–Pb–Ca–Sr–Cu–O system from metal acetate solutions with an addition of

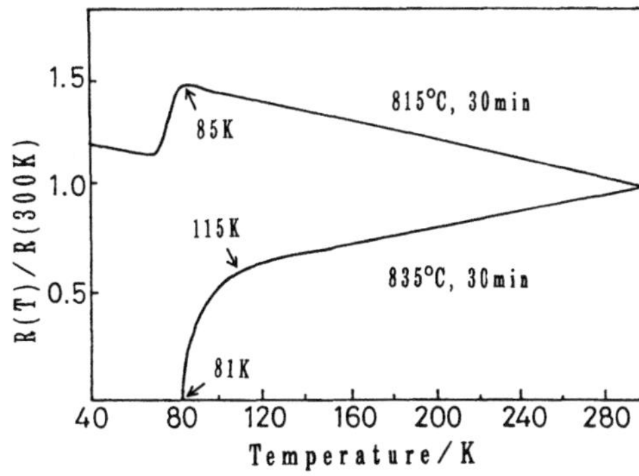

Fig. 17. DC electrical resistance of Bi–Ca–Sr–Cu–O coating film as a function of temperature [25]

tartaric acid. Resistivity-temperature curves are shown in Fig. 17. T_c(end) is found at 81 K, when the film is heat-treated at 835 °C for 30 min.

3.2.5 Magnetic Coating Films

Magnetic thin films are important materials for magnetic recording devices. So far, several attempts have been made to form magnetic films of Fe_2O_{3-x} [83], $NiFe_2O_4$ [84] and Fe_3O_4 [85] using the sol–gel method. The Fe_3O_4 film prepared from iron and nickel acetylacetonates on silica glass substrate showed a magnetization curve with hysteresis loop which is characteristic of ferromagnetism as shown in Fig. 18 [85]. The coercive force was estimated to be 230 Oe

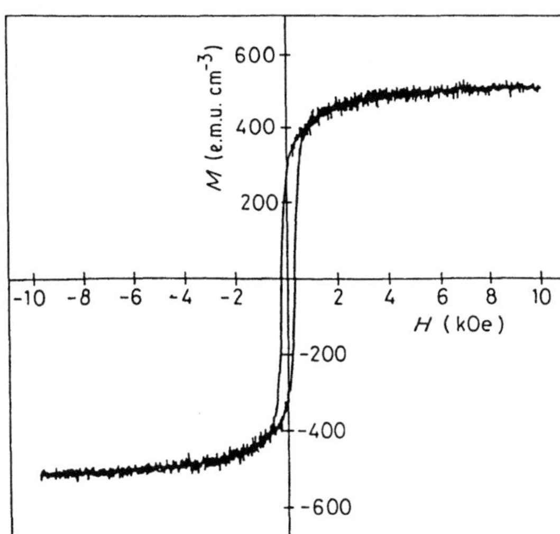

Fig. 18. Magnetization curve of sol–gel-derived Fe_3O_4 film; after Tanaka et al. [85]

which is larger than that of bulk Fe_3O_4. This is ascribed to the smaller Fe_3O_4 crystallite size of about 50 nm in the sol–gel-derived film.

Taking into account that the sol–gel-derived oxide films are usually transparent due to smaller crystallite size, they are promising materials as recording media of large size for magnetooptical devices. From this point of view, ferrite films may be one of the most important candidates for magnetooptical recording device.

3.3 Sol–gel Coating Films with Chemical Functions

Chemical functions of the coating films include photoelectrochemical effect, catalyst support, and chemical protection of the substrate and the second coating film. Table 6 shows the lists of sol–gel coating films with chemical functions.

3.3.1 Photoelectrochemical Catalyst Films

Utilization of solar energy is necessary for meeting future energy demands. One of the methods is to convert solar energy directly into electrical or chemical energy using a semiconductor electrode [86, 87]. Yoko et al. [88–90] applied the sol–gel derived TiO_2 films to a photoanode to decompose water into oxygen and hydrogen. Figure 19 shows the photocurrent-bias potential curves obtained with the TiO_2 film electrodes of various thicknesses. It is found that the TiO_2 film electrodes heated at 500 °C for 20 min show the saturated photocurrent of 14 mA cm^{-2} which is comparable to or even better than that of single crystal TiO_2 [87]. Since the sol–gel-derived oxide films are porous in nature and characterized by a large surface area at which electrode reaction takes place under illumination [89], they are very much suitable for the photoelectrode.

The photoelectrochemical properties of $SrTiO_3$ and $BaTiO_3$ prepared by the sol–gel method are also investigated [91]. However, these oxide films do not

Table 6. Sol–gel coating films with chemical functions

Effect	Application	Examples of composition
Photoelectrochemical catalyst	Hydrogen generation by decomposition of water	TiO_2 [88–90], $SrTiO_3$, $BaTiO_3$ [91], α-Fe_2O_3 [92], $NiFe_2O_4$ [93]
Catalyst support	Catalyst	Al_2O_3: Pt [94], TiO_2: Pd [95], SiO_2 (TiO_2 or Al_2O_3): Fe, Co or Ni [96–98]
Chemical protection	Prevention of corrosion	SiO_2 [11, 99], B_2O_3–SiO_2 [100], ZrO_2 [101, 102], CeO_2–TiO_2 [32]
	Passivation of semiconductor	Al_2O_3 [103], Si–O–N, Si_3N_4 [104, 105]
	Prevention of alkali diffusion	SiO_2 [65]

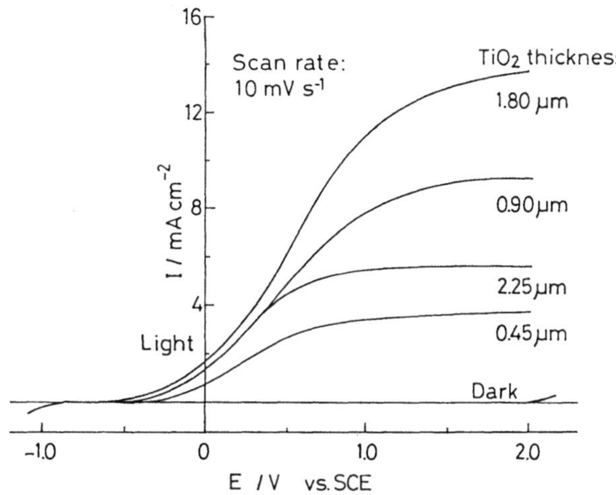

Fig. 19. Photocurrent-bias potential curves for TiO_2 film electrodes with different film thicknesses; after Yoko et al. [88, 89]

show so large photocurrent as TiO_2. The same workers also reported the preparation and photoelectrochemical properties of the sol–gel-derived α-Fe_2O_3 [92] and $NiFe_2O_4$ [93] film electrodes. These materials have a direct optical band gap at about 2.6 eV which is slightly smaller than that of TiO_2, $SrTiO_3$, and $BaTiO_3$, suggesting that they might show a higher conversion efficiency.

The future subjects in this field are to develop photoelectrochemically stable *n*- and *p*-type semiconductor film electrode materials with an optical bandgap at around 1.6 eV and high photoelectrochemical stability.

3.3.2 Coating Films as Catalyst Support

Pt and Pd metals exhibit various catalytic abilities. Al_2O_3 [98] and TiO_2 [95] coating films containing these metals have been made to catalyze the reduction or oxidation reaction of inorganic compounds. For making these catalysts, glass beads are dipped in aluminum or titanium alkoxide solutions, accompanied by drying to change the coating films to Al_2O_3 or TiO_2 gel film, and then the solutions dissolving Pt or Pd compounds are infiltrated into the gel and heated to precipitate the metal particles dispersed in oxide.

In order to obtain fine particle catalysts of iron, nickel or cobalt held in porous SiO_2, TiO_2, or Al_2O_3 supports, nickel or cobalt metal dissolved in ethylene oxide is mixed in silicon, titanium, or aluminum alkoxide and changed to an oxide via sol–gel processing [96]. The reduction of the product by heating in hydrogen atmosphere provides fine iron, nickel, or cobalt metallic particles dispersed in SiO_2, TiO_2 or Al_2O_3 catalyst carriers. The metallic particles have a sharp size distribution, showing a selective catalytic action depending on the size. Thus, these catalyst systems are used as model catalyst or practical catalyst for hydrogenation or oxidation reactions [97]. These catalyst systems are assumed to be easily applied to coating films.

3.3.3 Coating Films for Chemical Protection of Substrates

It is easily assumed that chemically stable coating films will chemically protect metal, glass and plastic substrates from oxidation and corrosion.

SiO_2 [11, 99] and B_2O_3–SiO_2 [100] glass films made by sol–gel coating can be used for the protection of metal substrates from oxidation and attack by acid. With SiO_2 coating film on stainless steel sheets, De Sanctis et al. [100] indicated that silica coatings effectively protect metallic substrates, increasing their oxidation and corrosion resistance markedly, as shown in Table 7. ZrO_2 [101] and ZrO_2–SiO_2 [102] coatings proved to increase the alkali-resistance of the glass substrates. Transparent, yellow-colored coating of TiO_2–CeO_2 film [32] obtained by sol–gel coating is highly resistant to acid and alkali and so is applied to chemical protecting films.

Passivation of silicon semiconductors can be made by SiO_2 and Al_2O_3 coatings [103], and introduction of nitrogen into SiO_2 film [104, 105] increases the protection ability of SiO_2 films. These films also suppress alkali diffusion from glass substrates, serving in preventing the functional coating film from attack by alkalis.

In order to give the coating film a certain ductility which is a property required for bending the coated steel substrates after the film formation, it is reasonable to use $CH_3Si(OC_2H_5)_3$ as a starting material for SiO_2 coating film [103–106]. Besides these inorganic coatings, coatings by organic–inorganic composite films are attempted to make [107]. Also SiO_2 coating films modified by polyfluoroalkoxysilane [108, 109] are very unique in that they repel both water and oil. As an example [109], such coating film is prepared from

Table 7. Tests of static oxidation and acid attack on coated and uncoated stainless steel plates (de Sanctis et al. [99])

AISI	Static oxidation C (g/m^2)		Acid attack* Re (g/m^2 h)	
	coated	uncoated	coated	uncoated
304	0.11	1.26	1.27	1.93
316	0.09	1.08	1.02	1.58
310	0.08	1.85	—	—
316**	—	—	0.07	1.58

* Boiled in 65% HNO_3 for 36 h
** 316 corresponds to AISI 316 samples heat-treated before coating
Coating film: SiO_2 film 0.4 μm thick

$CF_3CF_2CF_2C(CF_3)_2(CH_2)_3Si(OC_2H_5)_3$–$Si(OC_2H_5)_4$–$H_2O$–HCl solution [109].

3.4 Coating Films with Mechanical Functions

It has been shown in Sect. 2 and Table 2 that only very thin coating films of the thickness in the order of 0.1 μm can be made in one coating procedure in brittle films. Even repeating of the coating procedure can achieve the thickness of only less than 1–3 μm. It is suspected that such thin films are useful for increasing mechanical resistance of the substrate. Certain improvement of mechanical resistance of the substrate is reported, however. Table 8 shows coating films with mechanical functions.

In order to have a certain improvement in hardness, the coating films have to have the hardness value that is expected for the substance of coating film. Fabes and Oliver [110] measured the hardness of tetraethoxysilane-derived SiO_2 coating films prepared on sapphire (Al_2O_3) substrate by an indentation technique, showing that the hardness of the coating film surface increases with heating temperature, that is, with densification of the gel film and becomes comparable to the value of silica glass on heating at 1000 °C. This may indicate that an improvement of hardness, and accordingly scratch resistance of soft metals may be possible by sol–gel oxide coatings using SiO_2, Al_2O_3, ZrO_2, and so on.

Inorganic–organic composite materials [111] based on a ceramic oxide, such as SiO_2, TiO_2, ZrO_2, and Al_2O_3, and organic groups, such as epoxy and methacrylate groups, have a combination of properties shown by those two types of substances. Thus, hard coatings have been made on polycarbonate [112] and polyethyleneterephthalate [113]. It is noted that the coating films have flexibility and improve the scratch resistance of the polymer substrates.

Strengthening of glass by the sol–gel SiO_2 coatings has been reported by several research groups [114–116]. The breaking strength of silica glass rod drastically increases to as much as about two times that of uncoated glass by coating with tetraethoxysilane solutions. For instance, the breaking strength of as-received silica glass increases from 15 ksi to 22 ksi by heating the coated glass to 900 °C. Heating to 800–900 °C is necessary to densify the gel film. The increase in strength has no significant dependence on coating thickness in the thickness

Table 8. Sol–gel coating films with mechanical function

Effect	Application	Examples of composition
Mechanical protection	Increase of hardness of metal substrates	SiO_2 [110], ZrO_2 [101, 102]
Strengthening	Strengthening of glass	SiO_2 [114–117]

range of 0.2–1.0 μm. As mechanisms of strengthening, formation of a barrier to moisture, formation of a compressive layer at the surface, and infiltration and partial healing of flaws have been proposed. The strengthening due to infiltration and flaw healing has been proposed by Fabes and Uhlmann [116]. Fabes and Berry [117] have shown by model experiments using vickers indentation that tetraethoxysilane solutions infiltrate into flaws and partially heal flaws in silica glass and that flaws less than 35 nm wide appear healed. It is believed that strengthening is not due to formation of surface compression layer, but due to filling-in of the flaws.

4 References

1. Sakka S (1982) In: Tomozawa M, Doremus RH (eds) Treatise on materials science and technology, Vol 22, Glass III. Academic Press, New York, p 129
2. Sakka S (1988) Science of sol–gel method. Agne-Shofu-Sha, Tokyo, pp 221 (in Japanese)
3. Brinker CJ, Scherer GW (1990) Sol–gel science, Academic Press, San Diego, pp 908
4. Klein LC (ed) (1988) Sol–gel technology for thin films, fibers, preforms, electronics, and specialty shapes. Noyes Publications, Parkridge, pp 407
5. Sakka S (1985) J Non-Cryst Solids 73: 651
6. Dislich H, Hussmann E (1981) Thin Solid Films 77: 129
7. Schroeder H (1969) Physics of thin films, Vol 5. Academic Press, New York, p 87
8. Sakka S (1989) In: Aegerter MA, Jafelicci Jr, M, Souza ED, Zanotto ED (eds) Sol–gel science and technology. World Scientific, Singapore, p 346
9. Brinker CJ, Hurd AJ, Frye GC, Ward KJ, Ashley CS (1990) J Non-Cryst Solids 121: 294
10. Brinker CJ, Scherer GW (1990) Sol–gel science. Academic Press, San Diego, p 788
11. Sakka S, Kamiya Makita K, Yamamoto Y (1984) J Non-Cryst Solids 63: 223
12. Monde T, Kozuka H, Sakka S (1988) Chemistry Letters, 287
13. Sakka S, Kokubo T (1983) Jpn J Appl Phys, Vol 22, Supplement 22-2, 3
14. Makishima A, Asami M, Wada K (1990) J Non-Cryst Solids 121: 310
15 Yoko T, Kamiya K, Sakka S (1987) J Ceram Soc Jpn, 95: 150
16. Landau LD, Levich VG (1942) Acta Physica Chim USSR 17: 41; cf. Kamiya K, Yoko T (1986) Surface 24: 131 (in Japanese)
17. Yang CC, Josefowicz JY, Alexandru L (1980) Thin Solid Films 74: 117
18. Strawbridge I, James PF (1986) J Non-Cryst Solids 82: 366
19. Strawbridge I, James PF (1986) J Non-Cryst Solids 86: 381
20. Guglielmi M, Zenezini S (1990) J Non-Cryst Solids 121: 303
21. Mukherjee SP (1985) In: Wright AF, Dupuy J (eds) Applied Science, No 92, Glass—Current Issues, Martinus Nijhoff Publ, Dordrecht, p 232
22. Hirano S, Kato K (1988) Adv Ceram Mater 3: 503
23. Okuwada K, Imai M, Kakuno K (1989) Jpn J Appl Phys 28: L271
24. Yamashita H, Yoko T, Sakka S (1990) J Ceram Soc Jpn 98: 913
25. Zhuang HR, Kozuka H, Sakka S (1990) Jpn Appl Phys 29: in press
26. La Serra ER, Charbouillot Y, Baudry P, Aegerter MA (1990) J Non-Cryst Solids 121: 323
27. Hermann PP, Wildman D (1983) IEEE J Quantum Electron QE-19: 1735
28. Hewak DW, Lit JWY (1988) Appl Opt 27: 4562
29. Geotti-Bianchini F, Guglielmi M, Polato P, Soraru GD (1984) J Non-Cryst Solids 63: 251
30. Yamamoto Y, Makita K, Kamiya K, Sakka S (1983) J Ceram Soc Jpn 91: 222
31. Duran A, Fernandez Navarro JM, Casariega P, Joglar A (1986) J Non-Cryst Solids 82: 391
32. Makishima A, Kubo H, Wada K, Kitami Y, Shimohira T (1986) J Am Ceram Soc 69: C127
33. Mohallen NDS, Aegerter MA (1988) J Non-Cryst Solids 100: 526
34. Sainz MA, Duran A, Fernandez Navaro JM (1990) J Non-Cryst Solids 121: 315

35. Hou LS, Sakka S (1989) J Non-Cryst Solids 112: 424
36. Sakka S, Kamiya K, Yoko T (1988) In: Zeldin M, Wynne KJ, Allcock HR (eds) ACS Symposium Series 360, American Chemical Society, Washington DC, p 345
37. Arfsten NJ, Kaufmann R, Dislich H (1984) In: Hench LL, Ulrich DR (eds) Ultrastructure processing of ceramics, glasses, and composites, Wiley, New York, p 189
38. Hutter F, Schmidt H, Scholze H (1986) J Non-Cryst Solids 82: 373
39. Gurkovich SR, Blum JB (1984) In: Hench LL, Ulrich DR (eds) Ultrastructure processing of ceramics, glasses, and composites, Wiley, New York, p 152
40. Hattori A, Makita K, Okabayashi S (preprint) Abst Meeting of the Society of Photo-Optical Instrumentation Engineers, USA, Aug 1989
41. Mukherjee SP, Lowdermilk WH (1982) J Non-Cryst Solids 48: 177
42. Yoldas BE, Partlow DP (1985) Thin Solid Films 129: 1
43. Hinz P, Dislich H (1986) J Non-Cryst Solids 82: 411
44. Dislich H (private communication)
45. Avnir D, Kaufman VR, Reisfeld R (1985) J Non-Cryst Solids 74: 395
46. Kobayashi Y, Kurokawa Y, Imai Y, Muto S (1988) ibid 105: 198; Tanaka H, Takahashi J, Tsuchiya J, Kobayashi Y, Kurokawa Y (1989) 109: 164; Kurokawa Y (1989) Kagaku Kogyo (Chemical Industry) 40: 937 (in Japanese)
47. Yoko T, Yoshida Y, Kozuka H, Sakka S (Submitted to J Non-Cryst Solids)
48. Xu Y, Chen CJ, Mackenzie JD, Proc SPIE's 1990 Symp, Vol 1328: p 428, p 441
49. Reisfeld R (1989) In: Aegerter MA, Jafelicci Jr M, Souza ED, Zanotto ED (eds) Sol–gel—science and technology. World Scientific, Singapore, p 323
50. Reisfeld R, Eyal M, Chernyak V, Zusman R (1988) Solar energy materials, 17: 439
51. Livage J (1983) Solid state chemistry 1982. Elsevier, Amsterdam, p 17
52. Unuma H, Tonooka K, Suzuki Y, Furusaki T, Kodaira K, Matsushita T (1986) J Mater Sci Lett 5: 1248
53. Baudry P, Rodrigues ACM, Aegerter MA (1990) J Non-Cryst Solids 121: 319
54. Tohge N, Matsuda A, Minami T, Matsuno Y, Katayama S, Ikeda Y (1988) J Non-Cryst Solids 100: 501
55. Yanovskaya MI, Turova NY, Turevskaya EP, Novoselova AV, Venevtsev YN, Sagitov SI (1981) Inorg Mater 17: 221
56. Budd KD, Dey SK, Payne DA (1985) Brit Ceram Proc 36: 107
57. Budd KD, Dey SK, Payne DA (1986) Mat Res Soc Symp 73: 711
58. Yi G, Wu Z, Sayer M (1988) J Appl Phys 64: 2717
59. Dey SK, Zuleeg R (1989) Proc 1st Symp Integrated Ferroelectrics, Colorado Microelectronics Conf, Colorado Spring, Mar 30–31
60. Silverman LA, Teowee G, Uhlmann DR (1986) Mat Res Soc Symp Proc 73: 725
61. Ling HC, Yan MF, Rhodes WW (1984) In: Hench LL, Ulrich DR (eds) Ultrastructure processing of ceramics, glasses, and composites, Wiley, New York, p 285
62. Kang J, Yoko T, Sakka S (submitted to J Am Ceram Soc)
63. Eichorest DJ, Payne DA (1988) Mat Res Soc Symp 121: 773
64. Yamashita H, Yoko T, Sakka S (submitted to J Am Ceram Soc)
65. Ogiwara S, Kinugawa K (1982) Yogyo-Kyokai-Shi 90: 157 (in Japanese)
66. Gonzalez-Oliver CJR, Kato I (1986) J Non-Cryst Solids 82: 400
67. Dislich H (1983) J Non-Cryst Solids 57: 371
68. Maddalena A, Dal Maschino R, Dire S, Raccanelli A (1990) J Non-Cryst Solids 121: 365
69. Bullot J, Cordier P, Gallais O, Gauthier M, Livage J (1984) J Non-Cryst Solids 68: 123
70. Yoldas BE, Partlow DP (1980) Am Ceram Soc Bull 59: 640
71. Tatsumisago M, Minami T (1987) Nippon Kagakukaishi, 1958 (in Japanese)
72. Perthuis H, Velasco G, Colomban Ph (1987) Jpn J Appl Phys 23: 534
73. Perthuis H, Colomban Ph (1985) J Mater Sci Lett 4: 344
74. Ravaine D, Seminel A, Charbouillot Y, Vincens M (1986) J Non-Cryst Solids 82: 210
75. Monde T, Kozuka H, Sakka S (1988) Chem Lett 287
76. Kumagai T, Yokota H, Kawaguchi K, Kondo W, Mizuta S (1987) Chem Lett 1645
77. Tatsumisago M, Sato H, Minami T (1988) Chem Express 3: 311
78. Kumagai T, Manabe T, Kondo W, Minamiue H, Mizuta S (1990) Jpn J Appl Phys, Part 2, 29: L940
79. Agostinelli JA, Paz-Pujalt GR, Mehrotra AK (1988) Physics C 156: 208
80. Nasu H, Makida S, Ibara Y, Kato T, Imura T, Osaka Y (1988) Jpn J Appl Phys, Part 2, 27: L536

81. Hirano S, Hayashi T, Tomonaga H (1990) Jpn J Appl Phys, Part 2, 29: L40
82. Tohge N, Tatsumisago M, Minami T (1990) J Non-Cryst Solids 121: 443
83. Kordas G (1986) Mat Res Soc Symp Proc 73: 685
84. Chen KC, Janah A, Mackenzie JD (1986) Mat Res Soc Symp Proc 73: 731
85. Tanaka K, Yoko T, Atarashi M, Kamiya K (1989) J Mater Sci Lett 8: 83
86. Fujishima A, Honda K (1972) Nature 238: 37
87. Finklea HO (ed) (1988) Semiconductor electrodes, Elsevier, Amsterdam
88. Yoko T, Kamiya K, Sakka S (1986) Denki Kagaku 54: 284
89. Yoko T, Kamiya K, Yuasa A, Tanaka K, Sakka S (1988) J Non-Cryst Solids 100: 483
90. Yoko T, Yuasa A, Kamiya K, Sakka S (in press) J Electrochem Soc
91. Yoko T, Yuasa A, Kamiya K, Tanaka K, Sakka S (1987) Res Rep Fac Eng Mie Univ 12: 41
92. Yoko T, Kamiya K, Tanaka K, Sakka S (1990) Res Rep Bull Chem Res Kyoto Univ 67: 249
93. Yoko T, Inagaki Y, Sakka S (1990) Rep Asahi Glass Foundation Ind Techn 13: 56
94. Cairns JD, Segal DL, Woodhead JL (1984) Mat Res Soc Symp Proc 32: 135
95. Carturan G, Facchin G, Navazio G, Gottardi V, Cocco G (1984) In: Hench LL, Ulrich DR (eds) Ultrastructure processing of ceramics, glasses, and composites, Wiley, New York, p 197
96. Ueno A, Suzuki H, Kotera Y (1983) J Chem Soc, Faraday Trans I 79: 127
97. Ida T, Tsuiki H, Ueno A, Tohji K, Udagawa Y, Imai K, Sano T (1987) J Catal 106: 428
98. Sakka S (1990) Catalysts 32: 2 (in Japanese)
99. De Sanctis O, Gomez L, Pellegri N, Parodi C, Marajofsky A, Duran A (1990) J Non-Cryst Solids 121: 338
100. Tohge N, Matsuda A, Minami T (1987) Chem Express 2: 141
101. Ganguli D, Kundu D (1984) J Mater Sci Lett 3: 503
102. Tohge N, Matsuda A, Minami T (1987) J Am Ceram Soc 70: C13
103. Schlichting J, Neumann S (1982) J Non-Cryst Solids 48: 185
104. Martinsen J, Figat RA, Shafer MW (1984) Mat Res Soc Symp Proc 32: 145
105. Brow RK, Pantano CG (1984) Mat Res Soc Symp Proc 32: 361
106. Murakami M, Izumi K, Deguchi T, Morita A, Tohge N, Minami T (1989) J Ceram Soc Jpn 97: 91 (in Japanese)
107. Schmidt H, Phlipp G, Patzelt H, Scholze H (1986) Collected Papers, XIV Intl Congr on Glass, Indian Ceramic Society, Vol 2, p 429
108. Izumi K, Tanaka H, Murakami M, Deguchi T, Morita A, Tohge N, Minami T (1990) J Non-Cryst Solids 121: 344
109. Monde T, Yamamoto M, Konakabara T, Sakka S, to be published
110. Fabes BD, Oliver WC (1990) J Non-Cryst Solids 121: 348
111. Schmidt H (1988) J Non-Cryst Solids 100: 51
112. Schmidt H, Scholze H, Tunkes G (1989) J Non-Cryst Solids 80: 557
113. Nass R, Arpac E, Glaubitt W, Schmidt H (1990) J Non-Cryst Solids 121: 370
114. Maddalena A, Guglielmi M, Raccanelli A, Colombo P (1988) J Non-Cryst Solids 100: 461
115. Fabes BD, Doyle WF, Zelinski BJJ, Silverman LA, Uhlmann DR (1986) J Non-Cryst Solids 82: 349
116. Fabes BD, Uhlmann DR (1990) J Am Ceram Soc 73: 978
117. Fabes BD, Berry GD (1990) J Non-Cryst Solids 121: 357

Thin Films, the Chemical Processing up to Gelation

Helmut Schmidt

Institut für Neue Materialien, Universität des Saarlandes, Gebäude 43, Im Stadtwald, 6600 Saarbrücken, FRG

The sol–gel process is an interesting method of synthesizing coating materials and applying them in the form of thin films. The techniques are similar to those of conventional coating techniques such as dip, spin or spray coatings. These techniques allow one to prepare coatings with smooth surfaces or planarization effects. Based on the chemistry of the synthesis, a large variety of materials can be obtained: glasses, ceramic materials and composites. One of the most interesting features is the possibility of synthesizing multicomponent compositions or composites. For multicomponent materials, the stoichiometry can be controlled very exactly and composites can include glass–ceramic, ceramic–ceramic and ceramic–organic compositions. Especially, the composites open a field of coatings which is difficult to achieve by other than sol–gel techniques. The exploitation of the potential will be an interesting future task.

1 Introduction

The sol–gel process has gained importance in science as well as in industry during the last twenty years. The process describes, in general, the formation of solid materials, mainly inorganic non-metallic materials from solutions. This can be a solution of monomeric, oligomeric, polymeric or colloidal precursors. In nature, for example, minerals like agate or chalcedony have been formed by a polycondensation reaction from aqueous siliceous solutions [1]. Similar reactions are commercially used for the preparation of silica from aqueous solutions of alkali silicates [2]. In these reactions, the "solid" forming step is the condensation reaction of ≡SiOH groups, which forms ≡Si–O–Si≡ bonds by elimination of H_2O, Eq. (1).

$$\equiv\text{Si–OH} + \text{HO–Si}\equiv \;\rightarrow\; \equiv\text{Si–O–Si}\equiv \; + H_2O \tag{1}$$

By this reaction an inorganic backbone is built up, leading to a solid material, the so-called gel. The gel state is a not very well defined state but it is widely accepted to consider a gel as a material which, at least, is diphasic (where pores can be considered as one phase). Due to the ability of Si to crosslink three-dimensionally, an almost infinite number of structures becomes possible.

Condensation reactions are very common in organic as well as inorganic chemistry, and the progress of chemistry made new precursors available for many synthetic products, e.g. ceramic precursors like alkoxides. Although they have been known since the second half of the last century [3], they were considered as significant for material synthesis at this time. Other condensation step containing reactions, e.g. the precipitation of hydroxides are widely used in industry for the synthesis of oxidic powders but are not considered as a sol–gel process. The breakthrough of sol–gel technologies occurred from 1970 to 1980 and was based on the pilot work of Geffcken [4], Roy [5] and Dislich [6]. Widespread interest developed in the material science community. The idea of synthesizing real glasses far below melting temperatures and especially the formation of thin films by this method encouraged a great number of material scientists to investigate this route. A similar development took place in the ceramic community and the idea of lowering the sintering temperature or the increase of the sinter activity of new ceramic powders was an important driving force.

However, the limits of the sol–gel processing soon became very clear: The chemistry behind the reactions turned out to be much more complicated than expected and material tailoring and reproducibility of properties required a very thorough knowledge of the parameters influencing reaction mechanisms, structure and properties. Moreover, the processing of monoliths turned out to be extremely difficult due to the unavoidable gel step, characterized by a low density as a result of pores. The pores are formed by a three-dimensional crosslinking of oligomeric units (colloidal particles or macromolecules) leading to a rigid network, the gel. Due to the relative weakness of the gel and the building up of high stresses as a result of the rigidness, the gel tends to crack due to an anisotropic shrinkage during drying and heating [7]. Thus, the idea of monolithic component processing was abandoned with a few exceptions. For example, as shown by Hench et al. [8] or Yamane [9], the preparation of high quality silica or GRIN materials requires special methods such as drying aids.

The sol–gel process includes several steps as the following schematics indicates:

starting material		sol		gel		final material
A	→	B	→	C	→	D
molecular		colloidal		solid (porous)		solid (dense)

Since the B phase can be stabilized in most cases, it can also be used as starting material. As already mentioned B → C and C → D are critical for technical application, if large three-dimensional parts are considered. This leads to the conclusion, that the fabrication of products with one or two very small dimensions should be the preferable way for sol–gel processes. The step A → B is important as a formation step for intermediates. If viscosity can be controlled, shaping techniques such as fiber, film or foil formation become possible [10]. Therefore, the investigation of sol–gel techniques for the formation of thin films has gained much interest, especially since with the work of Dislich [6] the industrial application of sol–gel films has become popular.

In order to illustrate the important features of sol–gel technology for film formation, one has to realize the specific features of the different processing steps, which include a large variety of parameters.

2 The Basis of Sol–Gel Chemistry

As shown in Ref. [11], the majority of the elements of the periodic table have already been introduced into sol–gel processes. The question whether an element is suitable for sol–gel processes or not cannot be answered very simply. However, there are some rules which can be drawn from the basics of the process. In order to form an inorganic network from a solution (pure or colloidal solution) a network-forming step is required. This step mainly depends on the structure of the sol and can roughly be divided into two alternative mechanisms. The first is a mechanism which is based on the growth of molecules, leading to macromolecules which then grow together to an infinite network. This mechanism leads to the so-called polymerized gels and is very common in the acid-catalysed hydrolysis and condensation of tetraalkyl silicates [2, 12]. The other type is based on the aggregation of colloidal particles from a so-called colloidal sol and requires a fairly stable sol as intermediate. Otherwise the whole procedure would end up in a precipitation process with no sol phase to be identified. The network-forming step in these sols is the aggregation of particles to an infinite network. In the case of acid-catalyzed silica from alkoxides the polymerization process can be simply controlled by the limitation of water which leads to stable sols if enough unhydrolysed $\equiv$SiOR groups can be maintained to keep the average molecular weight small.

The influence of protons and water on the structure of SiO_2 sols has been intensively studied by Sakka [10, 13] who showed by rheological analysis, that various structures of viscous sols can be obtained just by varying H^+ and H_2O concentration in the starting solution.

The stabilization of the colloidal sol includes the control of surface chemistry and can be successfully performed by charging the surface electrically (control of the ξ potential) or by the adsorption of neutral molecules acting as steric barriers [14]. The stabilization of sols is of high importance in sol–gel processing because it defines very strongly their processing properties which are of special importance for film formation. These are, for instance, the rheology, the maximum solid content and the particle size and distribution. Rheology is a complex parameter and depends on particle shape, temperature, solvent, concentration and particle interaction. The reduced viscosity η/c of non-interacting particles does not depend on concentration [15, 16] (Eq. 2).

$$\frac{\eta}{c} = \frac{k}{\rho} \tag{2}$$

k represents a constant and ρ the density of the particles. For an organic polymer solution, the intrinsic viscosity η_i is related to the average molecular weight M by Eq. (3),

$$\eta_i = k \cdot M^\alpha \tag{3}$$

where k is a constant depending on temperature, solvent and the chemistry of the polymer and α represents a parameter depending mainly on the polymer structure [16, 17]. Thus, a rough distinction can be made between "basic macromolecular forms", for example, $\alpha = 0$ for rigid spherical particles; $\alpha = 0.5 - 1.0$ for flexible chainlike molecules, and for rigid, rod-like molecules α becomes 1.0–2.0. The determination of α allows one to tailor processing properties, if the mechanisms of particle shaping can be controlled. An interesting example was given by Sakka [13, 18, 19] who was able to demonstrate that in the hydrolysis and condensation process of tetraethylorthosilicate, α can be selected for optimal fiber spinning. The process has been industrialized by Asahi Glass Co. for the production of high quality SiO_2 fibers [20]. In this case hydrolysis under acid conditions leads to chainlike flexible polymers suitable for fiber drawing.

In contrast to this, according to Stöber [21], at high pH values, particulate sols are obtained. These sols are not useful for fiber drawing or spinning. The basic ability of film formation, however, does not depend so strongly on the sol structure parameters, but the film properties can be affected strongly again, as pointed out by Frye and Brinker [22, 23]. They compared acid- and base-catalyzed SiO_2 films after dip coating by N_2 adsorption. Whereas from acid-catalyzed sols pretty dense structures were observed (Type II isotherms, refractive index $n_D \approx 1.43$), in the base catalyzed systems, particulate sols ($\approx$ 20 nm SiO_2 particle diameter) were obtained leading to porous film structures (type IV isotherms, 45% porosity, $n_D \approx 1.24$) (Fig. 1). This demonstrates clearly the influence of the sol structure on the coating properties.

The SiO_2 system has received the highest scientific interest up to now. This may be due to the fact that SiO_2 precursors in the form of alkylortho silicates have been readily available for almost 150 years, and, compared to almost all other common alkoxides they are relatively insensitive to moisture which means in this case, that the hydrolysis and condensation to gels take place pretty slowly (depending on concentration, type of alkoxides, and solvent, between hours and weeks). Therefore, it is possible to study the reaction kinetics and structure forming mechanisms of sols by condensation in details relatively conveniently compared to other systems. In the recently published book by Brinker and Scherer [24], the reaction of tetraethyl orthosilicates as interesting precursors in sol–gel reactions are intensively discussed.

The sol structure also influences the maximum film thickness. As shown in [25], for given systems, the "cracking thickness" of coatings is defined by the system parameters only. It is shown that films can be transformed to dense layers by temperature treatment only if the thickness of them does not exceed certain limits defined by the intrinsic system parameters. In general, the film thicknesses to be obtained in a one-step coating process (spin or dip coating) does not exceed some tenths of a μm. If the viscosity of a system is adjusted to obtain thicker films, cracks occur during drying.

In the work of Dislich [6], who developed an industrial coating technology for plate glass, the question of film thickness does not play an important role since he was able to prove that the dip-coating process employed leads to dense chemically very stable thin films of SiO_2, TiO_2 or combinations. Therefore, two requirements had to be fulfilled: the films had to be very perfect (pinhole-free) and they had to be very dense. Since these films on plate glass as substrate have never been fired to temperatures around Tg, the densification process has to be different from the "normal" viscous flow sintering of glasses.

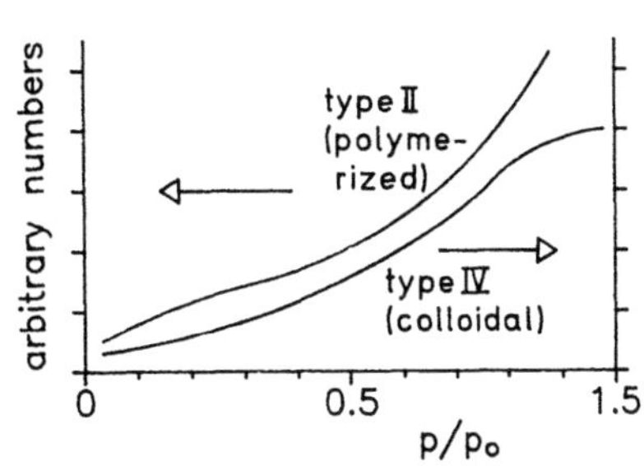

Fig. 1. Influence of the structure of two different sols on the N_2 adsorption of dry films (data after 22 and 23)

The observation of Brinker et al. that weakly branched SiO_2 polymers from hydrolysis of TEOS or highly branched borosilicate polymers lead to very dense films after dip coating by simply drying them [less than 5% porosity, type II isotherms] is explained by a special particle packing process taking place at the drainage and drying front of a dip coating process (Fig. 2).

The maximum obtainable thickness of the films does not seem to depend strongly on the sol structure. Films up to 0.2 and 0.3 μm can be obtained from most sols. With increasing thickness, the "bulk" behavior gains more and more importance, as shown in the model according to Fig. 3.

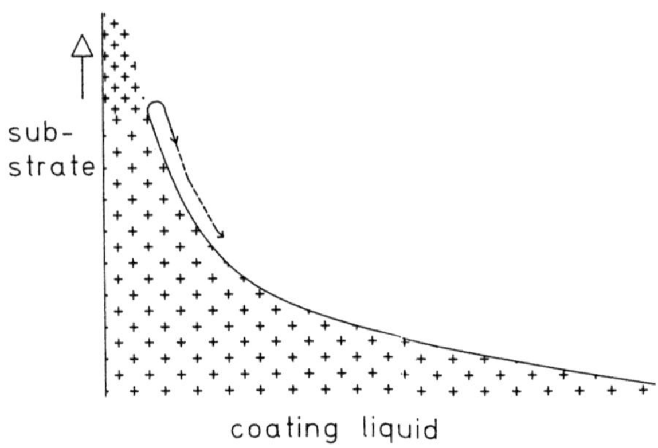

Fig. 2. Model of the densification process during the dip coating step from a sol

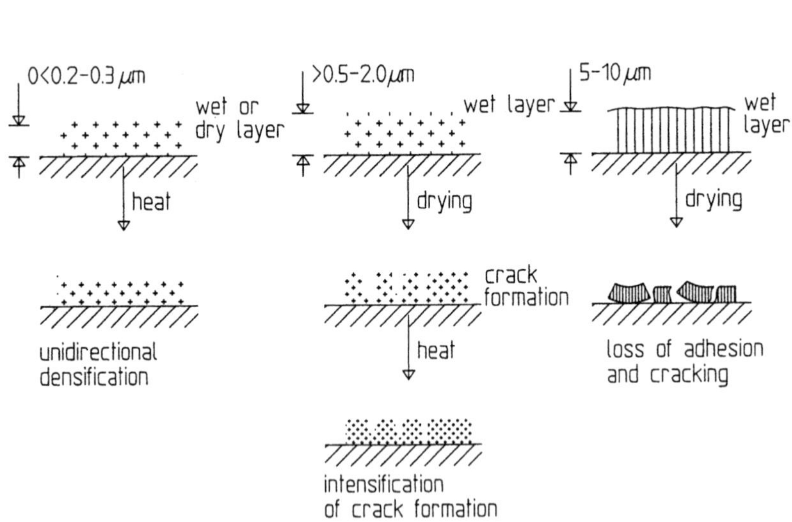

Fig. 3. Model of the effect of thickness on the drying/densification behavior of sol–gel coatings

There exist some qualitative rules, which can act as guidelines to influence film thickness:

- If the interaction of the sol particles is high, the densification process will be stopped at a very porous gel structure with a low-density brittle system leading to crack formation during drying and sintering.
- Densely packing particles and easily relaxing networks can help to obtain dense films even at increased thicknesses.

Unfortunately, there have been no systematic investigations on the influence of solvents other than alcohols and other systems than SiO_2 on film thickness depending on processing parameters. The most important progress in understanding the film formation from sols is shown in [24] and it is amazing, that these investigations were carried out about 20 years later than the first successful industrial application of sol–gel coatings had been performed by Schott Glaswerke, Mainz. The findings of Dislich and coworkers that they could succeed in coating large dimensioned plate glass and densify the films at $T \ll Tg$ had found a late explanation.

3 Chemistry as a Tool For Sol Tailoring

As mentioned in Sect. 2, it is highly important to control particle size and properties of sols in order to obtain optimum coatings. In the case of SiO_2, the reaction control is rather easy since the reactivity of SiO_2 precursors can be adjusted by the pH and the H_2O addition over a wide range. These methods cannot be employed in highly reactive systems such as Ti, Zr, Al alkoxides or glass forming systems containing alkaline and earth alkaline components since the addition of water leads to insoluble precipitates in the majority of cases and no homogeneous sols suitable for coating processes are obtained. In these cases, it is necessary to control the reaction by other means, e.g. by complex formation, as demonstrated by Livage et al. in the system TiO_2/acetic acid [26]. In this case, the hydrolysis is not completed with acetic acid present due to the formation of an intermediate complex (Fig. 4). The first reaction step is the substitution of an OR group by an acetic acid molecule. By water addition, a complex where two OR groups are substituted by OH groups is obtained, which is stable against further hydrolysis and which does not form insoluble precipitates. As shown by LaCourse, the viscosity of TiO_2 sols can be controlled by

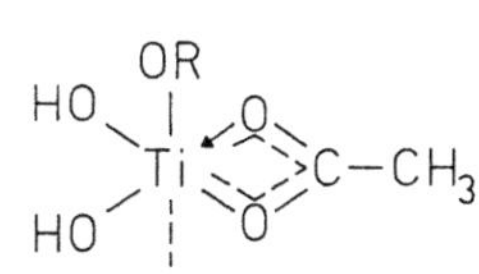

Fig. 4. Acetic acid complex with titania

complexing Ti(O^{i}prop)$_4$ by addition of β-diketones, e.g. bis acetyl acetonate [27, 28].

This type of precursor is suitable for fiber drawing but no data on the film-formation properties of these sols are available. But there seems to be a good chance of adjusting the viscosity for optimum film forming conditions.

In the case of alumina, investigations into the film forming properties of β-diketone-modified sols have been carried out by Naß et al. [29, 30]. In these investigations, two types of Al containing sols (with and without a complex former) have been compared with respect to their film forming properties. It could be shown that Al propoxides complexed with acetyl acetonate (acac) or acetoacetates form clear sols with nanosized particles (2–5 μm depending on the complex former concentration). The solid content of these sols can be increased up to 25 wt % without gelation (≈ 5 wt % in the uncomplexed case).

The film formation ability of the complexed sol is substantially better than that of the non-complexed one. In Fig. 5a and 5b two SEM micrographs with

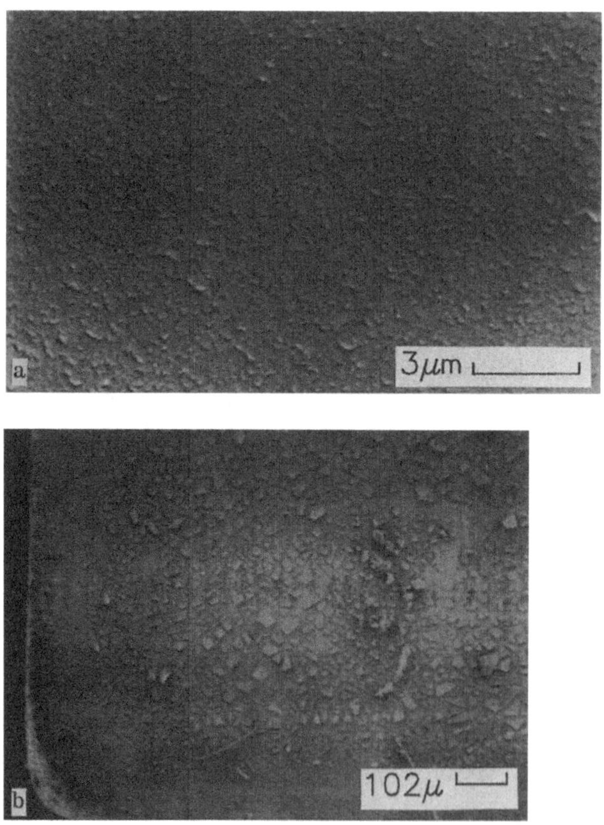

Fig. 5a, b. SEM micrograph of a film obtained from a non-complexed sol with intensive crack formation. SEM micrograph of a film obtained from a complex sol. Coating thickness 1 μm

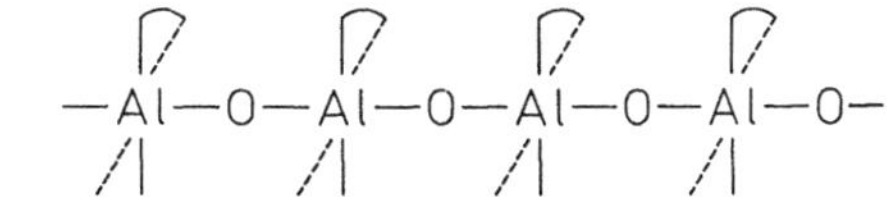

Fig. 6. Structural model of complexed Al sols obtained by NMR data according to [29]

films of the two sols are shown. NMR and IR analysis show that chainlike molecules are obtained (Fig. 6). The coatings obtained from complexed precursors can be densified at 600 °C to clear transparent layers. These sols can be used for fiber-drawing processes, too. Bulk gels are amorphous up to 800 °C and then turn directly into α-Al_2O_3 [31].

The problem of the homogeneity of multicomponent systems in sol–gel processing becomes obvious if one considers the large hydrolysis and condensation rate differences and solubilities of the various precursors. We can distinguish between several types of basic properties important for sol–gel processing:

- slow reacting components like ethylortho silicates
- fast reacting components like $Ti(OR)_4$, $Zr(OR)_4$, $Al(OR)_3$, $P(OR)_{3/5}$, halides, alkaline or earth alkaline alkoxides (alcoholates), $B(OR)_3$
- soluble species (especially in alcohols) like $B(OH)_3$, alkaline and earth alkaline hydroxides, P_2O_5.

In order to obtain homogenous sols and gels one has to obey some rules: It is necessary to introduce at least one good gel former into the system in order to obtain homogenous sols and gels [32]. The additional components can be either linked to the gel network or gel particles or entrapped in pores (Fig. 7).

In the case of the PZT gel, for example, it depends strongly on the processing conditions whether the homogeneity can be maintained during the drying and heating step.

If the network formers are very different in hydrolysis or condensation rates one has to take care that no phase separation occurs as a consequence of these reaction rate differences. To overcome these difficulties in the case of $Ti(OR)_4/Si(OR)_4$ mixtures, Yoldas suggested that prehydrolyzed $Si(OR)_4$ with

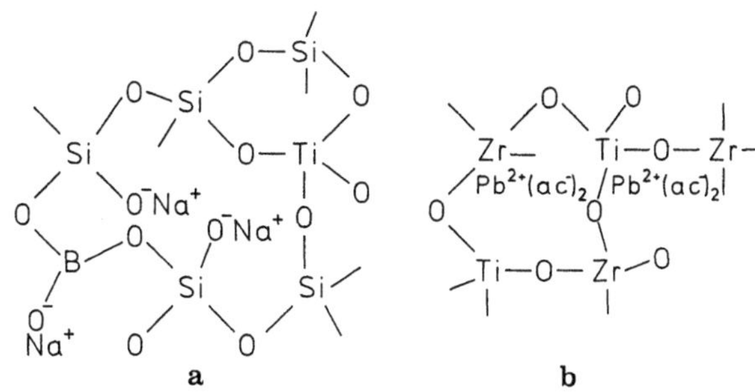

Fig. 7. Sodium ions linked to a glass network by ionic bonds (a) and lead acetate entrapped within a PZT gel (b)

high ≡SiOH concentrations should lead to the formation of ≡Si–O–Ti≡ bonds by (Eq. (4)) [33]

$$\equiv SiOH + RO\,Ti(OR)_3 \rightarrow \equiv Si{-}O{-}Ti\equiv \quad (4)$$

Closer investigations of this hypothesis by Basil and Lin [34–36] showed that the number of ≡SiOH groups is drastically reduced by the addition of $Ti(OR)_4$ to a prehydrolysed $Si(OR)_4$ but no ≡Si–O–Ti≡ bonds could be detected by NMR. The authors conclude that $Ti(OR)_4$ acts as a condensation catalyst leading to an increase of the condensation rate including ring structures but with no incorporation of the $Ti(OR)_4$ into the silica network taking place. The authors, however, did not test their hypothesis by adding water to the system but by just looking to see whether the precipitation behavior of the $Ti(OR)_4$ had changed or not.

This test was made by Schmidt et al. [37] with another, more complicated system in connection with the development of a sol–gel coating system for plastic eye glasses. It was found out, that TiO_2 in combination with SiO_2 and an organically modified silicic acid ester leads to very hard coatings being densified at surprisingly low temperatures (90–110 °C) suitable to be employed on plastics. This system was prepared from $Si(OMe)_4$ (I), $Ti(OEt)_4$ (II) and $(MeO)_3Si(CH_2)_3{-}O{-}CH_2{-}CH{-}CH_2{-}O$ (III) as precursor. The epoxide polymerizes partially to polyethylene oxide chains. The hydrolysis reaction rate at neutral pH decreases in the order $k_I < k_{III} \ll k_{II}$. Direct addition of water to the precursor mixture (I : II : III = 5 : 25 : 70, molar ratio) leads to the precipitation of

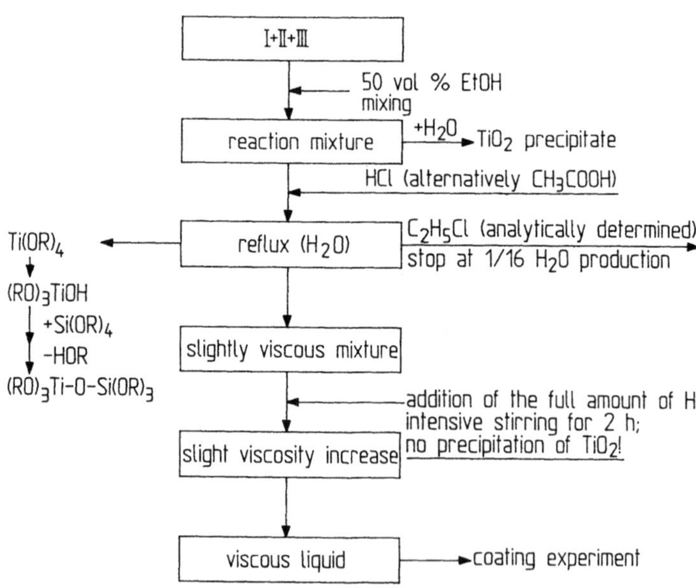

Fig. 8. Schematics of the process of preparing Ti containing hard coatings

TiO_2. H_2O, however, can be generated homogeneously within the reaction mixture (ethanolic solution), e.g. by adding HCl and refluxing the system (Eq. 5).

$$C_2H_5OH + HCl \rightarrow C_2H_5Cl + H_2O \tag{5}$$

In the system I + II + III, it is sufficient to generate about 1/16 of the amount of water necessary to hydrolyze the OR-groups of the system completely to prevent any precipitate after adding the full amount of H_2O. In the precondensation step, $Ti(OR)_4$ was linked to the silicate precursors, thus being prevented from precipitating.

Figure 8 shows the process schematically for easier understanding. The findings can only be interpreted by the formation of a ≡Si–O–Ti≡ bond. In this case, the reaction runs the opposite way to that in Yoldas' experiment. Due to the higher reactivity, $(RO)_3TiOH$ is formed. Of course, the statistic consideration leads to the fact that an already hydrolysed $(RO)_3TiOH$ will be hydrolysed a second time at the expense of an unhydrolysed one, but the probability is low compared to the probability of the consumption of the ≡TiOH groups by ROSi≡. The resulting prepolymers can be used for coating experiments as shown later.

The establishment of a suitable viscosity is an important means for coating experiments. It is important to know whether it is possible to predict film thicknesses by physical data of the coating solution. As shown in [38] and [39] independently, the relationship derived by Landau and Levich [40] (Eq. 6), which describes the coating thickness h as a function of various parameters, fits fairly well for sol–gel derived coatings (Figs. 9a and 9b).

$$h = 0.94(\eta \cdot U/\gamma)^{1/6} (\eta \cdot U/\rho \cdot g)^{1/2} \tag{6}$$

(γ = surface tension, η = dynamic viscosity, ρ = density, g = gravity constant, U = withdrawal speed)

Both findings represent slow withdrawal rates (up to some cm/min) where the dependence of the viscosity on shear rates can be neglected.

A general schema for the different steps includes precursor chemistry, reaction to coating liquids (e.g. by viscosity and molecular weight control), the coating step, post treatment of the coating (e.g. curing, densification, patterning) is given in the following.

The schema of course, can only represent a selection of important or interesting techniques. In special cases, others are used. But it shows the large variety of possibilities.

4 Precursor Chemistry

There is no general difference in the precursor chemistry between the preparation of intermediates for coatings or for gels. The problem with precursor chemistry is its complexity once the first hydrolyzed species are formed which are able to react along a large variety of reaction paths.

Alkoxides are used as precursors in most cases. The well-established field of alkoxide chemistry considered hydrolysis as an undesired side reaction for a long time [41, 42]. It is only recently that Mehrotra emphasized the significance of alkoxide chemistry for sol-gel processing [43–44]. The effect of the structure of alkoxide solutions on hydrolysis and condensation reaction is not very well investigated in most cases, but there are some examples that clearly demonstrate processing advantages if the chemistry is controlled. Hirano [45] showed that $LiNbO_3$ films could be prepared much more homogeneously and crystallized at lower temperatures if the alkoxide precursors were reacted to a double alkoxide

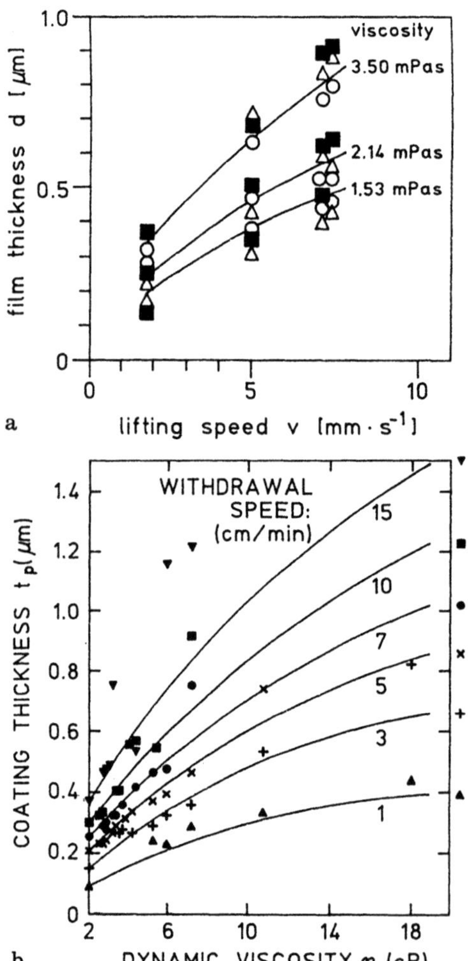

Fig. 9a, b. Dependence of dynamic viscosity η on coating thickness of acid catalyzed solutions of TEOS according to [37]. Dependence of coating thickness on withdrawal speed at various viscosities of a solution representing a sodium borosilicate glass system according to [39]

complex before hydrolysis. Results indicating the important role of double alkoxides were reported by Pouxviel [46] by NMR analysis of the $Si(OR)_4/Al(OR)_3$ two component system. The problem of all these investigations seems to be the fact, that structural analysis, especially on the short range level can only be carried out in homogeneous media and only give clear results up to complexes or aggregates of only a few monomeric units. There is a gap between the "oligomeric" or even macromolecular state on the one side and the gelified (infinite molecular size) state on the other side which is very difficult to characterize by structural analysis. More recently, efforts have been made to

PRECURSORS
- Alkoxides
  Si.., Al.., Ti.., Zr..
  $Si{-}CH_2$...
- colloids (boehmite, colloidal silica ..)

→

REACTION TO COATING LIQUID
- controlled hydrolysis and condensation
- controlled aggregation
- stabilization by ξ potential control or sterical surface modification

COATING TECHNIQUES	IMPORTANT PARAMETERS
• dip coating:	viscosity, solid content, withdrawal speed, contact angle to substrate (wetting).
• spin coating:	viscosity, shear rate dependence of viscosity, shear rate, contact angle (wetting), surface tension (smooth surfaces), additives for surface tension control.
• spray coating:	spray parameters (e.g. nozzle size, pressure, ...), droplet size, evaporation velocity of solvents (viscosity increase), surface tension, additives for surface tension control, drying versus surfaces smoothing velocity.
• electrophoretic coating:	particle or droplet size and charge (ξ potential), electric field strength, viscosity of liquid media, conductivity of the deposited layer.

↓

POST TREATMENT

• heat treatment:	IR, furnace, LASER
• photo curing:	UV with initiators, if reacting groups are present e.g. $X{-}X + \xrightarrow{h\cdot\nu} X\bullet\bullet X$ (radial mechanism) e.g. for –Si– (olefinic polymerization in composites) or $X{-}H \xrightarrow{h\cdot\nu} H^+ + X^-$ (ionic mechanism) for $-SiOR \xrightarrow{H^+/H_2O}$ hydrolysis and condensation.
• patterning:	UV, LASER writing.

"construct" complex oligomers with well-defined crystal structures acting as "molecular building block precursors". The significance of this concept was shown by Payne [47] on $BaTiO_3$ clusters prepared by understoichiometric addition of water to Ba and Ti alkoxides.

As a general rule, it seems to be suitable to tailor the oligomeric state very carefully for coating. This is valid, too, for fiber drawing, as shown in [19] for SiO_2. Similar behavior was found by LaCourse for "linearized" TiO_2 precursors [26]. Naß [48] found that Al-alkoxides complexed with β-diketones lead to precursors from which excellent coatings can be obtained compared to uncomplexed prehydrolyzed precursors. These few examples demonstrate the importance of sufficient knowledge of the precursor chemistry for sol tailoring. This is valid for precursors leading to inorganic compositions (glasses, ceramics) as well as for precursors containing chemically bonded organics.

There are various means of controlling precursor chemistry with respect to tailor sol structures. The majority of investigations concerns SiO_2 as a one-component system [49–50]. In this system, the pH dependence of the sol structure is obvious. At pH values from about 1 to 5 the macromolecular types of sols with more or less linear molecules are obtained whereas at higher pH values sols containing spherical particles are obtained. From the work of Thomas [51] it is known that the particulate type of sols can be obtained in the system Al_2O_3/SiO_2 which was advantageously used for preparing laser damage protective coatings with excellent optical properties. If the optical properties of the coatings are satisfactory they do not necessarily have to be fully densified if the requirements for chemical stability are not critical [51]. The sols used in this case are colloidal sols with nanometer-sized particles.

For sol property tailoring, precursor chemistry is the key issue. In contrast to its tremendous importance, the state of the art is rather low. It is possible, however, to formulate some general rules helpful for sol development.

4.1 Homogeneity

Homogeneity or controlled heterogeneity affects the properties of films in various ways. The "molecular" homogeneity of precursor solutions can be changed by different processes as shown in [52], e.g. different rates of hydrolysis and condensation, phase separation during ageing or transport phenomena in wet gels. Therefore, the control of reaction rates or avoiding the effect of different reaction rates becomes an important objective. Two different routes are of interest: First one can try to slow down the rate of faster reacting species by chemical modification of precursors, by increasing the chain length of alkoxides, e.g. $\equiv MeO(CH_2)_nCH_3$ with n increasing. In this case, the hydrolysis rate decreases with increasing n and the overall condensation rate decreases by keeping the concentration of condensable species low. The effectiveness of this method is very limited and can only effectively be applied to silicon alkoxides.

To overcome reaction rate differences of very different metal alkoxides as Si and Ti seems rather unlikely.

Another means is to block reactive sites by complex formation which can be used for adapting rates as well as for molecular design of sols. As shown by Yamane [53], the formation of $acac_2$-Ti-alkoxides leads to a sol without precipitation through water addition (Fig. 10). b is stable against hydrolysis but can be used for further synthesis with reactive alkoxides. In the case of SiO_2–TiO_2 systems [53] the most homogeneous glasses have been obtained with precursors with similar reaction rates. As shown in [29] in the case of ß-diketone complexed Al alkoxides there is no significant difference in the H_2O consumption rate to the uncomplexed alkoxide, but the complexed form leads to a clear solution with polymeric unit sizes in the range of 2–5 nm, indicating that by complexation the condensation process is slowed down drastically. Keeping particle sizes in the nano range in sols restricts the maximum heterogeneity dimension in multicomponent sols, even if complete phase separation of the different components occurs during the growth process. It depends on further processing whether this inhomogeneity dimension is disturbing or not. In case of consecutive high temperature processes e.g. for glass or ceramic phase formation nano-range inhomogeneities as a rule, are not serious. If the particle size can be controlled by chemical means, e.g. by desactivating the surface by reaction with organics as complex formation, polymeric coating or adsorption of the tenside type of molecules onto the surface, the independent preparation of sols with different composition and mixing after preparation seems to be a comfortable way for the preparation of multicomponent sols as shown in [54] for the preparation of semiconductor cluster containing glasses by the sol–gel process.

For the preparation of particle size controlled sols, which can be used for coatings, the control of surface chemistry of colloids by surface active additives is the most important point. In this case, the maximum heterogeneity is defined by the particle size distribution of the colloidal phase. Examples for sol processing by proper colloid chemistry are given by Messing [55] and Aksay [14]. Aksay processed 20 nm SiO_2 sols by surface treatments controlling the hydrodynamic radius (Fig. 11), thus being able to end up with densely packed (close to theoretical hexagonal or cubic packing) nano sols. These findings are

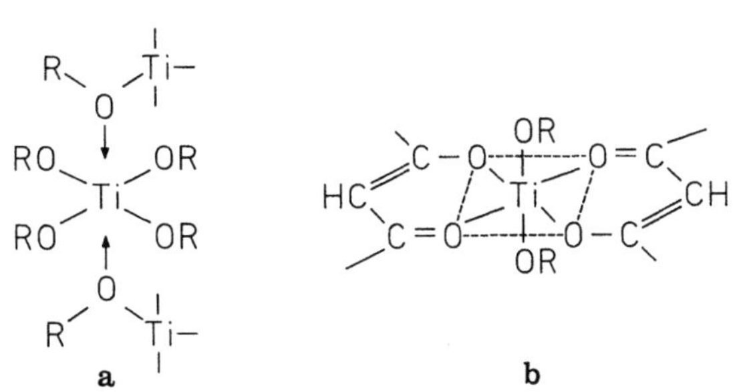

Fig. 10a, b.
Structures of Ti alkoxides

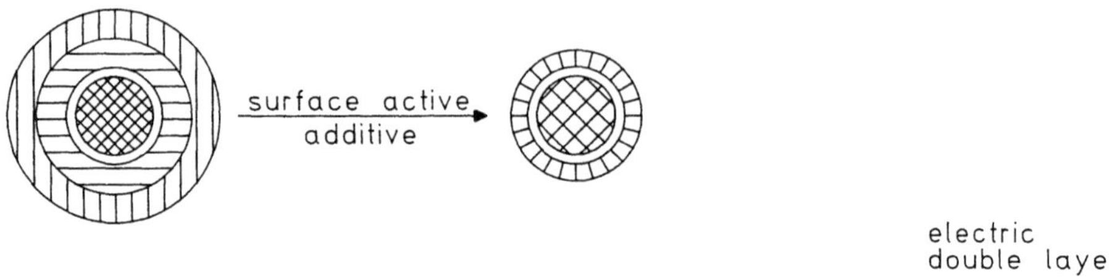

Fig. 11. Reduction of the hydrodynamic radius by surface treatment

highly important, if the preparation of coatings is considered, but details regarding this step and sol structure to coating performance relationships have been published. Compared to conventional ceramic processing, the homogeneity gain using nanosized sols is up to three orders of magnitude. The benefits of colloidal sol processing for ceramic coating has been shown by a variety of authors [56–60] who could show a substantial reduction of sintering temperatures compared to conventional processing. The standard coating process for sols is the dip coating process. The main interest is focussed on the film

Material	Characterisation		next reaction step	properties to be affected
precursor	IR, UV//VIS, GC, GC/MS, phys. data (m.p.; b.p.), NMR	ΔΔΔ	I purification	purity level, all properties depending on purity
mixed precursor	IR, UV/VIS, NMR reaction kinetics	ΔΔ	II hydrolysis/ condensation	solution or sol structure, rheology, particle size, ξ-potential
colloidal macromolecular type sol	light scattering, SAXS, ξ-potential, viscosity (NMR), SAX, LAXS	Δ	III gelation	gel structure, porosity, pore size distribution, homogeneity (phase distribution)
wet gel	porosity, density, mechanical properties, thermal analysis, thermoporometry, NMR, IR, SAXS	Δ •	IV drying ageing	homogeneity, pore distribution, gel structure, further processing properties
dry gel	porosity (adsorption, isotherms), densification mechanisms, thermal analysis, crystallization, mechanical properties, X-ray	• • •	V processing to materials	crystallization, phase formation, density, mechanical, optical, electrical, chemical or other properties

• bulk Δ thin films

ΔΔΔ well investigated for silica and a few other systems [24 and literature cited herein]

ΔΔ only a few systematic data available, mainly on silica

Δ no systematic data and only a few others available

properties, but not to the influence of sol structure or coating parameters on film properties. Meanwhile, coatings are prepared from almost all compositions to be obtained by sol–gel processing, reviewed in [61–79], but the deficiency in the wet-step mechanisms is growing rather than decreasing. Two main reasons can be identified. First, it is very difficult to characterize sol structures and second, the structural characterization of thin films is also very difficult. Examples of investigations to correlate film properties to sol properties are given in [12, 18, 24, 80–82].

The complex processing and the interdependencies are schematically overviewed in the schema on page 00. Step II is extremely important since it represents the homogeneity determining step for sol preparation and thus will affect strongly the coating properties. This is essential since especially for thin films characterized by one very small dimension homogeneity is very difficult to achieve by thermal diffusion after coating.

4.2 Molecular Design

Therefore, it seems to be necessary to take care of sufficient homogeneity during step II. As already mentioned, if the sol particle size can be controlled on a nanometer-sized level, homogeneity can be sufficient for a variety of applications, even for optical purposes as shown by Zarzycki [83] and Nogami [84] with sol–gel glasses containing semiconductor crystallites such as CdS or CdSe. To obtain dense glasses from particulate sols as well as from polymerized sols the densification temperature has to be high enough, probably around Tg. There are no exact data available, since it is extremely difficult to measure the low porosities of thin films. In the case of crystalline coatings, the development of crystalline micro-structures is a complicated process, where nucleation and crystal growth play an important role [85]. But it can be generalized that only in sols with nanometer-phased micro-structures resulting in optically transparent coatings. According to suggestions from Payne [86] the microstructure of crystalline films can be tailored by molecular design of the precursors where the different components already have positions in a crystal close to their final structure. As shown in $BaTiO_3$, thin films, this can lead to nanocrystalline films with domains in the nanometer range but already showing good electrical properties otherwise observed only on much larger domain dimensions.

Other interesting examples has been given by Hirano [45] who was able to demonstrate the effect of the formation of a double alkoxide on the sintering temperature of $LiNbO_3$ and its optical performance. By use of the double alkoxide (Fig. 12), the crystallization temperature of thin films could be reduced to less than 300 °C. The results can be interpreted with the reduced diffusion paths within a super amorphous solid and the high level of free energy. In the early papers of Dislich [6] it was demonstrated that the addition of a β-diketone and subsequent refluxing leads to much better coatings. This can be attributed to the formation of precursor complexes in which the several compounds are

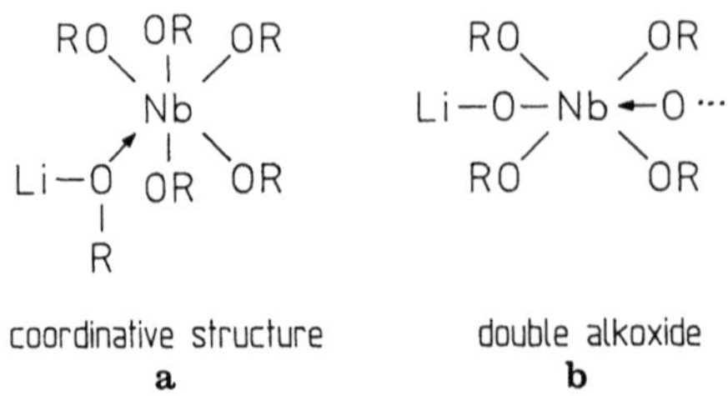

Fig. 12a, b. Mixture of lithium-niobium alkoxide and double alkoxide

already immobilized on a molecular level. Livage and coworkers pointed out in an extensive review [25] the possibilities of molecular tailoring of precursors. His considerations are based on an electrostatic reaction control model and on the role of complex formation with complexing acids or β-diketones. In [87] they try to give some first interpretations of correlations between precursor and the structure of the resulting product.

The control of colloidal particle size and agglomeration cannot of course, be considered as "molecular design". Molecular design has to take into consideration the formation of precursors representing desired structural elements, the formation or maintainance of such an element by reaction control and the structural control during the various processing steps.

4.3 *Reaction Control*

The possibilities of controlling reaction rates with respects to oligomer, macromolecule or colloid tailoring are limited as already pointed out. An interesting alternative is to avoid the undesired effects of rate differences. This was first suggested by Yoldas [32] in the system SiO_2/TiO_2, as already mentioned. The general idea is to prehydrolyse the slower reacting precursor in order to supply it with an "advantage on the time scale" (Fig. 13).

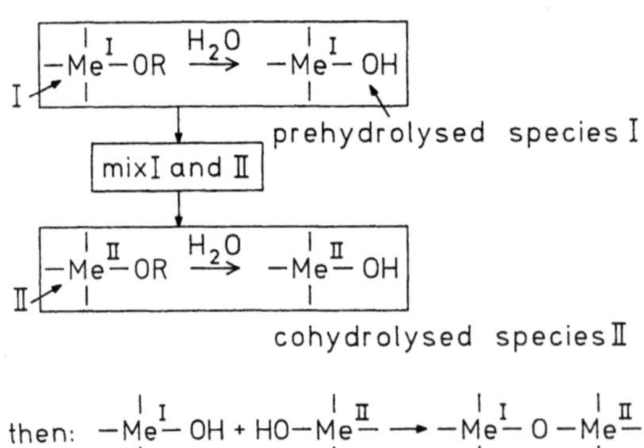

Fig. 13. Prehydrolysation of slower reacting precursors for improvement of homogeneity

But this model can be only employed if the $\equiv Me^{I}OH + HOMe^{II}\equiv$ condensation step is fast. This is, as pointed out by Livage [26], not the case in the presence of an $\equiv$MeOR bond at the same time. Another route was suggested by Schmidt [88] who proposed to prehydrolyse the fast reacting species by controlled generation of water in a reaction mixture (Fig. 14). In this case, the $\equiv MeOH_{fast}$ can react with all types of $\equiv$MeOR and would lead to a homogeneous immobilization of the fast reacting species to other precursor molecules in accordance with Livage's theory. The effectiveness of this method was demonstrated on the eye glass lens coatings and contact lens materials. The use of this route allows the synthesis of optical materials from multi-component systems with very different reaction rates [88, 89, 90].

The usefulness of prehydrolysed precursors for homogeneous distribution can be demonstrated very clearly with "stable hydroxides" such as $(C_6H_5)_2Si(OH)_2$. Transparent coatings can be prepared by reacting $(C_6H_5)_2Si(OH)_2$, which is a solid non-sensitive to moisture solvent, e.g. toluene with $Ti(OR)_4$. The reaction can be followed by monitoring the OH vibration by IR [91] (Fig. 15).

The condensation reaction takes place at room temperature and the $\equiv$SiOH groups are consumed almost completely by condensation with $\equiv$TiOR groups. A viscous liquid results from this process and it can be used for coatings. The refractive index of this coating is between 1.65 and 1.7. The reaction can be interpreted by the formation of polymers with alternating $-O-Si(C_6H_5)_2O-$ and –O–Ti–O– units, as shown in Fig. 16.

These sols (Ti can also be substituted by Si) can be densified at low temperatures (150 °C) and fired to organic-free materials at surprisingly low temperatures, as shown in [92].

These examples show that control of precursor chemistry is an adequate means for property tailoring by design of the sol especially if homogeneity plays an important role.

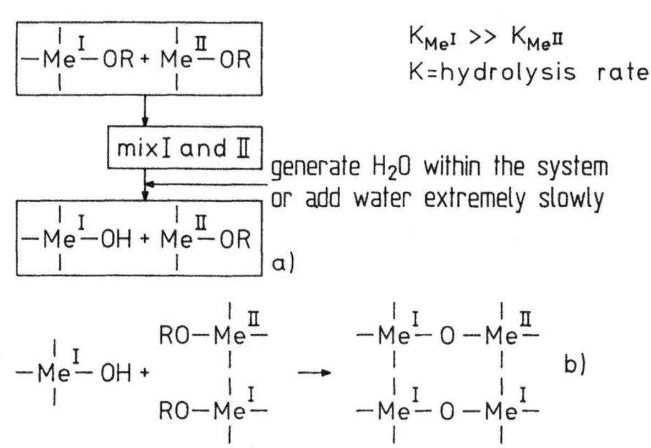

Fig. 14. Prehydrolysation of faster reacting precursors for optimum homogeneity

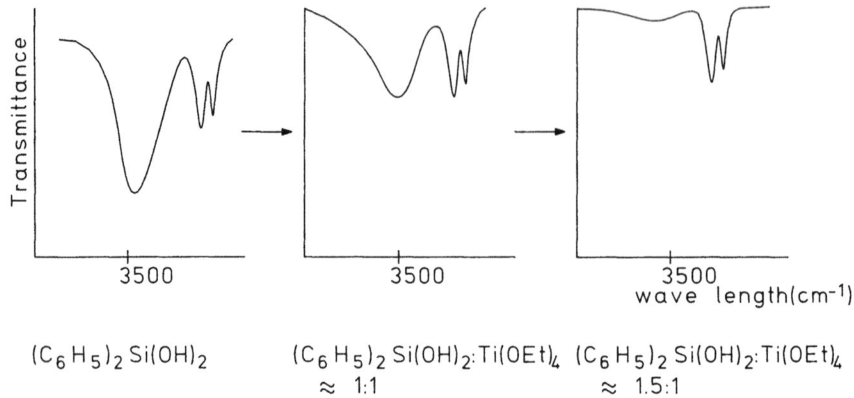

Fig. 15. IR controlled reaction of $(C_6H_5)_2Si(OH)_2$ in toluene with $Ti(OEt)_3$ in toluene as solvent

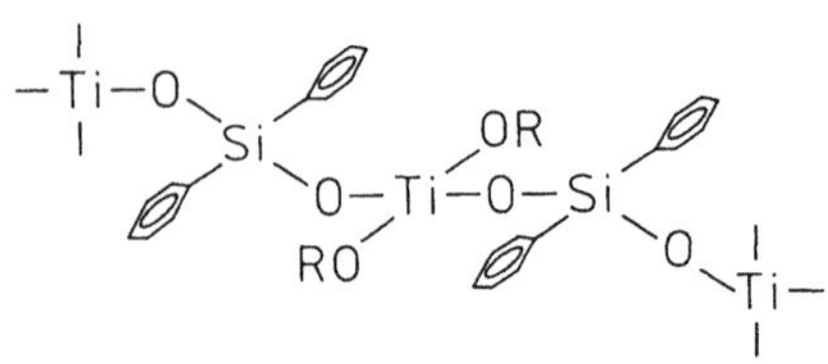

Fig. 16. Suggested structures of $(C_6H_5)_2$–Si–O–Ti-sols

The results from Sanchez and Livage [26] and also from Naß [28] point out that the control of precursor chemistry by the proper design of intermediates or oligomers can be used for increasing homogeneity, but since organic groups chemically bound to inorganic precursors are used this means at the same time that a homogeneous distribution of organic components within an inorganic network occurs. For example, it is possible to disperse more than 50 mole % of ZrO_2 within a polymeric system using complex formers [93]. In this case, the methacrylic acid (MMA) was used as "link" as shown in Fig. 17 [94, 95].

The possibility of molecular design of sols can be extended to almost innumerable variations, but, therefore, possibilities of organic chemistry, organometallic chemistry and coordination chemistry are exploited for these objectives. Especially the latter one is only at its infancy and a few examples can show the importance of these routes. As shown by Schubert and coworkers [96], a homogeneous distribution of metal ions is possible by the formation of complexes with sol–gel precursors (Fig. 18a). Cu can be incorporation by complexation with $NH_2(CH_2)_3Si(OR)_3$. The viscosity can be adjusted by hydrolysis and condensation control, films and foils can be received and the colour intensity can be varied by the Cu content [97] (Fig. 18b). Another type of complex formation was suggested by Charbouillot [98] and Ravaine [99]

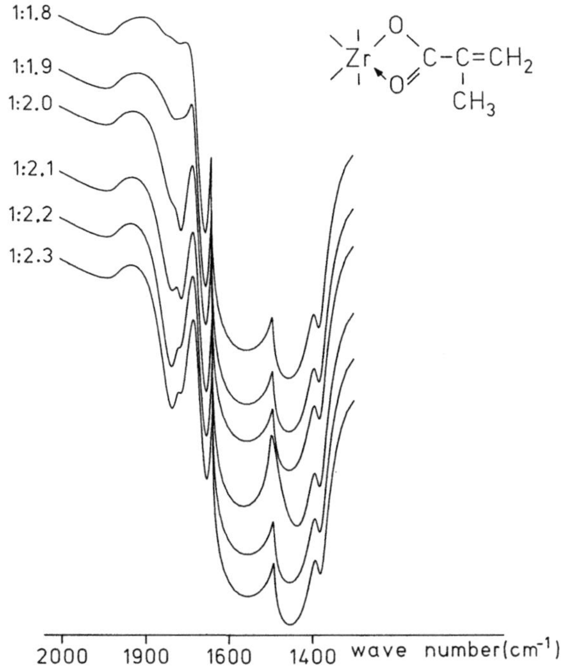

Fig. 17. Complex formation of Zr alkoxides with MMA and IR analysis of increasing MMA to Zr ratio

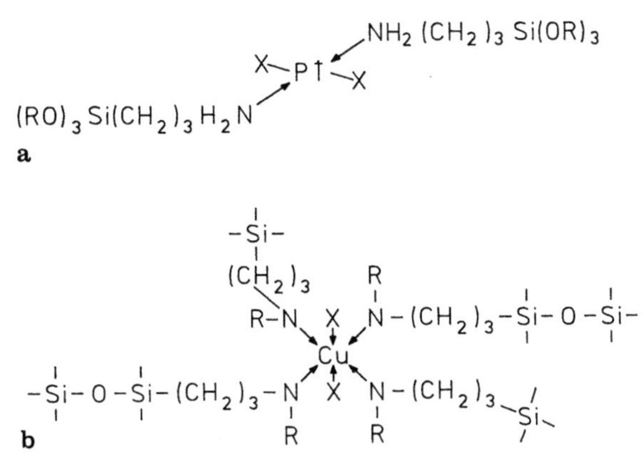

Fig. 18a, b. Structure of a soluble Pt complex and of a film containing a copper tetrammine complex [96]

who prepared proton conducting films by doping condensates from $H_2O(CH_2)_3Si(OMe)_3$ with acids such as $HClO_4$ or CF_3SO_3H postulating a structure stabilized by an anionic complex within the gel network (Fig. 19) and protonic transport mechanism by a chain movement step.

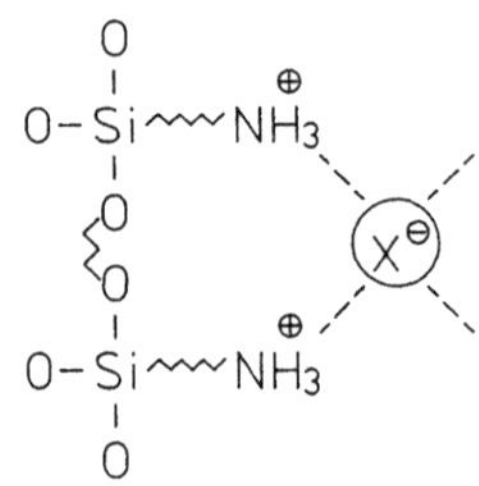

Fig. 19. Structure of a proton conducting anionic complex

5 Material Developments For Coating Application

The key issue for tailored material development is structure tailoring in connection with correct structure analysis, the knowledge of the mechanisms of the reactions and the correlation of these data to material properties. One has to distinguish between two basic features when considering property tailoring: the tailoring of processing properties and/or the tailoring of final product properties. It depends mainly on the product type and application to which of them has to be given priority. A very important point of view is the question of to what extent is it possible to maintain properties, designed by sophisticated precursor chemistry through all the processing steps up to the final product. To illustrate this, Hirano's and Payne's results [45, 87, 100] can be taken as an example. To obtain high quality $LiNbO_3$ films, various sol–gel routes can be employed, e.g. mixing the precursors or synthesize double alkoxides. The main benefit of the double alkoxide route in this case is to reduce processing temperature which opens the application to a variety of substrates and devices as well as the film properties. Another benefit, which is based on low temperature processing, is the slow crystal growth which can lead to extremely small crystal sizes down to nanosized microstructures. But despite this, to maintain the precursor structure is not the objective. If we think of doping silica or oxide glasses with lanthanides, the precursor structure may play an important role if a special coordination structure can be maintained as demonstrated by the preparation of complexed Eu containing films [101, 102] or SiO_2 sol–gel glass films which are heat densified at low temperatures.

The lower the processing temperatures are, the more likely structures generated by chemistry in the sol can be maintained in the final product or in other words, if one wants to maintain structures, one has to use low processing temperatures adapted to the temperature sensitivity of these structures.

5.1 *Glass Coatings*

The development of glass coating materials during the last decade was devoted to fundamental questions, mainly related to hydrolysis condensation and rheology. Models for sol–gel film formation have been developed by Bornside and Macosko [103, 104]. Basic methods to determine rheological parameters are summarized and related to structural models.

Glass coatings were developed for coating plate glasses very early [6]. Since this time only a few large scale applications have been developed, for example for head-up dislays in cars. The only well understood system up to now is the SiO_2 system but this system has only a limited application potential. The SiO_2 films have become interesting for integrated optic systems. Since high grade SiO_2 exhibits extremely low losses, it is a potential material for planar wave guides in the form of passive waveguides or doped with lanthanides for active wave guides. Due to the high densification temperature of SiO_2 gels, direct laser densification including laser patterning was investigated [105]. The same process was used for Ta_2O_5 and SiO_2/TiO_2 films. The problem of this technique is the heat transfer by a laser beam to the surface and the heat dissipation into the non-irradiated gel or sol. It seems to be difficult to obtain smooth surfaces of the patterned channels, which are necessary for low losses for integrated optics. The structure of the sol, the resulting gel and the sol–gel transition by prebaking should be of high relevance for this interesting process. This is true for another interesting technique developed to pattern gel glass films: the embossing step. The prerequirement therefore is, that the gel transition is not completed. The viscosity has to be low enough to allow the embossing of patterns in the range of some μ or less (Fig. 20). This can be obtained by condensation control of the sol

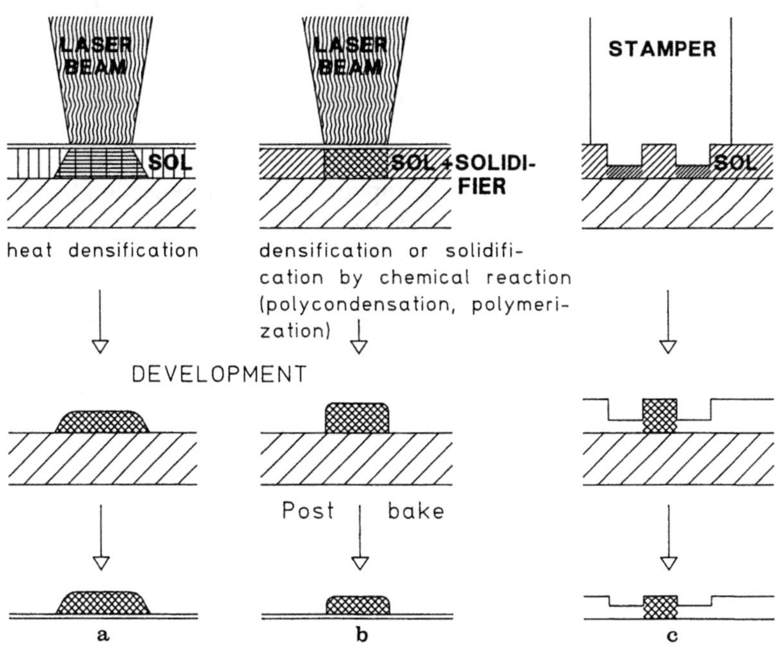

Fig. 20a–c. Patterning processes to produce waveguide channels; *a*) direct laser densification; *b*) chemically processed material e.g. photoinitiator assisted crosslinking; *c*) embossing

or, as shown by Matsuda and coworkers [106, 107] by addition of an organic additive, in this case polyethylene glycol (PEG), acting as a softener. With PEG it was possible to emboss a gel layer on a glass disc for a CD–ROM.

The development of direct densification of sol layers is at its beginning and it is not yet possible to estimate the potential of this method. The chemical-reaction-assisted densification route which is very well developed for organic polymer patterning in microelectronic processes, is an interesting way for patterning ceramics, too. It can be advantageously used in the form of a composite (Fig. 21a) or in the form of an inorganic–organic hybrid material (ceramer, polyceram or ORMOCER) (Fig. 21b) [108–109]. This route which includes, in general, a development step (e.g. dissolving the non-irradiated areas) has an interesting potential for the development of patterned ceramics as well as for composites to be used in optical as well as in electronic application. Therefore, it is necessary to investigate the reaction mechanisms and structure formation in detail. For the exploitation of this potential it is necessary to close this gap in our knowledge. The embossing process seems to be rather simple compared to the described processes. If the sol preparation can be controlled in a way that the viscosity allows the embossing step, the method is very promising for optics. Whereas laser writing is limited by the laser focus, heat dissipation, and reaction control (some µm resolution up to now) by embossing, patterns in the sub-µm range could be obtained, as shown in [110]. Titania–silica was used preferably since it allows one to adjust the refractive index between 1.46 and 2.60. In [111–112] the embossing of thin films of TiO_2–SiO_2 to produce complex gratings with 1200 lines/mm is shown. These gratings were used for sensors [113–114] and thermo-optical switches [115]. With ORMOCERs containing polymerizable organic ligands, the embossing step can be combined with a polymerizing step leading to high quality patterns in a quick process [116].

Another interesting field of applications is antireflection (AR) and reflective coatings. Of course, it is possible to produce antireflection coatings by sol–gel techniques. There are two approaches for AR coatings: Coatings which fulfill the quarter wave length conditions and coatings with or based on the index discontinuity of the interface. Interesting recent developments are AR coatings for solar cells [117–119] for TV screens or computer terminals [120–121] for the near infrared region [122] or for increasing the damage threshold for glass lasers

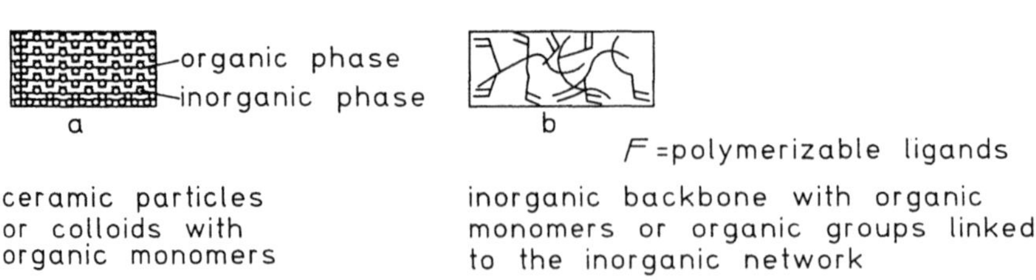

Fig. 21a, b. Various types of composites for patterning

[123]. Two routes have been used for these approaches. The one is to evaluate sol properties and coating conditions to obtain the desired film thickness after densification, the second to obtain desired n variations by adjusting the film density (or porosity). For both approaches it is desirable to tailor the sol properties and to work out the correlation of sol structure and film properties, which is only at its beginning.

The sol–gel method which can be very promising even with respect to costs will be successful in future if these basic questions can be answered. Therefore, more investigations are to be carried out in this respect.

Another interesting question is how far sol–gel coatings are able to improve mechanical strength or chemical stability of glass surfaces. As shown in [39] sol–gel coatings can act in various ways for improving strength either by flaw healing or by building up compressive strain or by combination of both (Fig. 22).

Compressive stress can be built up by use of a coating with a TEC α smaller than that of the substrate glass. In [39] rheological analysis led to the conclusion that particulate sols which follow the Landau–Levich expression exactly can be used to calculate the coating thickness in dip coating experiments. Whereas with flaw healing a bending strength increase up to 70% was observed [124], with sol–gel coatings including the formation or surface compressive stress an increase by a factor of four was achieved [39].

Another interesting development is electrochromic films based on WO_3, MoO_3 and V_2O_5 or combinations of them [125–133]. Although a variety of systems with electrochromic properties have been prepared, the structure to property relationships are poorly understood and the influence of the chemistry on the electrochromic performance is not known.

5.2 Ceramic Coatings

These coatings have become very interesting for optical application in the last decade and a special emphasis can be observed on the synthesis of electrooptic materials such as PZT, PLZT, SBN, PMN, $LiNbO_3$, $Li_2B_4O_7$ TiO_2/B_2O_3 TiO_2/CeO_2 and others [134–136]. The objectives of these investigations are, in general, to obtain thin films with high electrooptic coefficients [137–144]. Surveys on electrooptic sol–gel films are given in [145, 146]. The comparison of

Fig. 22. Sol–gel layers effective in flaw healing and compressive strain formation α_1, α_2 = TCE; $\alpha_1 > \alpha_2$

these films with bulk materials generally show lower linear and quadratic electrooptic coefficients. The mechanisms of the influence of sol structure to crystallization behavior is not well investigated and poorly understood and it is not clear whether it caused the crystallization effects (e.g. pyrochlore formation), interface phases or defects in the films (pores, microcracks). The film thicknesses which can be realized in a one-step process are in the range of 0.1 to 0.5 μm. In [147, 148] it is reported that planar wave guides have been prepared with a 1 μm thickness by using as many as 15 dip coating steps.

Hutter et al. [149] synthesized 50 nm particulate sols from WCl_6, Ti- and V-alkoxides which were used for the coating of cordierite honey comb structures to form a catalytically active coating for NO_X conversion in waste gases of power plants. These coatings were built up within the pores and on the surface of the cordierite ceramic body [150].

5.3 Composite Films with Organics

The sol process as a low temperature synthesis process for inorganic networks has the unique potential to incorporate organic components into the inorganic structure. There are a variety of possibilities of linking organic with inorganic components. A survey of some important mechanisms is shown in Fig. 23a–e.

The chemistry of the synthesis of these structures is very complicated and only a few structures have been proposed [151–154]. The basic advantage of the sol–gel process is its ability to form inorganic structures at low temperatures which allow organics to survive the processing. Organics can be organic groupings chemically linked to the inorganic backbone, organic polymers or entrapped organic molecules such as molecules with special electrical or optical properties (e.g. NLO, lasing dyes).

For the incorporation of dyes two routes have been investigated, the adsorption of dyes in pores or the condensation of dye-containing sols. In the first case the sol–gel process is not influenced by the dye molecule, in the second case, the dye molecule can either interfere with the reacting species, e.g. as catalysts or can “form” its own tailormade environment during the gelation step. This interesting reaction which allows one to control interaction and the properties of the dye molecule has not been investigated in these preparations.

The films prepared by these novel methods have been developed for solid-state dye lasers, solar light collectors and optical memory devices [155–158] where SiO_2 has been used as the matrix. In Al_2O_3 or Al_2O_3/SiO_2 films, rhodamine 6G shows lower degrees of aggregation and a remarkable increase in stability [159, 160]. New χ^3 materials have been prepared by Prasad using an inorganic backbone stabilizing a χ^3 organic polymer [161]. In [160] the use of an ORMOCER type as described in [88] was used as a matrix for incorporating lasing dyes showing a remarkable stabilization of rhodamine 6G against laser shots.

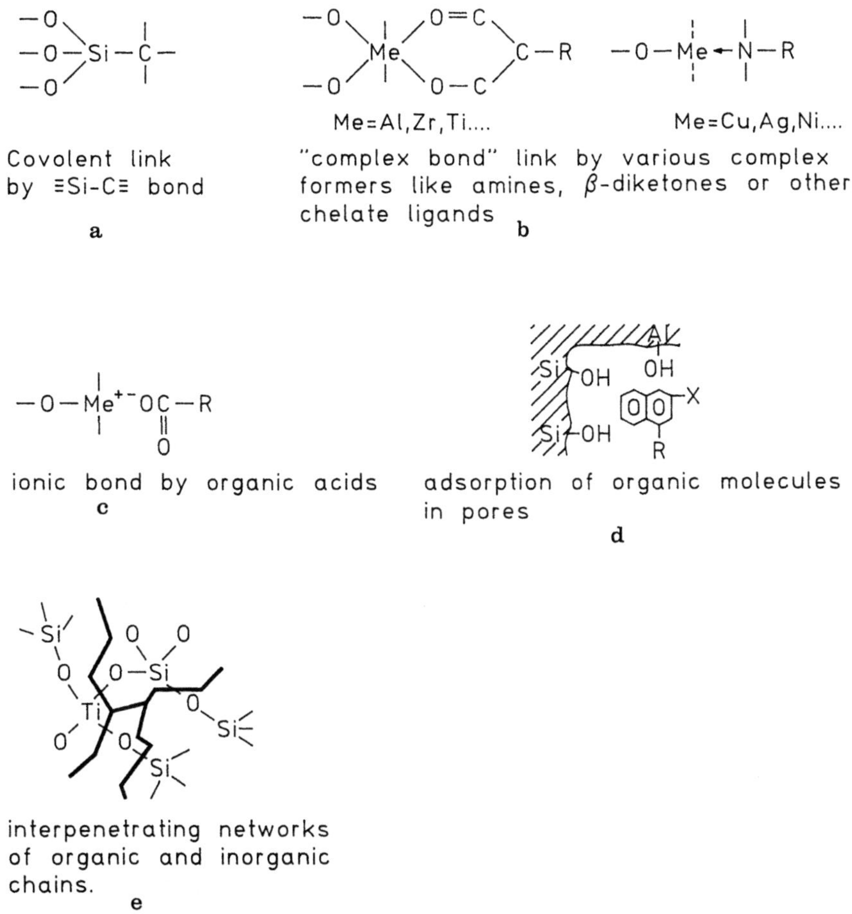

Fig. 23a–e. Suggested connections of inorganic with organic units

Further developments of inorganic–organic composites have been carried out by Teowee and coworkers [162], using butadienes, polyurethanes and polyethyleneimines as matrices and modifying them with SiO_2 for low DK coatings.

Another development of inorganic–organic hybrid films has been carried out by Schmidt [163–166] who was able to show that some of these films exhibit an abrasion resistance close to inorganic coatings [153]. The interesting feature of these coatings is, first, the fact that due to the organic groups they can be densified at remarkably low temperatures (70–150 °C) and thus, can be applied to plastic surfaces, and second, that the Young's modulus can be varied over a wide range without affecting the abrasion resistance too much. The modulus adaption becomes important for plastic surface protection.

The moduli difference has to be low enough to overcome the TCE mismatch between the coating and the substrate and to keep the internal stresses low. The high homogeneity of this type of material was achieved by exact reaction control during the synthesis as demonstrated in Sect. 4.3, and structural models were developed in [152]. These coatings are used for mechanical protection on CR 39 polymers for eye-glass lenses and additionally act as a "hard" interface for sputtered inorganic antireflection coatings, which are otherwise difficult to apply to polymeric surfaces for reasons shown in Fig. 23. By careful adjustment of the chemistry, especially the formation of the organic chain in epoxy polymerization, it is possible to create surprising elasticity within such coatings. These types of hard coatings can not only be used on optical polymers but also on metals where in addition to their scratch protection they can also provide corrosion protection [167]. Another type of inorganic–organic film forming material has been developed for electronic purposes. It is based on the $-(C_6H_5)_2SiO-$ chain forming unit which allows one to control the structure in combination with three-dimensional crosslinking units [e.g. SiO_4^{4-}, TiO_4^{4-}, ZrO_4^{4-}] and organic chain forming units [e.g. $-CH_3Si(CH{=}CH_2)-$ or –Si–methacryl]. Thus, coatings on Al for adhesion purposes [168–169], for corrosion protection on glass surfaces [170], for electronic application as low dielectric constant material with high thermal and chemical resistance [163, 171, 172] have been developed. Variations of this type have been tailored for direct laser writing to generate patterns for microelectronic application [108].

There are almost unlimited material variations to be synthesized just by varying the organic component, the inorganic backbone and the sequence of crosslinking. This opens the interesting potential of multifunctional coatings, where basic function, such as mechanical strength are used as a starting point and additional functions are added by structure tailoring. First results have already been obtained e.g. with respect to low permeation for small molecules [167], special chemical properties to link biological compounds [173], to incorporate functional transition metal ions [174] or to incorporate dyes [97].

6 Conclusions

Sol–gel processing covers a wide range of compositions to produce thin films on a variety of substrates. The basic advantages of sol–gel processing are low costs, obtaining precisely the desired stoichiometries in multicomponent systems, the homogeneous incorporation of low concentrations of dopants, the use of conventional wet coating techniques and the synthesis of inorganic organic composites on a molecular level. In order to make a breakthrough with this technology, a better understanding of the structure to property correlation has to be worked out with special focus on the structures of the coating materials, which, in general, are oligomeric or colloidal solutions.

7 References

1. Ramdohr P, Strunz H (eds) (1967) Klockmann's Lehrbuch der Mineralogie, 15th edn, Enke, Stuttgart
2. Iler RK (1979) The chemistry of silica. John Wiley, New York
3. Ebelmen M (1846) Ann Chem 57: 319
4. Geffcken W, Berger E (1983) DRP-No. 736: 411
5. Roy R (1969) J Amer Cer Soc 52: 344
6. Dislich H (1971) Angew Chem 83: 428
7. Scherer GW (1988) J Non-Cryst Solids 100: 77
8. Hench LL (1984) Mat Res Soc Symp Proc 32: 101
9. Yamane M, Yasumori A, Iwasaki M, Hayashi K (1990) in: SPIE Proc 1328, p 133
10. Sakka S (1988) In: Klein L (ed) Sol–gel technology for thin films, fibers, preforms, electronics and speciality shapes. Noyes, Park Ridge, New Jersey, p 140
11. Dislich H (1988) In: Klein L (ed) Sol–gel technology for thin films, fibers, preforms, electronics and speciality shapes. Noyes, Park Ridge, New Jersey, p 50
12. Brinker CJ, Drotning WD, Scherer GW (1984) Mat Res Soc Symp Proc 32: 25
13. Sakka S, Kamiya K (1982) J Non-Cryst Solids 48: 31
14. Tarasevich BJ, Liu J, Sarikaya M, Aksay IA (1988) Mat Res Soc Symp Proc 121: 225
15. Kamiya K, Yoko T, Suzuki H (1987) J Non-Cryst Solids 93: 407
16. Einstein A (1906) Ann Phys 19: 289
17. Abe Y, Misono T (1943) Polym Chem 21: 41
18. Sakka S (1984) Mat Res Soc Symp Proc 32: 91
19. Sakka S, Kamiya K, Makita K, Yamamoto, Y (1984) J Non-Cryst Solids 63: 223
20. Ohta H, In: Proceedings of the First International Symposium on New Glass, December 1–2, 1987, Tokyo, Japan, The Association of New Glass Industries, Tokyo (ed)
21. Stöber W, Fink A, Bohn E (1968) J Colloid Interface Sci 26: 62
22. Frye GC, Rco AJ, Martin SJ, Brinker CJ (1988) Mat Res Soc Symp Proc 121: 349
23. Brinker CJ, Hurd AJ, Ward KJ (1988) In: Mackenzie JD, Ulrich DR (eds) Ultrastructure Processing of advanced ceramics. John Wiley, New York, p 223
24. Brinker JC, Scherer GW (eds) Sol–gel science (1990) Academic, London
25. Lange F, In: Proc. 4th International Conference on Ceramic Powder Processing Science, Nagoya, March 1991. American Ceramic Society (ed) in print
26. Livage J, Henry M, Sanchez C (1988) Progress in Solid State Chemistry 18: 259
27. LaCourse WC (1988) In: Klein L (ed) Sol–gel technology for thin films, fibers, preforms, electronics and speciality shapes. Noyes Publications, Park Ridge, New Jersey, p 184
28. Kim S (1986) Ph.D. Thesis, NYS College of Ceramics, Alfred University, Alfred
29. Naß R, Schmidt H (1989) In: Hausner H, Messing GL, Hirano S (eds) Ceramic Powder Processing. Deutsche Keramische Gesellschaft, Köln, p 69
30. Sporn D, Naß R, Schmidt H, Brehler KP, Private communication
31. Schmidt H, In: Proc. 4th International Conference on Ceramic Powder Processing Science, Nagoya, March 1991. American Ceramic Society (ed) in print
32. Schmidt H (1988) J Non-Cryst Solids 100: 51
33. Yoldas BE (1980) J Non-Cryst Solids 38: 81
34. Lin CC, Basil JD (1986) Mat Res Soc Symp Proc 73: 173
35. Basil JD, Lin CC (1988) In: Mackenzie JD, Ulrich DR (eds) Ultrastructure Processing of advanced ceramics. John Wiley, New York, p 783
36. Basil JD, Lin CC (1988) Mat Res Soc Symp Proc 121: 49
37. Schmidt H, Seiferling B (1986) Mat Res Soc Symp Proc 73: 739
38. Strawbridge I, James PF (1986) J Non-Cryst Solids 82: 366
39. Strehlow P, Schmidt H, Oral presentation at S.G.T. Symposium on Glass Strength, Harrogate, UK, May 1988
40. Landau LD, Levich BG (1942) Acta Physiochim URSS 17: 42
41. Bradley DC, Mehrotra RC, Gaur DP (1978) Metal Alkoxides, Academic Press, New York
42. Mehrotra RC (1988) J Non-Cryst Solids 100: 1
43. Mehrotra RC (1989) In: Aegerter MA, Jafelicci Jr M, Souza DF, Zanotto ED (eds) Sol–gel science and technology. World Scientific Publishing Co. PTE LTD, Singapore, p 17
44. Mehrotra RC (1990) J Non-Cryst Solids 121: 1
45. Hirano S, Kikuta K, Collection of Abstracts of The First International Conference on

Advanced Ceramics (ICAC-1)—Molecular Design of Ceramics by Sol–Gel Processing, November 1990, Kyoto, Japan, p 25
46. Pouxviel OC, Boilet JP, Beloeil JC, Lallemand JY (1987) J Non-Cryst Solids 89: 345
47. Payne DA (1991) In: Proceedings International Symposium on Molecular Level Designing of Ceramics, Nagoya, March 1991. Team on the NEDO International Joint Research Project (eds) in print
48. Naß R, Schmidt H (1990) J Non-Cryst Solids 121: 329
49. Brinker CJ, Drotning WD, Scherer GW (1984) Mat Res Soc Symp Proc 32: 25
50. Brinker CJ (1988) J Non-Cryst Solids 100: 31
51. Floch HG, Priotton JJ, Thomas IM (1990) In: Mackenzie JD, Ulrich DR (eds) SPIE Vol. 1328 Sol–Gel Optics, p 307
52. Schmidt H (1989) In: Proceedings Seminar Nano-Hybridization of Ceramic and Creation of New Functions, February 1989, Oiso, Japan. The Association for the Progress of New Chemistry, Tokyo, Japan (ed) p 1
53. Yamane M, Inove S, Nakazawa K (1982) J Non-Cryst Solids 48: 153
54. Spanhel L, Anderson MA, In: IS & T conference proceedings, St. Paul Symposium on electronic and ionic properties of silver halides, 1991, in print
55. Wu M, Fujiu T, Messing GL (1990) J Non-Cryst Solids 121: 407
56. Chen KC, Mackenzie JD (1990) Mat Res Soc Symp Proc 180: 663
57. Budd KD, Payne DA (1990) In: Mackenzie JD, Ulrich DR (eds) SPIE Vol. 1328 Sol–Gel Optics, p 450
58. Ostertag R, Rinn G, Tünker G, Schmidt H (1989) In: Moulson AJ, Binner J, Morrell R (eds) British Ceramic Proceedings No. 41, Electroceramics. The Institute of Ceramics, Stoke-on-Trent, UK p 11
59. Tomandl G, Stiegelschmitt A, Boehner R (1988) Sci Ceram 14: 305
60. Vilminot S, Rehspringer JL (1989) In: Proceedings of Eurogel '89, December 1989, Colmar, France, p 125
61. Brinker CJ, Clark DE, Ulrich DR (eds) Mat Res Soc Symp Proc 32. Materials Research Society, Pittsburgh
62. Brinker CJ, Clark DE, Ulrich DR (eds) Mat Res Soc Symp Proc 73. Materials Research Society, Pittsburgh
63. Brinker CJ, Clark DE, Ulrich DR (eds) Mat Res Soc Symp Proc 121. Materials Research Society, Pittsburgh
64. Zelinski BJJ, Brinker CJ, Clark DE, Ulrich DR (eds) Mat Res Soc Symp Proc 180. Materials Research Society, Pittsburgh
65. Hench HL, Ulrich DR (eds) Proceedings of the International Congress on Ultrastructure Processing of Ceramics, Glasses and Composites. John Wiley, New York, 1984
66. Hench HL, Ulrich DR (eds) Proceedings of the International Congress on Science of Ceramic Chemical Processing, John Wiley, New York, 1986
67. Mackenzie JD, Ulrich DR (eds) Ultrastructure Processing of Advanced Ceramics. John Wiley, New York, 1988
68. Hench LL et al (eds) Chemical Processing of Advanced Materials, Proceedings 1991 Ultrastructure Conference, Orlando. John Wiley, in print
69. Gottardi V (ed) J Non-Cryst Solids 48, 1982. North-Holland Physics Publishers, Amsterdam
70. Scholze H (ed) J Non-Cryst Solids 63, 1984. North-Holland Physics Publishers, Amsterdam
71. Zarzycki J (ed) J Non-Cryst Solids 82, 1986. North-Holland Physics Publishers, Amsterdam
72. Sakka S (ed) J Non-Cryst Solids 100, 1988. North-Holland Physics Publishers, Amsterdam
73. Aegerter M (ed) J Non-Cryst Solids 121, 1990. North-Holland Physics Publishers, Amsterdam
74. The Association for the Progress of New Chemistry (ed) Proceedings of Seminar Nano-Hybridization of Ceramics and Creation of New Functions, February 7–10, 1989, Oiso, Japan
75. Mackenzie JD, Ulrich DR (eds) SPIE Proceedings Vol. 1328 Sol–Gel Optics, Bellingham, USA
76. Proceedings of Eurogel '89—Advanced Materials and Processes by Sol–Gel Techniques, European Conference on Sol–Gel-Technology, Colmar, France, December 4–5, 1989
77. Collection of Abstracts of the First International Conference on Advanced Ceramics (ICAC-1)—Molecular Design of Ceramics by Sol–Gel Processing, November 20–22, 1990, Kyoto, Japan
78. Klein LC (ed) Sol–gel Technology for Thin Films, Fibers, Preforms, Electronics and Speciality Shapes. Noyes Publications, Prk Ridge/New Jersey, 1988
79. Aegerter MA, Jafelicci Jr M, Souza DF, Zanotto Ed (eds) Sol–gel science and technology. World Scientific Publishing Co. PTE LTD, Singapore, 1989

80. Modak AR, Dey SK (1990) Cer Transact 15: 447
81. Dey SK, Budd KD, Payne DA (1986) In: Electronic Packaging and Materials Science. Materials Research Society, Palo Alto, California
82. Dey SK, Budd KD, Payne DA (1988) IEEE Trans. on Ultrasonics, Ferroelectrics and Frequency Control 35: 80
83. Bagnall CM, Zarzycki J (1990) In: Mackenzie JD, Ulrich DR (eds) SPIE Vol. 1328 Sol–Gel Optics, p 108
84. Nogami M, Nagasaka K, Kadono K, Kishimoto T (1988) J Non-Cryst Solids 100: 298
85. Uhlmann DR, Weinberg MC, Teowee G (1988) J Non-Cryst Solids 100: 154
86. Payne DA, In: Proceedings 4th International Conference on Ceramic Powder Processing Science, Nogoya, March 1991. American Ceramic Society, in print
87. Sanchez C, Livage J, Henry M, Babonneau F (1988) J Non-Cryst Solids 100: 65
88. Schmidt H, Philipp P (1985) In: Wright AF, Dupuy J (eds) Glass . . . Current Issues. Martinus Nijhoff, Dordrecht-Boston-Lancaster, p 580
89. Philipp G, Schmidt H (1984) J Non-Cryst Solids 63: 283
90. Schmidt H, Seiferling B, Philipp G, Deichmann K (1988) In: Mackenzie JD, Ulrich DR (eds) Ultrastructure Processing of Advanced Ceramics. John Wiley & Sons, New York, p 651
91. Schmidt H, Unpublished results
92. Strehlow P, Schmidt H, Birkhahn M (1988) Mat Res Soc Symp Proc 121: 743
93. Naß R, Schmidt H, Arpac E (1990) In: Mackenzie JD, Ulrich DR (eds) Spie Vol. 1328 Sol–Gel Optics, p 258
94. Schmidt H (1990) Mat Res Soc Symp Proc 180: 961
95. Arpac E, Schubert U, Glaubitt W, Naß R, Schmidt H, Unpublished results
96. Schubert U, Rose K, Schmidt H (1988) J Non-Cryst Solids 105: 165
97. Schmidt H, In: Mackenzie JD et al (eds) Chemical Processing of Advanced Materials, Proceedings 1991 Ultrastructure Conference, Orlando. John Wiley & Sons, in print
98. Charbouillot Y (1987) Ph.D. thesis. Institut National Polytechnique, Grenoble
99. Ravaine D, Seminel A, Charbouillot Y, Vincens M (1986) J Non-Cryst Solids 82: 210
100. Hirano S, Hayashi T, Kato C (1989) In: Hausner H, Messing GL, Hirano S (eds) Ceramic Powder Processing Science. Deutsche Keramische Gesellschaft, Köln, p 61
101. Lessard RB, Berglund KA, Nocera DG (1989) Mat Res Soc Symp Proc 155: 119
102. Dulebohn JI, Van Vlierberge S, Berglund KA, Lessard RB, Yu J, Nocera DG (1990) Mat Res Soc Symp Proc 180: 733
103. Bornside DE, Macosko CW, Scriven LE (1987) J Imaging Tech 13: 122
104. Macosko CW, McCartney ML, Scriven LE (1987) Mat Res Soc Symp Proc 190: 555
105. Fabes BD, Taylor DJ, Weisenbach LW, Stuppi MM, Klein DL, Raymond LJ, Zelinski BJ, Birnie DP (1990) In: Mackenzie JD, Ulrich DR (eds) SPIE Vol. 1328 Sol–Gel Optics, p 319
106. Matsuda A, Matsuno Y, Kataoka S, Katayama S, Tsuno T, Tohge N, Minami T (1990) In: Mackenzie JD, Ulrich DR (eds) SPIE Vol. 1328 Sol–Gel Optics, p 62
107. Matsuda A, Matsuno Y, Kataoka S, Katayama S, Tsuno T, Tohge N, Minami T (1990) In: Mackenzie JD, Ulrich DR (eds) SPIE Vol. 1328 Sol–Gel Optics, p 71
108. Popall M, Meyer H, Schmidt H, Schulz J (1990) Mat Res Soc Symp Proc 180: 995
109. Schmidt H, Popall M (1990) In: Mackenzie JD, Ulrich DR (eds) SPIE Vol. 1328 Sol–Gel Optics, p 249
110. Roncone RL, Weller-Brophy LA, Weisenbach L, Zelinski BJJ, submitted to the J Non-Cryst Solids for publication in 1990
111. Tiefenthaler K, Briguet V, Buser E, Horisberger M, Lukosz W (1983) Proc. SPIE 401: 165
112. Tiefenthaler K, Lukosz W (1983) IEE Conf Pub 227: 108
113. Lukosz W, Tiefenthaler K (1988) Sensors and Actuators 15: 273
114. Nellen PM, Tiefenthaler K, Lukosz W (1988) Sensors and Actuators 15: 285
115. Lukosz W, Briguet V (1985) Thin Solid Films 119: 197
116. Schmidt H (1991) In: Proceedings International Symposium on Molecular Level Designing of Ceramics, Nagoya, March 1991. Team on the NEDO International Joint Research Project (eds) in print
117. Pettit RB, Ashley CS, Reed ST, Brinker CJ (1988) In: Klein LC (ed) Sol–gel Technology for Thin Films, Fibers, Preforms, Electronics and Speciality Shapes. Noyes, Park Ridge, NJ, p 80
118. Yoldas BE, U.S. Patent 4 346 131, Aug 24, 1982
119. Pettit RB, Brinker CJ, Ashley CS (1985) Solar Cells 15: 267
120. Dislich H (1985) J Non-Cryst Solids 73: 599
121. Hinz P, Dislich H (1986) J Non-Cryst Solids 82: 411

122. Biswas PK, Kundu D, Ganguli D (1989) J Mat Sci Lett 8: 1436
123. Yoldas BE, Partlow DP, European Patent Application 0 130 801; June 29, 1983, assigned to Westinghouse Electric Corporation, USA
124. James PF (1988) J Non-Cryst Solids 100: 93
125. Covino J, McManis (1988) Mat Res Soc Symp Proc 121: 553
126. Chemseddine A, Babonneau F, Livage J (1987) J Non-Cryst Solids 91: 271
127. Chemseddine A, Morineau R, Livage J (1983) Solid-State Ionics 9–10: 357
128. Chemseddine A, Henry M, Livage J (1984) Rev Chim Miner 21: 487
129. Kato I, Ariizumi A, Kimura N, Jpn Kokai Tokkyo Koho JP 61/36292, Feb 20, 1986
130. Yoshino T, Baba N, Yasuda K (1988) Nippon Kagaku Kaishi 9: 1525
131. Navabi M, Doeuff S, Sanchez C, Livage J (1989) Mat Sci Eng B3: 203
132. Morineau R (1985) Vide Couches Minces 40: 281
133. Lynam NR, Moser FH, Hichwa BP (1987) Proc SPIE 823: 130
134. Aegerter MA, La Serra ER, Rodrigues ACM, Kordas G, Moore G (1990) In: Mackenzie JD, Ulrich DR (eds) SPIE Vol. 1328 Sol–Gel Optics, p 391
135. Yoko T, Yamashita H, Sakka S (1990) In: Mackenzie JD, Ulrich DR (eds) SPIE Vol. 1328 Sol–Gel Optics, p 416
136. Mackenzie JD, Ulrich DR (1990) In: Mackenzie JD, Ulrich DR (eds) SPIE Vol. 1328 Sol–Gel Optics, p 2
137. Budd KD, Dey SK, Payne SK (1986) Brit Cer Proc 36: 107
138. Vest RW, Xu J (1984) Ferroelectrics 93: 21
139. Chen CJ, Xu Y, Xu R, Mackenzie JD, paper presented at 1st Internat Cer Sci and Tech Cong, Anaheim, 1989
140. Hirano S, Kato K (1988) J Non-Cryst Solids 100: 538
141. Hirano S, Kato K (1988) Adv Cer Materials 3: 503
142. Hirano S, Kato K (1989) Solid State Ionics 32/33: 765
143. Hirano S, Kato K (1989) Bull Chem Soc Jpn 62: 429
144. Hirano S, Kato K (1989) Mat Res Soc Symp Proc 155: 181
145. Uhlmann DR, Boulton JM, Tewowee G, Weisenbach L, Zelinski BJJ (1990) In: Mackenzie JD, Ulrich DR (eds) SPIE Vol. 1328 Sol–Gel Optics, p 270
146. Uhlmann DR (1990) Mat Res Soc Symp Proc 180: 645
147. La Serra ER, Charbouillot Y, Baudry P, Aegerter MA, accepted by the J Non-Cryst Solids for publication, 1990
148. Aegerter MA, Charbouillot Y, Mohallem N, da Silva AA, de Godoy LH, presented at the February 1989 Ultrastructure Conference in Tucson, AZ, to be published in 1992
149. Schmidt H, Kaiser A, Kasemann R, In: Proceedings ENVICERAM '91, March 1991, Saarbrücken. Deutsche Keramische Gesellschaft, Köln, in print
150. Hutter F, Guldner G, Schmidt H, Koschlig HJ, Danowski F, In: Proceedings Fourth International Conference on Ultrastructure Processing of Ceramics, Glasses and Composites, Tucson, Arizona, February 1989. John Wiley & Sons, in print
151. Wilkes GL, Wang B, Brennan A, Rodrigues D, Huang H (1990) Mat Res Soc Symp Proc 171: 15
152. Gauthier-Luneau I (1988) Ph.D. Thesis, L'Université Paul Sabatier de Toulouse (Sciences)
153. Schmidt H (1984) Mat Res Soc Symp Proc 32: 327
154. Avnir D, Levy D, Reisfeld R (1984) J Phys Chem 88: 5956
155. Dunn B, Knobbe E, McKiernan JM, Pouxviel JC, Zink JI (1988) Mat Res Soc Symp Proc 121: 331
156. Reisfeld R (1989) In: Aegerter MA, Jafelicci Jr M, Souza DF, Zanotto ED (eds) Sol–gel science and technology. World Scientific Publishing Co. PTE LTD, Singapore, p 322
157. Levy D, Einhorn S, Avnir D (1989) J Non-Cryst Solids 113: 137
158. Makishima A, Tani T (1986) J Am Cer Soc 69: C72
159. Kobayashi Y, Imai Y, Kurokawa Y (1988) J Mat Sci Lett 7: 1148
160. Dunn B, In: Hench LL et al (eds) Chemical Processing of Advanced Materials, Proceedings 1991 Ultrastructure Conference, Orlando. John Wiley, in print
161. Prasad PN, In: Proceedings Fourth International Conference on Ultrastructure Processing of Ceramics, Glasses and Composites, Tucson, Arizona, February 1989. John Wiley, in print
162. Teowee G (1990) Mat Res Soc Symp Proc 180: 405
163. Schmidt H (1990) Mat Res Soc Symp Proc 171: 3
164. Schmidt H, In: Proceedings Fourth International Conference on Ultrastructure Processing of Ceramics, Glasses and Composites, Tucson, Arizona, February 1989. John Wiley, in print

165. Schmidt H (1990) Erzmetall 43: 75
166. Schmidt H (1988) ACS Symposium Series No. 360: 333
167. 1989 Annual Report of Fraunhofer-Institut für Silicat-forschung, Würzburg
168. Schmidt H, Scholze H, Tünker G (1986) J Non-Cryst Solids 80: 557
169. Schmidt H, Philipp G, Patzelt H, Scholze H (1986) Collected Papers XIV Intl. Congress on Glass, Vol. II, p 429
170. Schmidt H, Fuchs D (1991) Mat Res Soc Symp Proc 185: 239
171. Schmidt H (1990) J Non-Cryst Solids 121: 428
172. Wolter H, Schmidt H (1990) DVS-Berichte 129: 80
173. Schmidt H, von Stetten O, Kellermann G, Patzelt H, Naegele W (1982) IAEA-SM-259/67, Wien, p 111
174. Hutter F, Schmidt H, Scholze H (1986) J Non-Cryst Solids 83: 373

Aqueous Chemistry of Metal Cations: Hydrolysis, Condensation and Complexation

Marc Henry, Jean Pierre Jolivet, and Jacques Livage

Chimie de la Matière Condensée, Université Pierre et Marie Curie,
4, place Jussieu–75252, Paris, France

The sol–gel synthesis of metal oxides can be performed via the hydrolysis and condensation of metal cations in aqueous solutions. This inorganic route provides a cheap and reliable way for industrial uses. However, many hydrolyzed precursors can be present simultaneously in the precursor solution and very few data are available on the formation of condensed species. The so-called Partial Charge Model provides a useful guide to describe and predict hydrolysis and condensation reactions in aqueous solutions. A charge–pH diagram can be established. It shows how condensation can be initiated via acid–base or redox reactions. The two basic condensation processes, olation and oxolation, are then discussed in terms of a charge–electronegativity diagram. They can lead either to small solute condensed species (polyanions or polycations) or to the formation of an infinite network (colloids, gels, or precipitates). Complexation by anionic species is finally discussed in the frame of an electronegativity–pH diagram. It also plays an important role during the formation of solid phases.

1 Introduction

The use of sol–gel techniques to prepare glasses and ceramics has received an increasing interest during the last decade. Academic research is mainly directed towards metal–organic precursors and focusses on the hydrolysis and condensation of silicon alkoxides [1]. Sol–gel processing of thin film coatings from metal alkoxides has already been used with success for over thirty years by one glass company [2]. However, most industrial processes are based on inorganic precursors, i.e. metal salts in aqueous solutions. Such precursors are much cheaper and easier to handle than metal alkoxides in organic solvents.

Hydrolysis and condensation reactions occur in solution and the term "sol–gel processing" is often broadly used to describe the synthesis of metal oxides from solutions. It includes techniques such as precipitation, coprecipitation or forced hydrolysis. These techniques are widely used in industry for the preparation of ceramic powders from aqueous solutions of metal salts. The Bayer process is based on the precipitation of gibbsite $Al(OH)_3$ from aluminate solutions. TiO_2 pigments are made via the controlled hydrolysis of titanium chloride solutions. Stabilized zirconia can be coprecipitated from aqueous solutions of zirconium oxychloride and yttrium nitrate.

From a chemical point of view sol–gel chemistry is based on the hydrolysis and condensation of metal complexes [3]. Such reactions are rather well understood in the case of metal alkoxides. They can be described as nucleophilic substitutions as follows:

$$-M-OR + XOH \longrightarrow -M-OX + ROH \tag{1}$$

This reaction stands for hydrolysis (X = H), condensation (X = M) or even complexation (X = R′).

Aqueous chemistry of metal salts appears to be much more complicated owing to the occurrence of spontaneous hydrolysis and condensation reactions in the aqueous medium. These reactions depend on many parameters such as pH, concentration, or temperature.

Several monomeric or condensed solute species can exist simultaneously so that it is not obvious to know the real chemical nature of the precursor. Many data can be found in the literature for the hydrolysis of metal cations but very little is known about condensation reactions [4]. Moreover, the morphology, structure, and even chemical nature of the resulting solid phases strongly depend on the presence of anions even when these foreign species do not seem explicitly involved in the chemical process [5].

This paper presents a description of the main chemical processes (hydrolysis, condensation, and complexation) involved in the aqueous chemistry of metal cations. Their analysis is based on the so-called "Partial Charge Model" that provides a useful guideline for a rational explanation of these reactions.

2 The Partial Charge Model

The Partial Charge Model (PCM) is based on the electronegativity equalization principle stated by R. T. Sanderson [6] as follows: "when two or more atoms initially different in electronegativity combine, they adjust to the same intermediate electronegativity in the compound". According to L. Pauling electronegativity means the power of an atom in a molecule to attract electrons to itself [7]. Later on, Parr and co-workers showed that the electronegativity χ is related to the electronic chemical potential μ as follows $\chi = -\mu$ [8]. This provides a theoretical basis to Sanderson's assumption (cf Appendix A). The main consequence is that both the electronegativity χ_x of a given atom X and its partial charge δ_x vary when the atom is chemically combined. These two parameters must be related. A linear relationship is usually assumed as follows:

$$\chi_x = \chi_x^0 + \eta_x \delta_x \tag{2}$$

where η_x is the hardness of atom X as introduced by Pearson [9]. Hardness is related to the softness $\sigma_x = 1/\eta_x$ which provides a measure of the polarizability of the electronic cloud around X [10]. Softness increases with the size of the electronic cloud, i.e. with the radius r of X. Therefore, hardness varies as 1/r. According to the Allred-Rochow scale, electronegativity is proportional to Z_{eff}/r^2 [11]. Hardness may then be approximated as:

$$\eta = k\sqrt{\chi^0} \tag{3}$$

where k is a constant that depends on the electronegativity scale. It can be shown that k = 1.36 when Pauling electronegativities are expressed in the frame of Allred-Rochow's model (see Appendix A).

The total charge "z" of a given chemical species is equal to the sum of the partial charges of all individual atoms $z = \Sigma\delta_i$. This together with Eqs. (2) and (3) leads to the following expressions for:

the mean electronegativity: $$\chi = \frac{\Sigma_i\sqrt{\chi_i^0} + 1.36\,z}{\Sigma_i 1/\sqrt{\chi_i^0}} \tag{4}$$

and the partial charge: $$\delta_i = (\chi - \chi_i^0)/1.36\sqrt{\chi_i^0} \tag{5}$$

Equation (5) can also be written as: $$\delta_i = \sigma_i(\chi - \chi_i^0) \tag{6}$$

where $$\sigma_i = (1.36\sqrt{\chi_i^0})^{-1} \tag{7}$$

The Partial Charge Model provides an easy way to work out the mean electronegativity of chemical species and the charge distribution on each atom. In the case of the water molecule, for instance, Eqs. (4) and (5) lead to $\chi(H_2O) = 2.491$, $\delta(H) = +0.20$, and $\delta(O) = -0.40$. Less trivial applications of the partial charge model lead to characterize a chemical reaction rather than a mere partial charge distribution.

3 Hydrolysis of Cations

In an aqueous medium metal cations M^{z+} are solvated by dipolar water molecules giving rise to aquo-ions $[M(OH_2)_N]^{z+}$. Charge transfer occurs via the $M–OH_2$ σ bond. Electron density is transferred from the bonding $3a_1$ molecular orbital of coordinated water molecules towards empty orbitals of the metal cation. This results in a weakening of the O–H bond within the water molecule that becomes more acidic. Deprotonation then takes place as follows:

$$[M(OH_2)_N]^{z+} + hH_2O \longrightarrow [M(OH)_h(OH_2)_{N-h}]^{(z-h)+} + hH_3O^+ \quad (8)$$

The mean electronegativity χ_p of hydrolyzed precursors $[M(OH)_h(OH_2)_{N-h}]^{(z-h)+}$ depends on the metal cation M^{z+} and the hydrolysis ratio h. On the reverse, for a given metal cation M^{z+}, a relationship can be established between χ_p and h. It is thus possible to work out h for any value of χ_p.

3.1 *Electronegativity–pH Relationship*

In dilute solutions Eq. (8) leads to a whole set of more or less deprotonated monomeric species ranging from aquo-ions $[M(OH_2)_N]^{z+}$ to hydroxy species $[M(OH)_z]^0$ or even oxo-anions $[MO_m]^{(2m-z)-}$ when all protons have been removed from the coordination sphere of the metal cation. These equilibria depend on many parameters such as the pH of the solution, temperature or concentration. The question then arises whether it is possible to predict the value of the hydrolysis ratio h and which species can act as precursors for condensation at a given pH.

We can assume that proton exchange between the precursor and the solution proceeds until the mean electronegativity of the hydrolyzed precursor χ_p becomes equal to the mean electronegativity χ_w of the aqueous solution. The point now will be to find a relationship between χ_w and the pH.

As stated above, equilibrium conditions may be found by looking for which composition complex species can equalize their mean electronegativities to χ_w. However, as electronegativities are microscopic potentials, it follows that these conditions are completely insensitive to macroscopic parameters such as concentration, pH, or temperature. The only way to overcome this drawback is to introduce a transfer coefficient k between both sets of parameters. k will be determined empirically as illustrated below.

$[H_2O–H–OH_2]^+$ species have been identified in strongly acidic aqueous solutions (pH 0) by infrared spectroscopy. This $[H_5O_2]^+$ complex can then be taken as a microscopic reference state for the solvated proton. In water proton delocalization is expected to occur through the random network of hydrogen bonds. Owing to their quantum nature, protons cannot be distinguished from other hydrogen atoms (tunnel effect). This means that $[H_5O_2]^+$ complexes are able to share their proton with the surrounding water medium. In the same manner that electronegativities are equalized when atoms share electrons, equalization of mean electronegativities of solvated protons and bulk water is also expected to occur via fast proton exchange. Consequently, the mean electronegativity of an aqueous solution χ_w changes with pH. Such a variation is governed by the mean electronegativity of the solvated proton $\chi(H^+_{aq})$ which changes with the number of water molecules associated to the proton. At a macroscopic scale, the chemical potential of protons is given by:

$$\mu(H^+) = \mu°(H^+) + RT\,Ln(c_H) = \mu°(H^+) - \beta pH \tag{9}$$

where $\beta = 2.3RT$ and c_H is the proton concentration. At a microscopic scale and referring to the fact that electronegativities are also chemical potentials ($\chi = -\mu$), one may write:

$$\mu(H^+) = k\,\chi(H^+) \quad \text{and} \quad \mu°(H^+) = k\,\chi(H_5O_2^+) \tag{10}$$

i.e.

$$\chi(H^+) = \chi_w = \chi(H_5O_2^+) - 2.3kRT\,pH = 2.732 - 2.3kRT\,pH \tag{11}$$

The transfer coefficient k which links a macroscopic concentration scale to a microscopic electronegativity scale may be obtained by writing at pH 7 (T = 25 °C):

$$\chi(H^+) = \chi_w = \chi(H_2O) = 2.491 \tag{12}$$

This leads to the following expression:

$$\chi_w = 2.732 - 0.035\,pH \tag{13}$$

Then, the mean electronegativity of an aqueous solution decreases linearly with the pH, from 2.732 at pH 0 down to 2.242 at pH 14.

3.2 *The Charge–pH Diagram*

Following the Partial Charge Model, it can be assumed that proton exchange between the precursor $[M(OH)_h(OH_2)_{N-h}]^{(z-h)+}$ of mean electronegativity χ_h and the aqueous solution of mean electronegativity χ_w goes on until both values become equal:

$$\chi_h = \chi_w = 2.732 - 0.035\,pH \tag{14}$$

The hydrolysis ratio h of aquo species $[M(OH_2)_N]^{z+}$ can then be easily calculated at any pH value from Eqs. (4) and (14). The total charge $+(z-h)$ of the precursor must be equal to the sum of all partial charges. This leads to (cf Appendix B):

$$h = [z - n\delta_O - 2N\delta_H - \delta_M]/[1 - \delta_H] \tag{15}$$

where partial charges δ_i are given by:

$$\delta_i = (\chi_w - \chi_i^0)/1.36(\chi_i^0)^{1/2} \qquad (16)$$

These relations apply to any metal cation in an aqueous solution.

For Cr(VI) ions with a coordination number N = 4 they lead to $h \approx 6$ at pH 0 and $h \approx 8$ at pH 14. Three different solute species have then to be taken into account namely $CrO_2(OH)_2$ (h = 6), $CrO_3(OH)^-$ (h = 7) and CrO_4^{2-} (h = 8). Their relative abundance depends on the pH as shown below.

For Cr(III) in an octahedral coordination they lead to $0.4 < h < 5.1$ in the range pH 0–14. Six mononuclear species from $[Cr(OH_2)_6]^{3+}$ (h = 0) to $[Cr(OH)_5(OH_2)]^{2-}$ (h = 5) can then be found in aqueous solutions.

The hydrolysis ratio of Si(IV) species ranges from $[Si(OH)_3(OH_2)]^+$ (h = 3), to $[SiO_2(OH)_2]^{2-}$ (h = 6). This shows the complexity of inorganic salts compared to alkoxides for the sol–gel synthesis of metal oxides. A single molecular precursor $Si(OR)_4$ has to be taken into account in the parent alcohol while four different aqueous species can be found when silicates are dissolved into water.

According to Eqs. (15) and (16) the hydrolysis ratio h of a given precursor mainly depends on both the pH of the aqueous solution and the oxidation state z+ of the cation. A charge–pH diagram can then be drawn in which three domains are plotted (Fig. 1).

- The lower domain corresponds to aquo-cations $[M(OH_2)_N]^{z+}$
- The upper domain corresponds to oxo-anions $[MO_m]^{(2m-z)-}$
- An intermediate domain where hydroxo species can be found.

These domains are separated by two lines corresponding to h = 1 and h = 2m − 1, respectively. This analysis based on the Partial Charge Model

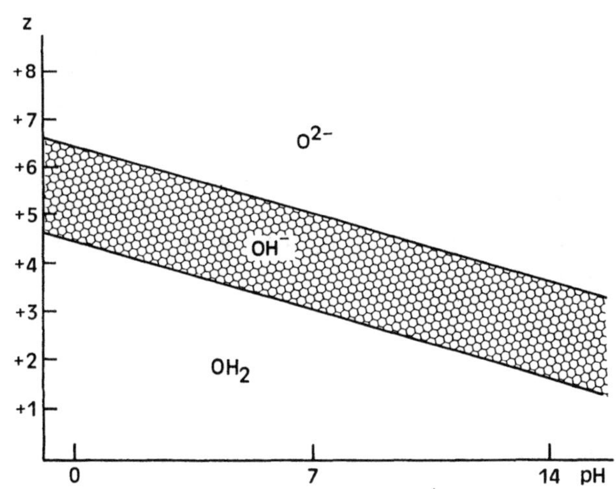

Fig. 1. The charge–pH diagram according to C. K. Jørgensen [12]. Condensation reactions are expected to occur when the system is brought from the aquo (low-valent cations) or oxo (high-valent cations) domains into the hydroxo one (tint middle domain)

provides a theoretical background for the charge–pH diagram established by C. K. Jørgensen [12].

The charge–pH diagram is a useful guide for sol–gel chemists. Condensation between mononuclear species in aqueous solutions becomes only possible when OH groups are present in the precursor i.e. inside the tint domain (Fig. 1). Therefore condensation can be initiated via pH variations, by adding a base to a low-valent metal ion (Cu^{2+}, Al^{3+}, Zr^{4+}) or an acid to a high-valent metal ion (W^{6+}, V^{5+}). It can also be initiated via redox reactions. MnO_2 gels, for instance, have been obtained upon reduction of permanganate ions $[MnO_4]^-$ [13].

4 Condensation

4.1 *Mechanisms*

Condensation of hydrolyzed metal ions follows two main mechanisms [3].

i) *Olation*: The nucleophilic attack of a negatively charged OH group onto a positively charged metal cation leads to the departure of an aquo ligand from the coordination sphere of the metal. This leads to the formation of "ol" bridges as follows:

$$\text{M–OH} + \text{M–OH}_2 \longrightarrow \text{M–OH–M} + \text{H}_2\text{O} \tag{17}$$

Such a process is only possible with aquo-precursors having their maximum coordination number.

ii) *Oxolation*: Two hydroxo groups react together forming a water molecule and an "oxo" bridge as follows:

$$\text{M–OH} + \text{HO–M} \longrightarrow \text{M–O–M} + \text{H}_2\text{O} \tag{18}$$

This reaction starts with the nucleophilic addition of OH groups onto metal ions. It is followed by a 1,3-proton transfer within the transition state (M–OH . . . M–OH) so that one water molecule can be removed easily.

In both cases, reacting species are OH groups so that condensation does not occur as long as stable hydroxylated precursors are not formed in the aqueous solution.

4.2 *The Charge–Electronegativity Diagram*

The stability of a M–OH bond in an aqueous medium can be analyzed in the frame of the Partial Charge Model (cf Appendix B). Two cases can be encountered:

i) $\chi^0_M > \chi^0_H$: M is more electronegative than hydrogen. Electron density is then withdrawn from the hydrogen atom that become more positively charged.

Hydroxyl groups exhibit an acidic behavior leading to their dissociation in water: $M\text{–}OH_{aq.} \rightarrow M\text{–}O^-_{aq.} + H^+_{aq.}$. This occurs mainly with electronegative elements at the upper right of the periodic table such as P(V), S(VI), or Cl(VII) and with highly charged "d" metal cations such as Mn(VII).

ii) $\chi^0_M < \chi^0_H$: Charge transfer goes towards the hydrogen atom that looses its acidic character. The M–OH bond remains stable as long as electron density in the bonding orbital of the M–OH bond remains large enough to avoid basic dissociation, $M\text{–}OH + H_2O \rightarrow M\text{–}OH_2^+ + OH^-$. This could occur with low valent large cations of groups 1 and 2 such as Na^+ or Ba^{++}. NaOH or $Ba(OH)_2$ are strong bases, completely dissociated in aqueous solutions.

An electronegativity range for metal cations can therefore be defined so that the M–OH bond remains stable in an aqueous medium. The critical electronegativities χ^*_+ and χ^*_- corresponding to acid and basic dissociation respectively can be calculated in the frame of the Partial Charge Model (cf Appendix B) [14]. When the M–OH bond is stable both against acid or basic ionization, condensation may occur if $\delta(OH) < 0$ through olation and/or oxolation. Two other critical electronegativities can then be defined χ^*_{OH} and χ^*_O ruling the nucleophilicity of the OH group and the occurrence of olation with respect to oxolation, respectively (cf Appendix B). A "charge–electronegativity" diagram (Fig. 2) can then be built from such considerations. Five groups of cations are then distinguished.

i) Those which undergo acid ionization ($\chi^0_M > \chi^*_+$). Such elements give rise to inorganic acids which are not able to condense in aqueous solutions.

ii) Those which undergo basic ionization ($\chi^0_M < \chi^*_-$). Such elements give rise to inorganic bases which again do not condense in aqueous solutions.

iii) Those which condense only through olation leading to hydroxide $M(OH)_z$ precipitation ($\chi^*_- < \chi^0_M < \chi^*_O$).

iv) Those which condense both through olation and oxolation leading to hydrous oxide $MO_{z/2} \cdot xH_2O$ precipitation ($\chi^*_O < \chi^0_M < \chi^*_{OH}$).

v) Finally, those which condense only through oxolation leading to molecular poly-acids formation as $\chi^*_{OH} < \chi^0_M < \chi^*_+$.

In the following sections, such a general behavior is illustrated for the last three classes which are of the utmost importance for sol–gel chemists. Examples mainly deal with transition metals which provide a wide range of behaviors related to their electronic configurations.

5 Oxolation and Polyanions

High-valent cations ($z > 4$) give rise to oxo-hydroxo anions $[MO_x(OH)_{m-x}]^{(m+x-z)-}$ in aqueous solutions. They are surrounded by oxo or hydroxo ligands only. Water molecules are no more coordinated to the metal

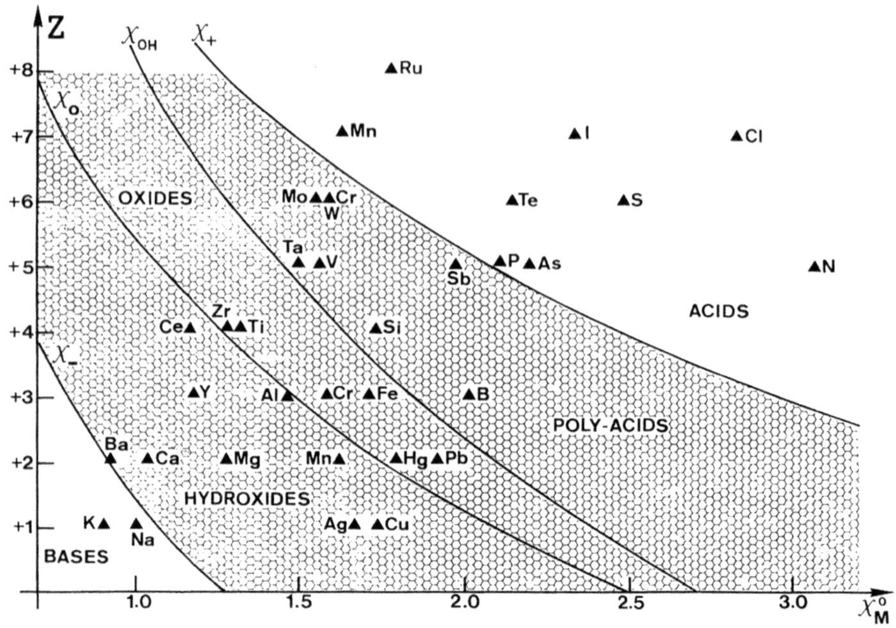

Fig. 2. The charge–electronegativity diagram according to Eqs. (75) and (77). Such a diagram tells, for a given neutral precursor $[MO_NH_{2N-z}]^0$, if the system remains soluble (class I and V) or if precipitation occurs (classes II–IV). Within the tint middle domains, a discrimination can be made between precursors undergoing only olation (class II) or oxolation (class IV) reactions, and those for whom both processes overlap (class III)

ion. Therefore, olation is not possible and condensation occurs only via oxolation. Protonation of anionic species is usually performed by adding an acid. Two basic mechanisms can be responsible for oxolation depending on whether coordination expansion of the metal ion is possible or not.

i) Nucleophilic addition via M–OH or M=O groups occurs in the first case increasing the coordination state [15]. No water molecule has to be removed so that chains or cycles are formed very rapidly. This mechanism leads to edge or face sharing $[MO_n]$ polyhedra.

ii) Nucleophilic substitution occurs when the metal atom already exhibits its maximum coordination state. It leads to corner-sharing polyhedra. The reaction starts with the nucleophilic addition of an OH group followed by a proton transfer towards the leaving ligand as follows:

$$\overset{\delta^-}{\text{M–OH}} + \overset{\delta^+}{\text{M–OH}} \longrightarrow \text{M–}\overset{\overset{\delta^+}{\text{H}}}{\underset{}{\text{O}}}\text{–}\overset{\delta^-}{\text{M–OH}} \longrightarrow \text{M–O–}\overset{\delta^+}{\text{M}}\text{–}\overset{\delta^+}{\text{OH}_2} \tag{19}$$

Water molecules have to be formed during the proton transfer so that oxolation should be slower than olation. However, it can be catalyzed by bases (first step) or acids (second step) and occurs over a wide range of pH [3].

At room temperature, highly condensed species are obtained around the point of zero charge. However, for oxolation to occur, OH groups have to be negatively charged ($\delta_{OH} < 0$). Otherwise the nucleophilic addition becomes no more possible and condensation stops giving rise to small oligomeric species called polyanions [16].

5.1 *Condensation of Hexavalent Cations*

As shown previously, two mononuclear protonated precursors have to be taken into account for Cr(VI) in aqueous solutions, namely $[CrO_2(OH)_2]^0$ and $[CrO_3(OH)]^-$. They correspond to h = 6 and 7, respectively. In both cases, OH groups have a negative partial charge so that oxolation is possible as follows:

- h = 7: $$2[CrO_3(OH)]^- \longrightarrow [Cr_2O_7]^{2-} + H_2O \qquad (20)$$

Condensation stops as hydroxy groups are no more available in the dimer.

- h = 6 $$2[CrO_2(OH)_2]^0 \longrightarrow [(HO)O_2Cr\text{–}O\text{–}CrO_2(OH)]^0 + H_2O \qquad (21)$$

OH groups are still available in the dimer but they become positively charged ($\delta_{OH} = +0.04$) meaning that condensation stops at this stage.

$$[Cr_2O_5(OH)_2] \longrightarrow [Cr_2O_7]^{2-} + 2H^+ \qquad (22)$$

Owing to the small size of Cr(VI), coordination expansion is not possible. Only dimeric bichromate species are therefore observed in aqueous solutions despite the negative partial charge of oxo groups.

Coordination expansion is possible with Mo(VI) and W(VI), leading to a more complex behavior in aqueous solutions [17, 18] (Fig. 3). According to R. H. Tytko [19], anionic precursors $[MO_4]^{2-}$ ($\delta_O = -0.57$) and $[MO_3(OH)]^-$ ($\delta_O = -0.42$) are tetrahedral while neutral ones are better formulated as $[MO_2(OH)_2(OH_2)_2]^0$ ($\delta_O = -0.35$). A comparison of oxygen partial charges in these precursors and in the water molecule ($\delta_O = -0.40$), leads to the following conclusions:

i) Tetrahedral $MO_3(OH)^-$ (h = 7) precursors are always better nucleophiles than water molecules. Therefore they tend to react with other anionic precursors (h = 8 or h = 7) in order to form polyanions (Fig. 3). Figure 3 shows that Mo(VI) mainly gives *cis*-dioxo polyanions while W(VI) forms mainly mono-oxo polyanions. Accordingly, ^{17}O- and ^{183}W-NMR studies [20] show that the only polytungstate having *cis*-dioxo groups (para A: $[W_7O_{24}]^{6-}$) transforms into a polytungstate with mono-oxo groups only (para B: $[H_2W_{12}O_{42}]^{10-}$). In the case of Mo(VI), ^{17}O and ^{95}Mo NMR show that $[Mo_7O_{24}]^{6-}$ and $[Mo_8O_{26}]^{4-}$ anions can undergo only isomerization reactions based upon concerted translations of close-packed fragments along molecular mirror planes [21]. Such a stabilization of *cis*-dioxo groups with Mo(VI) can be due to a smaller overlap between filled oxygen 2*p*-orbitals and the empty molybdenum

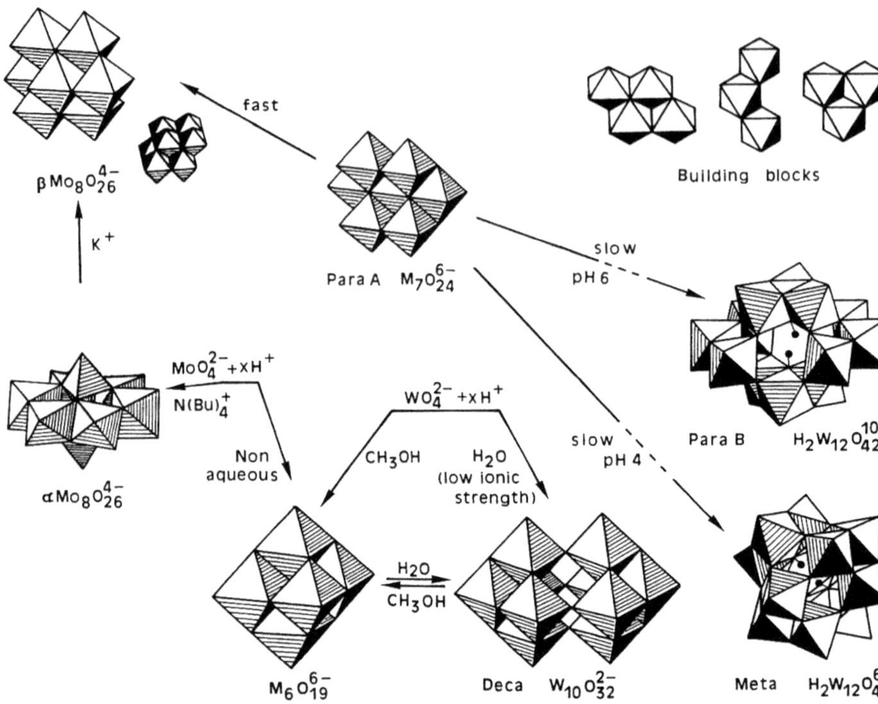

Fig. 3. The aqueous solution chemistry of Mo(VI) and W(VI) cations according to [16]. The structures of these polyanions can be obtained through oxolation between simple compact blocks shown at the upper left corner. The most prominent difference between both cations lies in the fact that Mo(VI) gives rise mainly to cis-dioxo polyanions while W(VI) leads mainly to mono-oxo ones

4*d*-orbitals [16]. For similar M–O bond length, 5*d*-orbitals of W(VI) give stronger σ(M–O) bonds than Mo which tends to compensate this weakness by forming π bonds with oxygen.

ii) Neutral metal precursors correspond to h = 6. These species are weaker nucleophiles than water. Addition of two water molecules then occurs leading to new molecular precursors in which the metal atom has a six-fold coordination [19]:

$$MO_2(OH)_2 + 2H_2O \longrightarrow MO_2(OH)_2(OH_2)_2 \tag{23}$$

At this stage, two behaviors are expected depending on the possible occurrence of intramolecular desoxolation:

$$MO_2(OH)_2(OH_2)_2 \leftrightarrow MO(OH)_4(OH_2) \tag{24}$$

Cis-dioxo Mo(VI) species are stabilized so that equilibrium of Eq. (24) is displaced towards the left. Polycondensation leads to solute molybdic polyacids $[Mo_nO_{2n+1}(OH)_2(OH_2)_x]$ and precipitation does not occur upon acidification

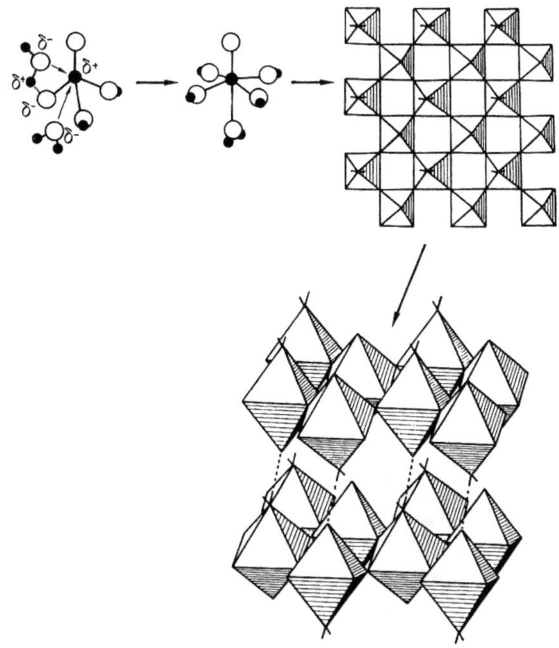

Fig. 4. A possible condensation pathway for the precipitation of tungstic hydrate $WO_3 \cdot H_2O$ from neutral h = 6 precursors $[MO_2(OH)_2]^0$ according to Ref. [3]. With Mo(VI) dissociation of the water molecule upon addition does not readily occur owing to the stabilization of cis-dioxo groups (cf Fig. 3). Such a mechanism could explain why Mo(VI) in contrast to W(VI) does not form precipitates upon acidification

[22, 23]. The reverse effect is observed for W(VI), mono-oxo species are more stable and the equilibrium is displaced towards the right. Lamellar tungstic hydrates are obtained upon acidification [24, 25] via an infinite bidimensional oxolation process (Fig. 4) [3].

5.2 Condensation of Pentavalent Cations

At high pH, V(V) species give vanadate ions $[VO_4]^{3-}$. Upon acidification they transform into more and more condensed anionic species, whose structures have been studied by ^{51}V and ^{17}O NMR [26] (Fig. 5). As the pH decreases, the negative charge per vanadium decreases progressively down to the point of zero charge (PZC = 2) corresponding to the slow precipitation of the hydrated pentoxide. Below that pH value, positively charged $[VO_2]^+$ monomeric solute species are formed.

The partial charge δ_O on oxygen corresponds to -0.35 for $VO(OH)_3$, -0.44 for $VO_2(OH)_2^-$ and -0.40 for H_2O. Therefore, $VO(OH)_3$ and

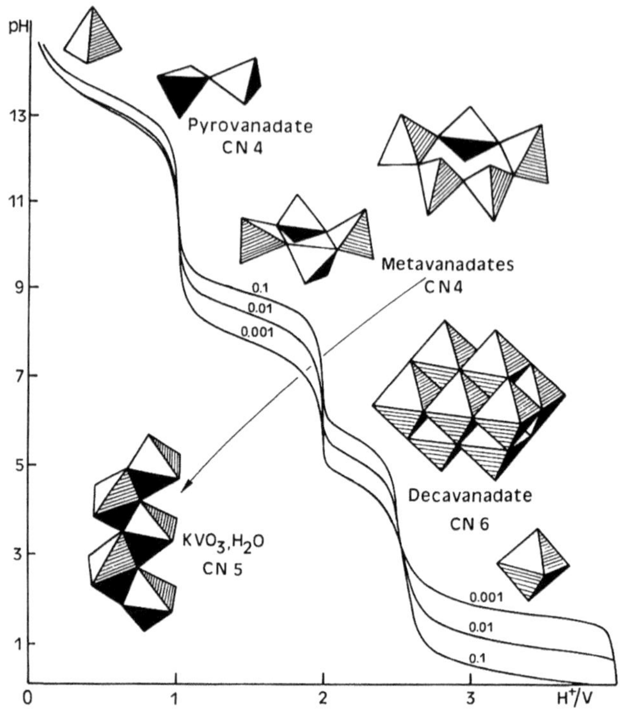

Fig. 5. The aqueous chemistry of V(V) according to Ref. [16] showing the transition from fourfold coordination ($H^+/V < 2$ or $pH > 7$) to sixfold coordination ($H^+/V > 2$ or $pH < 5$). Intermediate fivefold coordination can be observed only in the solid state when $H^+/V \approx 2$

$VO_2(OH)_2^-$ react together when pH > PZC leading to decavanadic acid (Fig. 5). Such a polyanion corresponds to a dead end for condensation and cannot lead to highly condensed species. Decavanadic acid has to be dissociated before V_2O_5, nH_2O gels are obtained. Water addition is expected around PZC:

$$VO(OH)_3 + 2H_2O \longrightarrow VO(OH)_3(OH_2)_2 \tag{25}$$

V–O bonds in this precursor are not equivalent so that olation occurs along one direction while oxolation occurs along another one [3]. Olation being faster than oxolation (no proton transfer) linear chains are rapidly formed accounting for the fibrous nature of V_2O_5 gels (Fig. 6).

Larger ions such as Nb(V) or Ta(V) are not able to form M=O bonds so that spontaneous desoxolation readily occurs:

$$MO(OH)_3(OH_2)_2 \longrightarrow M(OH)_5(OH)_2 \tag{26}$$

leading to the precipitation of hydrous oxides rather than gelation.

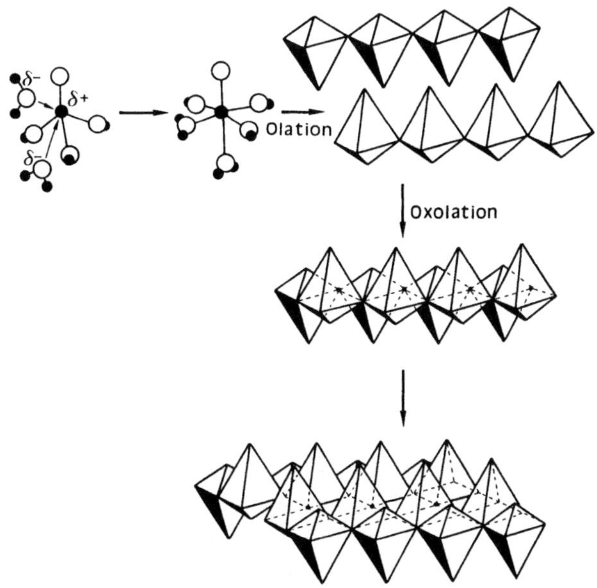

Fig. 6. A possible condensation pathway for the formation of vanadium(V) oxide gels from neutral h = 5 precursors $[VO(OH)_3]^0$ according to [3]. The presence of an aquo group in the equatorial plane introduces an intrinsic anisotropy which could help to explain the fibrous texture of these gels

6 Olation and Polycations

At least one water molecule has to be in the coordination sphere of the metal for olation to occur. Therefore, olation is observed with aquo-hydroxo precursors $[M(OH)_h(OH_2)_{N-h}]^{(z-h)+}$ where $h < n$ and $z < 5$. It leads to the formation of OH bridges as follows:

$$\overset{\delta^-}{M\text{–}OH} + \overset{\delta^+}{M\text{–}OH_2} \longrightarrow M\text{–}OH\text{–}M + H_2O \tag{27}$$

Actually it seems that the olation mechanism involves the formation of bridging $H_3O_2^-$ species [28]:

$$M\text{–}OH + H_2O\text{–}M \longrightarrow M\text{–}\overset{H}{O} \ldots\ldots \overset{H}{H\text{–}O}\text{–}M \tag{28}$$

A cleavage of the $M\text{–}OH_2$ bond then occurs following a SN_1 process [3]:

$$M\text{–}\overset{H}{O} \ldots\ldots \overset{H}{H\text{–}O}\text{–}M \longrightarrow M\text{–}OH\text{–}M + H_2O \tag{29}$$

In all cases H_2O ligands have to be removed from the coordination sphere of the metal so that olation rates depend on the lability of $M\text{–}OH_2$ bonds. They increase when the ionic radius increases and the oxidation state decreases. Crystal field stabilization effects (CFSE) have also to be taken into account with transition metal ions such as Cr^{3+} or Ni^{2+}. With such elements bridged intermediates have been isolated as single crystals providing a direct evidence of the bidentate nature of $H_3O_2^-$ ligands [28]. For other elements, bridged intermediates cannot be isolated and only the inner sphere bridge can be characterized.

6.1 Condensation of Trivalent Cations

Despite some analogy, trivalent metal cations exhibit very different behaviors towards hydrolysis and condensation in aqueous solutions. Cr^{3+} for instance forms stable polycations [32] or gels [33] while Fe^{3+} leads to very reactive polycations [34, 35] or gelatinous precipitates [36]. Al^{3+} exhibits an intermediate behavior with the well known "Al_{13}" polycation [37, 38], gels as well as gelatinous precipitates [39, 40]. Moreover $Al(OH)_3$ [41] and $Cr(OH)_3(OH_2)_3$ [42] hydroxides have been isolated while $Fe(OH)_3$ is still unknown [36]. All these elements form at least three different MOOH phases which transform upon heating into the M_2O_3 corundum structure [43–46]. An attempt is presented here to account for these observations in the frame of the Partial Charge Model.

The negative charge of the OH group in the $h = 1$ $[M(OH)(OH_2)_5]^{2+}$ precursor increases regularly from Fe^{3+} ($\delta_{OH} = -0.01$) to Cr^{3+} ($\delta_{OH} = -0.02$) and Al^{3+} ($\delta_{OH} = -0.03$). This may account for the regular increase of hydrolysis or dimerization constants (Table 1).

6.1.1 Small Oligomeric Species

Cr^{3+} complexes are known to be strongly inert. Therefore detailed structural studies of Cr^{3+} polycations in aqueous solutions have been published. Several chemical species are formed when a base is added to the aqueous solution. They

Table 1. Thermodynamics and kinetics data for three trivalent metals having quite close ionic radii in octahedral coordination (55, 62, and 53 pm, respectively). (*a*) Ref. [4] (*b*) Ref. [29] (*c*) Ref. [30] and (*d*) Ref. [31]

Ions	Fe^{3+}	Cr^{3+}	Al^{3+}
p*K*(h = 0)[a]	2.2	4.0	5.0
p*K*(dimer)[a]	3.0	5.1	7.7
p*K*(trimer)[a]	6.3	8.2	7.7
water lability (s^{-1})	$10^{2.2}$[b]	$10^{-5.627}$[c]	10^{-0R27}[d]
configuration	$3d^5$	$3d^3$	$3d^0$

Table 2. Equilibrium constants for various Cr(III) polycations according to Ref. [32]

Species	p*K*
$[Cr(OH_2)_6]^{3+}$	4.29
$[(H_2O)_5Cr(OH)Cr(OH_2)_5]^{5+}$	1.60
$[(H_2O)_4Cr(OH)_2Cr(OH_2)_4]^{4+}$	3.68
$[Cr_3(OH)_4(OH_2)_9]^{5+}$	4.35
$[Cr_4(OH)_6(OH_2)_{10}]^{6+}$	3.53
$[Cr_4O(OH)_5(OH_2)_{10}]^{5+}$	0.89

appear to be in slow equilibrium and are formed simultaneously owing to pH gradients during base addition (Table 2).

The first condensation step (Fig. 7) leads to the formation of an outer sphere complex involving two bridging $H_3O_2^-$ ligands stabilized through intramolecular hydrogen bonds. Elimination of one water molecule leads to a corner-sharing dimer which then transforms into the thermodynamically stable edge-sharing dimer [47, 48]. These dimerization reactions can be followed easily via color changes. The h = 0 monomer and the μ-hydroxo dimer are green while the μ_2-hydroxo compound is blue [49]. Hydroxo groups have a positive partial charge ($\delta = +0.01$) so that the blue dimer does not lead to further condensation. The molecular structure of such a dimer was determined by X-ray diffraction. Single crystals were obtained after dissolution of the "active hydroxide" into *p*-toluenesulfonic or mesitylene-2-sulfonic acids [50].

For condensation to proceed, hydrolysis must occur. This appears to be a quite easy process as dimers are more acidic than h = 0 monomeric precursors (Table 2). This may be attributed to the more positive partial charge of hydrogen atoms and the stabilization of deprotonated species through intramolecular hydrogen bonding (Fig. 7) [49].

Such a stabilization suggests that hydrolysis proceeds along the vertices of the dimer rather than in the equatorial plane. Another h = 1 monomeric precursor can then react at this position giving rise to trimeric $[Cr_3(OH)_4(OH_2)_{10}]^{5+}$ intermediates (Fig. 7). Intramolecular olation leads to the compact trimer $[Cr_3(OH)_4(OH_2)_9]^{5+}$ [51]. This polycation exhibits a cyclic structure based on three μ-hydroxo groups stabilized by a central μ_3-hydroxo group [32, 52]. A similar process occurs again upon further hydrolysis leading to a tetrameric intermediate $[Cr_4(OH)_6(OH_2)_{11}]^{6+}$ which can undergo two possible intramolecular reactions [52]:

i) Olation with a μ_2-hydroxo bridge leads to the planar tetramer $[Cr_4(OH)_6(OH_2)_{10}]^{6+}$ (structure II).

ii) Hydrolysis of the μ_3-hydroxo bridge leads to the skewed tetramer $[Cr_4O(OH)_5(OH_2)_{10}]^{5+}$ after substitution of one water molecule by the oxo group (structure III). This last reaction occurs at higher pH than the first one [53].

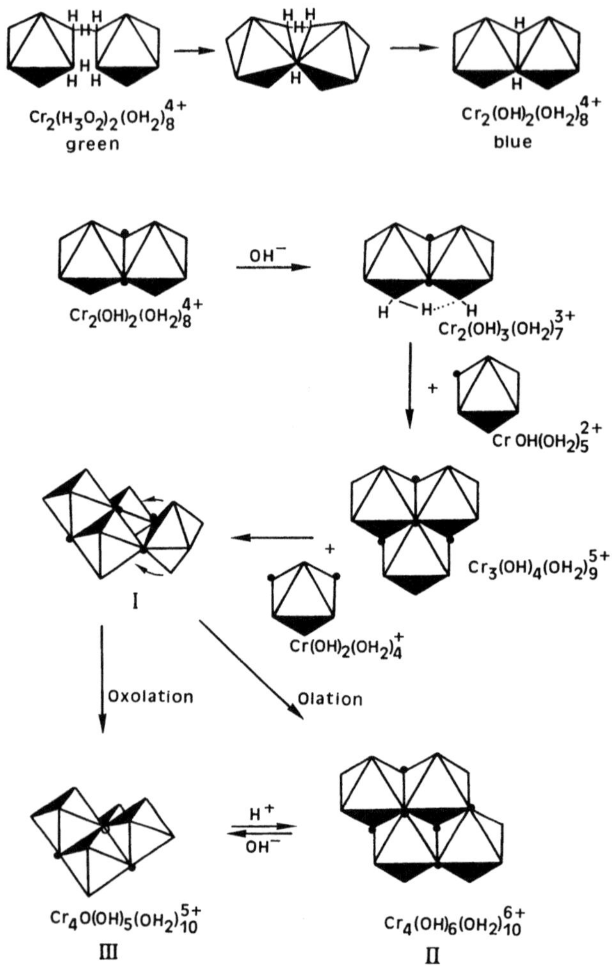

Fig. 7. The aqueous chemistry of Cr(III) according to Ref. [32]. Owing to its $3d^3$ electronic configuration, both olation and oxolation are rather slow processes for Cr^{3+} authorizing a detailed study as a function of the pH. At the present time only the μ_2-hydroxo dimer has been characterized in the solid state

This analysis is based on the fact that upon acidification ^{51}Cr labelled tetramers lead to a mixture of monomers and trimers among which ^{51}Cr is equally distributed [52]. Moreover kinetic studies [53] show that interconversion between tetrameric forms (I) (II) (III) occur in solution when the pH varies. Finally, owing to their compact structures, trimers and tetramers are the most inert forms of hydrolyzed Cr(III) species in aqueous solutions. Therefore, activation energies are too high for condensation to occur upon further hydrolysis.

Only the trimer $[Al_3(OH)_4(OH_2)_{10}]^{5+}$ was characterized by automated potentiometric titration [54]. ^1H-, ^{17}O-, and ^{27}Al-NMR measurements [37, 38] have recently confirmed that the di-μ_2-hydroxo cation

$[(H_2O)_4Al(OH)_2Al(OH_2)_4]^{4+}$ is not formed in detectable amounts upon neutralization. However, a sulfate dimer can be crystallized from sulfuric acid solutions [55]. X-ray diffraction data then provide accurate information on its structure. When such crystals are dissolved in pure water, ^{27}Al NMR shows that the dimer disproportionates giving $[Al(OH_2)_6]^{3+}$ monomers and a highly hydrolyzed (h ≈ 2.5–2.6) trimeric polycation of unknown structure [37]. As shown by dilution experiments such polycations transform directly and rapidly into $[Al_{13}O_4(OH)_{24}(OH_2)_{12}]^{7+}$ polycations. Fe^{3+} polycations are much less known than Al^{3+} ones owing to the intrinsic higher lability of Fe^{3+}. Dimeric species $[Fe_2(OH)_2]^{4+}$ and $[Fe_2O]^{4+}$ have been structurally characterized only as organic complexes [43]. The existence in aqueous solutions of both aquated forms ($[(H_2O)_4Fe(OH)_2Fe(OH_2)_4]^{4+}$ and $[(H_2O)_5FeOFe(OH_2)_5]^{4+}$) seems to be beyond doubt despite the lack of structural data.

6.1.2 Large Polycations

A quite different behavior is observed at higher pH (h = 2.5). Inert trimers or tetramers are formed with chromium. Aluminum gives the Keggin-like polycation $[Al_{13}O_4(OH)_{24}(OH_2)_{12}]^{7+}$ with twelve $[AlO_6]$ octahedra surrounding a central $[AlO_4]$ tetrahedron [37, 56]. For the same hydrolysis ratio Fe^{3+} does not form any polycation but leads to deep red sols made of amorphous colloidal particles about 2–4 nm in diameter [43, 58]. X-ray absorption measurements (EXAFS and XANES) show that Fe^{3+} ions are in a sixfold coordination with bridging oxo and hydroxo groups (corner and edge sharing) [59].

Such differences may be attributed to the electronic configuration of these trivalent cations. Let us assume that the rather stable compact trimer intermediate $[M_3(OH)_4(OH_2)_9]^{5+}$ is formed in all cases (M = Cr, Fe or Al) so that it can be considered as a precursor for further condensation. The μ_3-hydroxo bridge (Fig. 8) in which the oxygen atom is sp^3 hybridized makes four covalent bonds. Three O–M bonds are formed with metal atoms at the center of three $[MO_6]$ octahedra, the fourth one with a proton to give the OH bridge. Three different cases can now be described:

i) Al^{3+} ($1s^2 2s^2 2p^6$) is a small cation with empty *d*-orbitals. The oxygen electron cloud is strongly polarized towards Al^{3+} making the μ_3-OH bridge highly acidic. This favors hydrolysis reactions [60] as follows (Fig. 8):

$$[Al_3(OH)_4(OH_2)_9]^{5+} + H_2O \longrightarrow [Al_3O(OH)_3(OH_2)_9]^{4+} + H_3O^+ \quad (30)$$

Owing to this polarization, water molecules in a *trans*-position become much less acidic than water molecules in a *cis*-position (*trans*-influence). Therefore, half of these cis H_2O ligands are hydrolyzed upon mild conditions (h < 2.6) leading to:

$$[Al_3O(OH)_3(OH_2)_9]^{4+} + 3\,OH^- \longrightarrow [Al_3O(OH)_6(OH_2)_6]^+ + 3H_2O \quad (31)$$

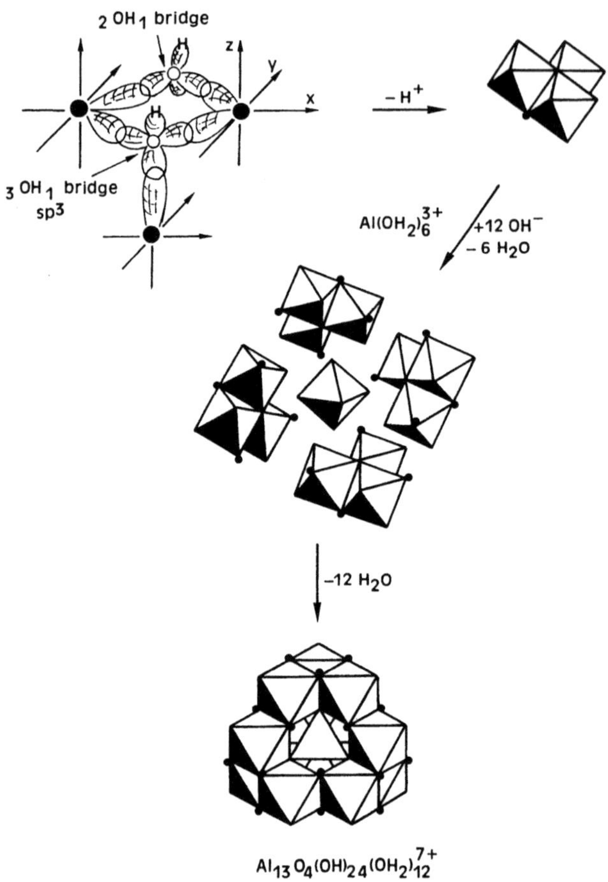

Fig. 8. A possible condensation pathway leading to the huge $[Al_{13}O_4(OH)_{24}(OH_2)_{12}]^{7+}$ polycation according to [60]. The prominent feature of the mechanism is the presence within an hydrolyzed trimer $[Al_3(OH)_7(OH_2)_6]^{2+}$ of a highly nucleophilic μ_3-hydroxo bridge, which could act as a ligand for a monomeric nonhydrolyzed Al^{3+} cation after deprotonation

This trimer has a highly nucleophilic oxygen atom and thus behaves as a ligand for h = 0 precursors leading to the formation of "Al_{13}" species:

$$[Al(OH_2)_6]^{3+} + 4[Al_3O(OH)_6(OH_2)_6]^+ \longrightarrow [Al\{Al_3O(OH)_6(OH_2)_6\}_4]^{7+} + 6\,H_2O \quad (32)$$

Trimers are rather large so that only four of them can accommodate around a small Al^{3+} ion [60]. Once they are bonded to a central aluminum, olation occurs via edge sharing between cis OH groups with elimination of the remaining cis water molecules leading to:

$$[Al\{Al_3O(OH)_6(OH_2)_6\}_4]^{7+} \longrightarrow [Al_{13}O_4(OH)_{24}(OH_2)_{12}]^{7+} + 12\,H_2O \quad (33)$$

Such a reaction path is in agreement with recent ^{1}H-, ^{17}O-, and ^{27}Al-NMR data [37]. These experiments show that the structure of this polycation remains the same in the aqueous solution as in crystalline sulfates or selenates [37, 38]. Moreover, they show that small oligomeric species (which may be trimers) directly condense upon hydrolysis into $[Al_{13}O_4(OH)_{24}(OH_2)_{12}]^{7+}$ with no detectable intermediates.

ii) The *d*-orbitals of Cr^{3+} ($3d^3$) ions are partially filled so that the μ_3-hydroxo bridge is less acidic. Moreover, such an electronic configuration is strongly stabilized in an octahedral coordination ($t_{2g}^3\, e_g^0$). Cr^{3+} cannot adopt a fourfold coordination. So that even if some trimers were hydrolyzed they could not react with the $h = 0$ precursor to form a "Cr_{13}" polycation. In such a case the skewed tetramer would be obtained.

iii) *d*-orbitals of Fe^{3+} ions ($3d^5$) are all singly occupied. The μ_3-hydroxo bridge is thus no more acidic. Olation occurs via μ_2-hydroxo bridges leading to the planar tetrameric polycation. Again discrete "Fe_{13}" polycations cannot be formed. No high activation energy prevents condensation between trimeric and/or tetrameric units. Amorphous colloidal particles are formed via the disordered condensation of such species through both olation and oxolation reactions.

6.1.3 Solid Phases

Neutralization of an aqueous solution of a M^{3+} salt ($h > 2.6$) mainly leads to neutral solute precursors $[M(OH)_3(OH_2)_3]^0$ from which $M(OH)_3$ solid phases are formed via rapid olation reactions:

$$[M(OH)_3(OH_2)_3]^0 \longrightarrow M(OH)_3 + 3\,H_2O \qquad (34)$$

Around room temperature ($T < 80\,°C$) crystalline hydroxides $Al(OH)_3$ precipitate from aqueous solutions either by increasing the pH of a nitrate solution or upon acidification of a sodium aluminate solution. These hydroxides exhibit lamellar structures based on hexagonal rings linked through μ_2-OH bridges (Fig. 9). Depending on the packing of these $Al(OH)_3$ sheets, gibbsite (cubic packing) or bayerite (hexagonal packing) structures are formed [41]. The formation of these planar sheets from aqueous precursors could be described following two different pathways (Fig. 9):

i) The first one occurs when a base is added to an Al^{3+} aqueous solution. It involves the octahedral $[Al(OH)_3(OH_2)_3]^0$ neutral precursor. This precursor can also be written as $[Al(H_3O_2)_3]^0$ where $H_3O_2^-$ groups behave as bidentate chelating ligands. Changing the coordination mode from chelating to bridging would first lead to cyclic oligomers. Because of the steric hindrance of $H_3O_2^-$, the smallest one would correspond to hexameric species with twelve $H_3O_2^-$ bridges. Tetramers would be too small for eight $H_3O_2^-$ bridges. Once formed, this embryo could easily collapse into a critical nucleus. Water elimination via

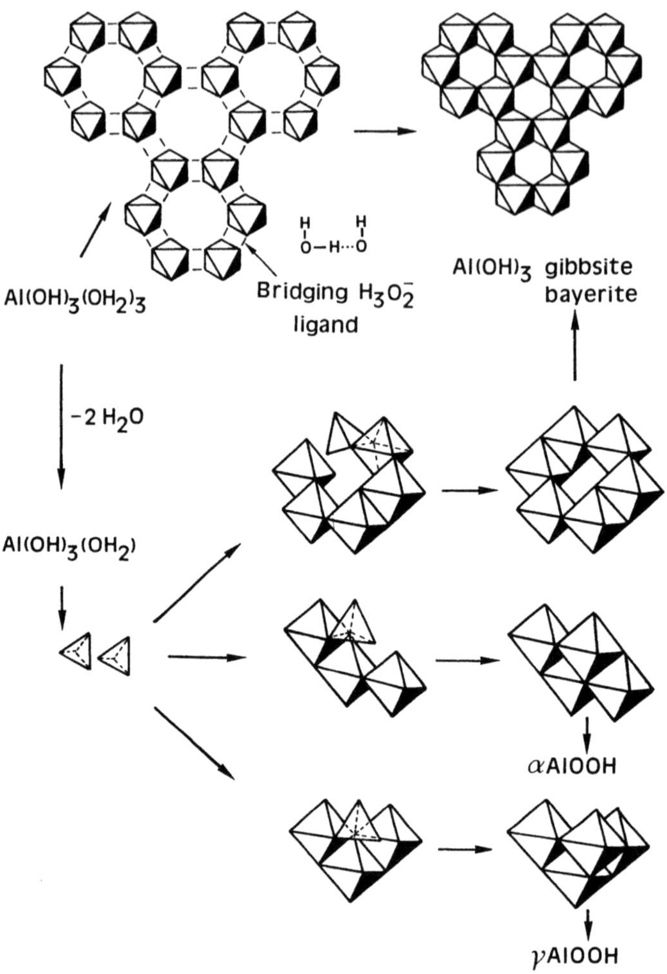

Fig. 9. Possible condensation pathways leading to aluminum hydroxides $Al(OH)_3$ or to aluminum oxy-hydroxides AlOOH. From acidic solutions olation involve nucleophilic substitution reactions with intermediate bridging $H_3O_2^-$ ligands. From basic solutions, olation involve nucleophilic addition reactions owing to the tetrahedral structure of the aluminate ion $Al(OH)_4^-$

olation leads to stronger μ_2-OH bridges. Condensation of $[Al(H_3O_2)_3]^0$ monomers or aggregation of hexameric units then leads to solid particles.

ii) The second pathway occurs upon acidification of an aluminate solution. Tetrahedral $[Al(OH)_3(OH_2)]^0$ neutral precursors are formed according to:

$$[Al(OH)_4]^- + H_2O \longrightarrow [Al(OH)_3(OH_2)]^0 + OH^- \qquad (35)$$

Tetrahedral monomeric aluminate ions have been clearly evidenced by ^{27}Al NMR [61]. After proton transfer from a water molecule, a $[Al(OH)_3(OH_2)]^0$ neutral precursor is formed. Owing to its low coordination number, it can undergo very fast addition reactions leading to octahedral oligomers such as trimers, tetramers, or hexamers. The hexameric oligomer is obviously a critical nucleus for bayerite or gibbsite. Smaller oligomers could act as critical nuclei for oxy-hydroxides phases.

At higher temperature ($T > 80\,^{\circ}C$) hydroxides $Al(OH)_3$ readily transform into oxy-hydroxides AlOOH, mainly boehmite γAlOOH [62, 63]. Diaspore αAlOOH is also known as a natural mineral and can be obtained via hydrothermal synthesis [46]. The preferential formation of boehmite rather than diaspore may well be due to the intrinsic acidity of Al_3OH bridges. Increasing the temperature probably destabilizes hexameric $[Al_6(H_3O_2)_{12}]^0$ oligomers leading to more compact $[Al_3(OH)_9(OH_2)_4]^0$ trimers. Moreover, temperature also enhances the intrinsic acidity of Al_3OH bridges. Therefore once a neutral monomer $[Al(OH)_3(OH_2)_3]^0$ has reacted to form a critical $[Al_4(OH)_{12}(OH_2)_5]^0$ nucleus, tilting occurs towards the Al_3O bridge as shown in Fig. 10. Such

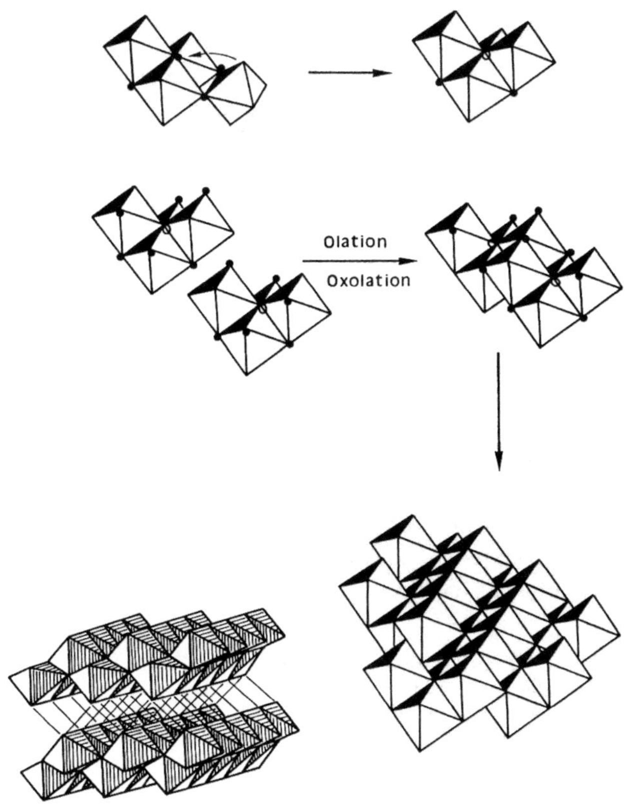

Fig. 10. Possible condensation pathway for nucleation and growth of the boehmite structure (γAlOOH) from aqueous solutions. The key point is as for Al_{13} formation the presence of a highly nucleophilic μ_3-OH bridge whose deprotonation orient nucleation towards a skewed tetramer. From such a tetramer the corrugated structure of boehmite sheets is easily derived

a nucleation process leads to the formation of skewed tetramers $[Al_4O(OH)_{10}(OH_2)_5]^0$. Polycondensation via olation and oxolation then leads to the characteristic corrugated sheets of the boehmite structure. Hydrogen bonding between these sheets leads finally to a three-dimensional rigid structure.

Fe^{3+} exhibits a quite different behavior. Crystalline ferric hydroxides $Fe(OH)_3$ have never been precipitated, but amorphous oxy-hydroxides with a variable water content (ferrihydrites) are rapidly formed [64]. Such a thermodynamic instability of $Fe(OH)_3$ can be explained by looking at the partial charge of the water molecules formed during solid-state oxolation as follows (Table 3):

$$M(OH)_3 \leftrightarrow MO(OH)(OH_2) \rightarrow MO(OH) + H_2O \tag{36}$$

Positively charged water molecules can be removed from the hydroxide when M = Fe or Cr. Their negative partial charge when M = Al, prevents such an elimination giving rise to the hydroxide phase. The occurrence of oxolation during the nucleation and growth of hydroxides also explains the very small solubility of ferrihydrite in alkaline solutions.

Amorphous solids formed via the neutralization of Fe^{3+} aqueous solutions are characterized by a disordered stacking of octahedral Fe^{3+} ions surrounded by OH^- and O^{2-} anions [64, 65]. Upon aging or gentle heating, crystalline oxy-hydroxides are generally obtained [43, 66, 67]. However, in contrast with Al^{3+}, goethite (αFeOOH) isostructural with diaspore is the stable phase [36]. Lepidocrocite (γFeOOH) isostructural with boehmite can only be obtained under very specific conditions (low pH, dilute Fe^{3+} solutions or Fe^{2+} oxidation) [36]. This could be due to the $3d^5$ electronic configuration of Fe^{3+} ions which leads to the weak acid character of the Fe_3OH bridge. Accordingly, as hydrolysis of this bridge does not readily occur, the fourth octahedron is tilted towards the Fe_2OH bridge (olation) (Fig. 11). Polycondensation of planar $[Fe_4(OH)_{12}(OH_2)_4]^0$ tetramers through olation would first lead to double rutile-type chains. Further association of these chains through oxolation then leads to the characteristic three dimensional structure of goethite. Tilting the other way towards the $Fe_3(OH)$ bridge (oxolation) leads to the skewed tetramer $[Fe_4O(OH)_{10}(OH_2)_5]^0$ precursor of γFeOOH. Other polymorphs with the rutile (εFeOOH) [68], αMnO_2 (βFeOOH) [69] or hexagonal (δFeOOH) [70] structures are also known. Finally, at high temperature and low pH, haematite (αFe_2O_3) can be precipitated directly from aqueous solutions [36, 43].

Table 3. Partial charges beared by water molecules and metal atoms within an hydroxide $M(OH)_3$ when M = Al, Fe and Cr

M	$\delta(OH_2)$	$\delta(M)$
Al	− 0.01	+ 0.62
Cr	+ 0.02	+ 0.56
Fe	+ 0.07	+ 0.46

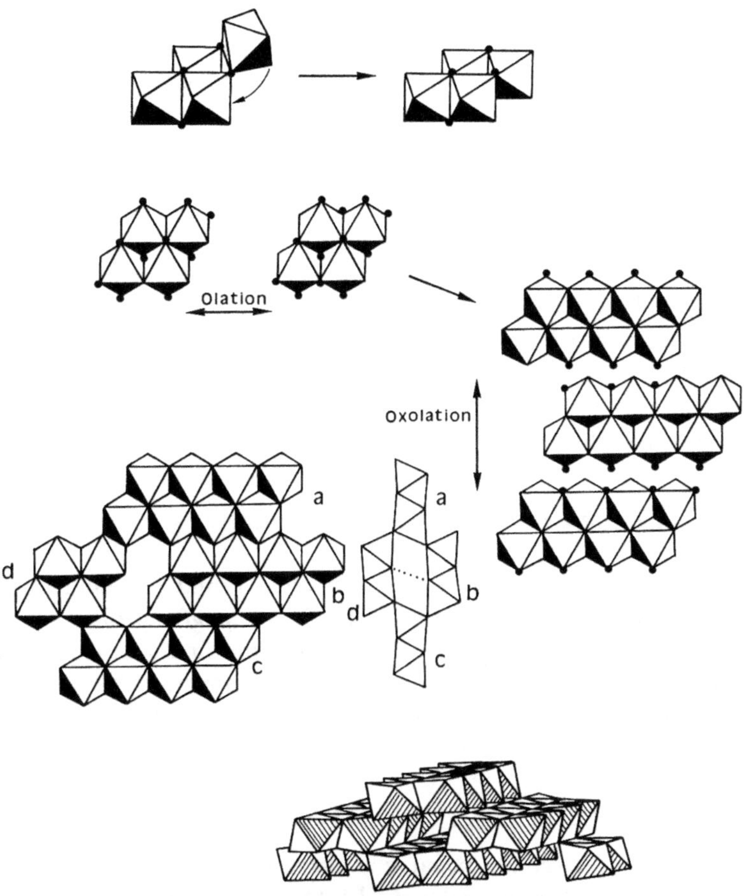

Fig. 11. Possible condensation pathway for nucleation and growth of the goethite (αFeOOH) structure from aqueous solutions. Here the non-acidic character of the μ_3-OH bridge orient nucleation towards the planar tetramer from which the double rutile chain characteristic of the goethite structure is readily obtained (compare Fig. 10)

As expected inert Cr^{3+} ($3d^3$) ions behave their own way. Amorphous gels which composition ranges from $Cr(OH)_3(OH_2)_3$ to Cr_2O_3 can be readily obtained when a base is added to the aqueous solution [33, 71]. The metastable and microcrystalline $Cr(OH)_3(OH_2)_3$ which is first obtained [42] probably corresponds to the kinetically stable intermediate $M(H_3O_2)_3$ postulated for $Al(OH)_3$ nucleation. Owing to the great inertness of Cr^{3+} ions, it is even possible to prepare a whole family of isomers having the same $Cr(OH)_3(OH_2)_3$ rough formula. Thus, besides the structure based on the monomer $Cr(OH)_3(OH_2)_3$ [43], dimeric $[Cr_2(\mu\text{-}OH)_2(OH)_4(OH_2)_4]\cdot 4H_2O$, and trimeric $[Cr_3(\mu\text{-}OH)_4(OH)_5(OH_2)_4]\cdot 4H_2O$ hydroxides have been obtained [72, 73]. These microcrystalline materials have been characterized by powder X-ray diffraction. Their structures are based on the various polycations studied above. All these crystals are unstable with respect to dehydration and polycondensation giving first oxy-hydroxides CrOOH then the oxide αCr_2O_3 [42, 45, 72, 73]. This behavior may be attributed to the positive partial charge of water molecules in $Cr(OH)_3(OH_2)_3$ (Table 3). Cr^{3+} ions form also upon neutralization green amorphous gels [33]. Dehydration of these gels leads to αCr_2O_3 with the possible formation of very special crystalline CrOOH phases [45]. Other oxy-hydroxides can also be crystallized via hydrothermal techniques. They are amorphous (the so-called γCrOOH) or isostructural with InOOH (βCrOOH) [74] and $Mg(OH)_2$ (αCrOOH) [75]. In this last case, the great inertness of planar $[Cr_4(OH)_{12}(OH_2)_4]^0$ tetramers towards both olation and oxolation is probably responsible for the orientation of polycondensation towards planar CdI_2-like octahedral sheets (Fig. 13) rather than linear double chains characteristic of goethite or diaspore structures.

6.2 Condensation of Divalent Cations

Divalent transition metal ions are often found in octahedral coordination. A regular decrease of the $M-OH_2$ bond length is observed from 0.220 nm (Mn^{2+}) down to 0.204 nm (Ni^{2+}) along the first series of divalent transition elements [76]. Condensation of divalent cations starts when the hydrolysis ratio h = 1. In that case, OH groups have a strong negative charge ($\delta_{OH} = -0.12$) so that olation goes beyond dimeric species [4]. Dimers are formed in a first step as follows:

$$2[M(OH)(OH_2)_5]^+ \longrightarrow [M_2(OH)_2(OH_2)_8]^{2+} + 2\,H_2O \qquad (37)$$

μ_2-OH bridges remain negatively charged ($\delta_{OH} \approx -0.10$) so that further condensation can proceed. All μ_2-OH bridges then transform into μ_3-OH bridges in order to give the most compact polycation. The simplest way to fulfill this condition is to add two edge-sharing dimers so that they form a right angle (Fig. 12). A compact and symmetrical tetramer $[M_4(OH)_4(OH_2)_{12}]^{4+}$ is thus

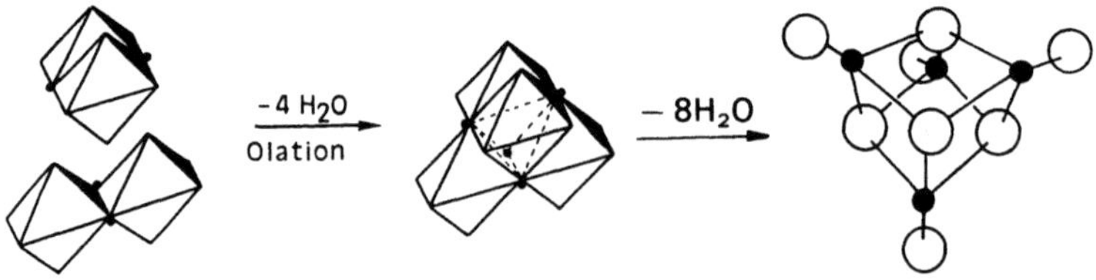

Fig. 12. Formation of the most compact polycation from h = 1 $[M(OH)(OH_2)_5]^+$ octahedral precursors. The resulting tetramer can be viewed as a tetrahedral cluster of metallic atoms surrounded by four μ_3-OH bridges above each face with aquo ligands completing the coordination sphere of the metals

obtained with four metal atoms at the vertices of a tetrahedron and four OH groups above each face. Such a structure was evidenced by kinetics measurements for Ni(II) [77] and X-ray diffraction for Pb(II) [78]. Similar polycations are probably formed with Co(II), Cd(II), and Mg(II) [4], although structural evidence is lacking.

Two main features can be pointed out concerning such tetramers:

i) Aquo and hydroxo ligands lie outside the cage cluster so that direct metal–metal bonds are possible. Magnetic coupling (Ni, Co) or lone pair interactions (Pb) are thus favored.

ii) The metal coordination can be modified by removing (Pb) or adding (Ni) two water molecules.

Cu^{2+} ions seem to form linear "core plus links" chains rather than compact tetramers [79]. This may be due to the Jahn-Teller effect typical of d^9 ions in an octahedral field. The solvated complex exhibits a tetragonal distortion with four equatorial water molecules at 0.194 nm and two axial ones at 0.238 nm [76]. Hydrolysis then occurs first in the equatorial plane. At low pH (pH < 3) a corner-sharing dimer $[Cu_2(OH)(OH_2)_6]^{3+}$ is probably formed [80] which transforms at higher pH into edge-sharing chain polymers $[Cu_n(OH)_{2n-2}(OH_2)_2]^{2+}$ (n = 2, 3) [79, 81].

Neutralization of an aqueous solution of a divalent metal around h = 2 leads to hydroxide nucleation and growth. Under such conditions solute species mainly correspond to the neutral precursor $[M(OH)_2(OH_2)_4]^0$ which can undergo infinite and rapid olation reactions:

$$[M(OH)_2(OH_2)_4]^0 \longrightarrow M(OH)_2 + 4\,H_2O \qquad (38)$$

The resulting solid hydroxide has a typical lamellar CdI_2 structure based on metal sheets having μ_3-OH groups on each side. Such a structure could be formed via the following condensation scheme (Fig. 13):

i) Formation of a compact and planar $[M_4(OH)_8(OH_2)_6]^0$ tetramer through olation reactions (critical nucleus).

ii) Bidimensional growth of these tetramers via olation along two perpendicular and equivalent directions.

Such a simple scheme appears to be followed by most divalent transition metal ions. In the case of Cu(II), the Jahn-Teller effect leads to $[Cu(OH)_2]_\infty$ edge-sharing chains formed from neutral square-planar $[Cu(OH)_2(OH_2)_2]^0$ precursors (Fig. 14). These chains can then be linked together to form a corrugated sheet isostructural of boehmite (γAlOOH) or lepidocrocite (γFeOOH). Such sheets exhibit unusual μ_4-OH bridging groups [82]. All four oxygen hybrid orbitals are involved in the formation of Cu–O bonds so that μ_4-OH bridges are expected to show some acidic character. This may account for the instability of copper hydroxide $Cu(OH)_2$ towards oxolation. Gentle heating or pH increase readily transforms the blue $Cu(OH)_2$ into black CuO, at least in the absence of counterions [83]. Other hydroxides having the CdI_2 structure

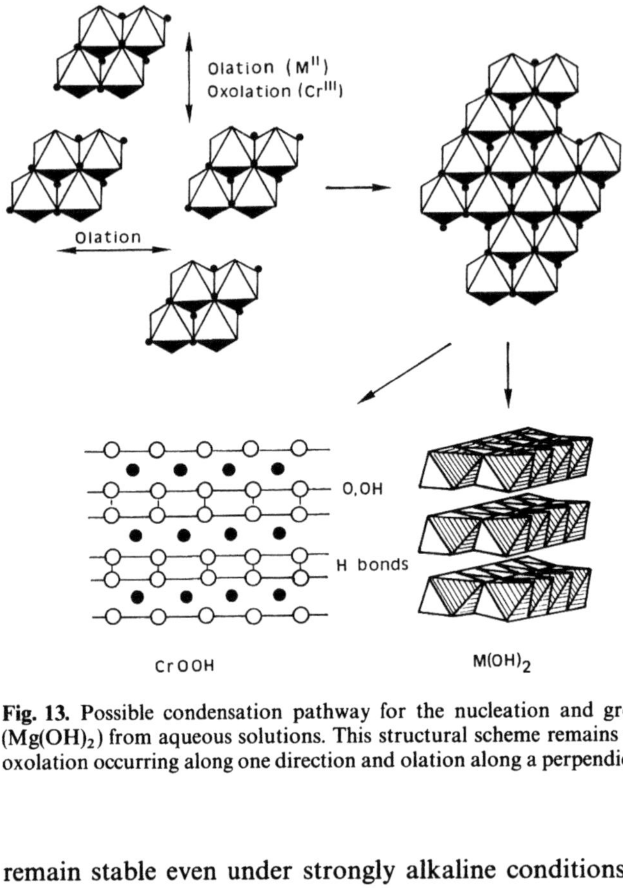

Fig. 13. Possible condensation pathway for the nucleation and growth of the brucite structure ($Mg(OH)_2$) from aqueous solutions. This structural scheme remains valid in the case of Cr^{3+} with oxolation occurring along one direction and olation along a perpendicular one (αCrOOH structure)

remain stable even under strongly alkaline conditions. They give rise to MO oxides only upon heating above 300 °C [84].

6.3 Condensation of Tetravalent Cations

According to the charge–pH diagram, hydrolysis of tetravalent cations already occurs at low pH. The first hydrolysis step (h = 1) leads to the formation of $[M(OH)(OH_2)_{N-1}]^{3+}$ species.

$Ti(OH)(OH_2)_5^{3+}$ precursors are stable only under strongly acidic conditions. They are not able to condense because of the positive charge of the hydroxo group ($\delta_{OH} = +0.06$). They then behave as acids leading to the h = 2 precursor as follows:

$$[Ti(OH)(OH_2)_5]^{3+} + H_2O \longrightarrow [Ti(OH)_2(OH_2)_4]^{2+} + H_3O^+ \qquad (39)$$

OH groups bear only a weakly negative charge ($\delta = -0.01$). This h = 2 precursor seems to remain monomeric [85, 86] as it probably undergoes a spontaneous intramolecular oxolation leading to the formation of titanyl ions

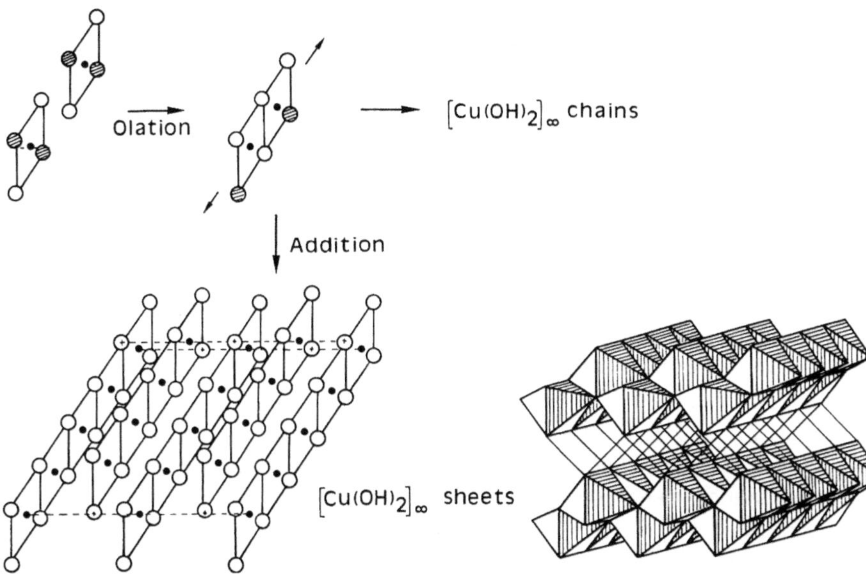

Fig. 14. Possible condensation pathway for the nucleation and growth of the $Cu(OH)_2$ structure (isostructural of boehmite) from aqueous solutions. The key-point is the square-planar coordination of Cu^{2+} ions induced by the Jahn-Teller effect occurring in the case of d^9 electronic configuration. This effect orients nucleation towards chain polymers whose association through nucleophilic addition (olation) leads to corrugated sheets characteristic of the boehmite structure

$[TiO]^{2+}$ as follows:

$$[Ti(OH)_2(OH_2)_4]^{2+} \longrightarrow [TiO(OH_2)_5]^{2+} \quad (40)$$

Zr^{4+} ions are bigger than Ti^{4+} so that $[Zr(OH)(OH_2)_7]^{3+}$ precursors are less acidic than $[Ti(OH)(OH_2)_5]^{3+}$. In the absence of complexing anion further hydrolysis occurs leading to h = 2 species.

$$[Zr(OH)(OH_2)_7]^{3+} + H_2O \longrightarrow [Zr(OH)_2(OH_2)_6]^{2+} + H_3O^+ \quad (41)$$

OH groups then have a rather high negative partial charge ($\delta_{OH} = -0.06$) so that condensation occurs leading to the formation of the well-known cyclic tetramers $[Zr_4(OH)_8(OH_2)_{16}]^{8+}$ [87, 88].

No evidence is available at the present time for highly polycondensed cationic species involving octahedral tetravalent elements. The only reported data is an octameric $[Ti_8O_8(OH)_{12}(OH_2)_x]^{4+}$ polycation of unknown structure [89, 90]. This may well be due to the high acidity of h = 3 based dimers (less negative OH groups). In that case hydrolysis must occur in the equatorial plane as axial positions are occupied by oxo and non-acidic *trans* aquo ligands ($O{=}M{-}OH_2$). This differs considerably from h = 2 based dimers of trivalent elements where axial hydrolysis is favored giving rise to compact trimers. Chain polycondensation then occurs preventing the formation of compact polycationic structures.

Tetravalent cations are too much acidic so that the nucleation of stable hydroxides $M(OH)_4$ cannot occur. Water molecules formed according to:

$$M(OH)_4 \leftrightarrow MO(OH)_2(OH_2) \leftrightarrow MO_2(OH_2)_2 \longrightarrow MO_2 + 2H_2O \qquad (42)$$

always bear a positive partial charge (Table 4). Therefore oxolation and olation can proceed simultaneously during nucleation and growth leading to amorphous hydrous oxides $MO_2 \cdot nH_2O$ where the number n of water molecules depends on experimental conditions.

Depending on experimental procedure, the precipitation of TiO_2 leads to rutile or anatase phases [91, 92]. The formation of such structures from aqueous molecular precursors could be described as follows (Fig. 15). When deoxolation ($O{=}M{-}OH_2 \rightarrow HO{-}M{-}OH$) does not occur during nucleation olation leads to a linear growth along one of the two equivalent directions in the equatorial plane of $[M_2O_2(OH)_4(OH_2)_4]^0$ dimers. Then, oxolation between the resulting $MO(OH)_2(OH_2)$ linear chains after an internal proton transfer leads to corner-sharing octahedral chains (M_3O bridges) characteristic of the rutile structure. The formation of rutile may then be associated to the metastability of apical M=O bonds within monomers or dimers. Now, if deoxolation occurs prior to olation, condensation can proceed along apical directions leading to skewed chains typical of the anatase structure. Such a deoxolation process seems to occur preferentially with titanium ($3d^0$) rather than with other transition metals (d^n with $n > 0$) for which the rutile structure appear to be favored.

Oxides of larger cations, such as Zr^{4+} or Hf^{4+} exhibit a distorted fluorite structure (CaF_2). It has been shown [88, 93] that such structures can be built from neutral $[M_4(OH)_{16}(OH_2)_8]^0$ tetramers. Upon neutralization condensation via olation rapidly occurs between such tetramers leading to an amorphous hydrous oxide (Fig. 16). Crystallization occurs upon heating leading to the monoclinic (stable), tetragonal or cubic (metastable) forms.

6.4 Condensation of Mixed Valence Systems

Electron exchange reactions in aqueous solutions may have strong effects upon the crystallization of solid phases. Typical examples are provided by the Fe(II)/Fe(III) and the Mn(III)/Mn(IV) couples. It may then be shown that the above simple structural schemes can overlap in a quite complex way to produce new original structures.

Table 4. Partial charges beared by water molecules and metal atoms within an hydroxide $M(OH)_4$ when M = Ti, Zr, and Hf

M	Ti	Zr	Hf
$\delta(OH_2)$	+ 0.01	+ 0.01	+ 0.03
$\delta(M)$	+ 0.76	+ 0.78	+ 0.73

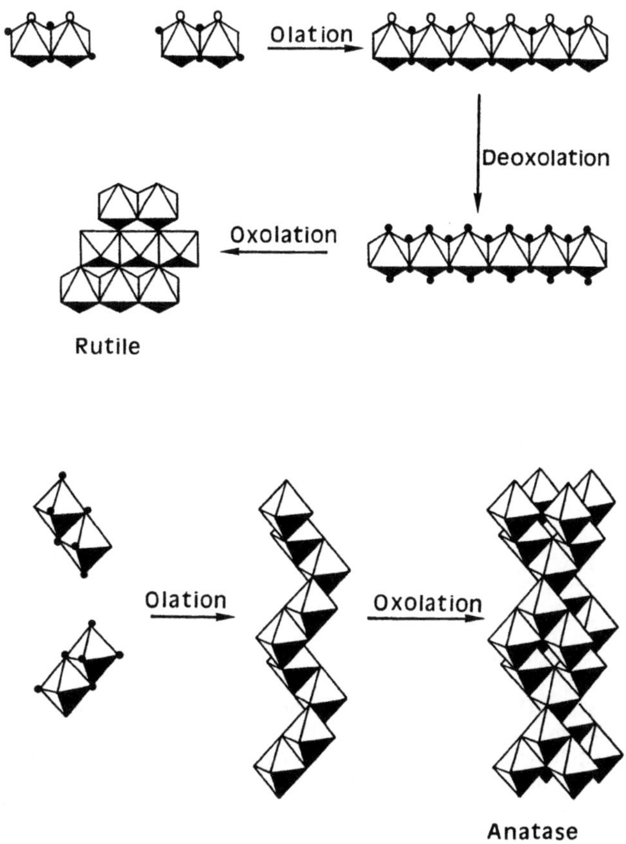

Fig. 15. Possible condensation pathway for the nucleation and growth of the rutile and anatase structures (TiO_2) from aqueous solutions. The key-point is the occurrence of the desoxolation reaction. If it occurs before nucleation, condensation may be oriented towards *cis*-skewed chains characteristic of the anatase structure. If it occurs after nucleation, the *trans*-linear chain characteristic of the rutile structure is rather formed

— Small amounts of Fe(II) within Fe(III) solutions (Fe(II)/Fe(III) > 0.1) prevent the formation of ferrihydrite. With stoichiometric Fe^{II}/Fe^{III} ratio equal to 0.5, magnetite (Fe_3O_4) is then obtained at high pH which may readily oxidize into maghemite (γFe_2O_3) at low pH [94]. Nucleation and growth of magnetite particles could be explained by the following structural scheme (Fig. 17):

i) Association at right angle through oxolation of two mixed Fe(II)/Fe(III) dimers $[Fe_2(OH)_5(OH_2)_5]^0$. This leads to a compact $[Fe_4O_4(OH)_2(OH_2)_{10}]^0$ critical nucleus having optimized magnetic couplings (cf Sect. 6.2). After olation and oxolation between such tetramers, electronic delocalization can occur inside the resulting chains and planes.

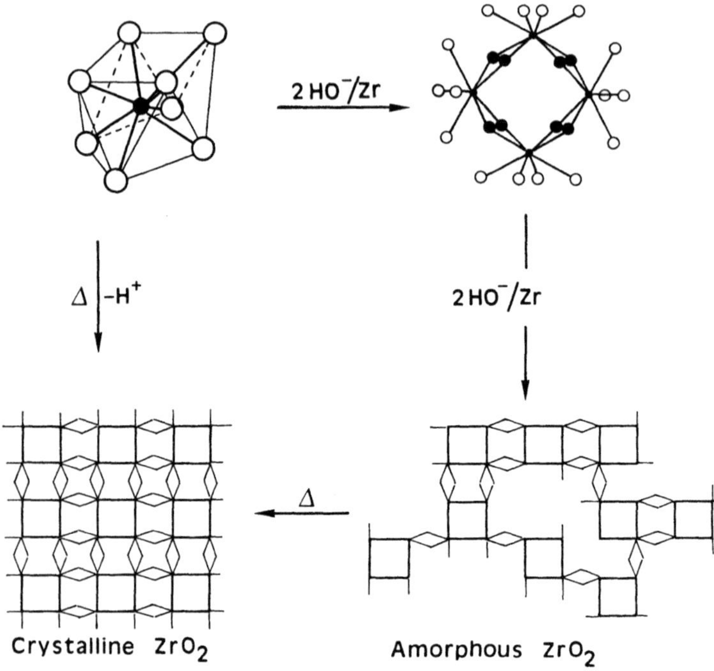

Fig. 16. The aqueous structural chemistry of Zr(IV), based on a eightfold coordination (square antiprism) according to Refs. [88] and [93]. Condensation through olation of four h = 2 monomers $[Zr(OH)_2(OH_2)_6]^{2+}$ leads to a characteristic tetrameric structure. Subsequent olation reactions induce the reticulation of such tetramers (gels) while heating promotes crystallization into three mains polymorphs (monoclinic, tetragonal and cubic) derived from the fluorite structure (CaF_2)

ii) Such chains and planes may then be further associated via monomeric $[Fe(OH)_3(OH_2)_3]^0$ precursors. These octahedral monomers become trapped between strongly magnetic planes and will then undergo strong steric interactions. Having no CFSE they can change their coordination from six to four in order to accommodate the tetrahedral sites characteristic of the cubic inverse spinel structure.

An indirect evidence favoring such a structural scheme is provided by the fact that this particular structure does not form at room temperature when Fe^{2+} ions are replaced by Co^{2+} ions. In such a case electron delocalization cannot occur at room temperature, but at a higher one. Accordingly, at $T > 100\,°C$ the spinel Fe_2CoO_4 is readily obtained.

— Manganese oxides are readily obtained through the reduction of permanganate MnO_4^- or the oxidation of manganous $[Mn(OH_2)_6]^{2+}$ ions [13, 95, 96]. When no Mn^{3+} or Mn^{2+} ions are present in the solution, the βMnO_2 polymorph having the rutile structure is obtained (Fig. 18). When only Mn^{3+} ions are present, the αMnOOH polymorph (groutite) having the goethite structure is obtained. This structure is conserved upon oxidation leading to the γMnO_2 modification. When Mn(II), Mn(III), and Mn(IV) are simultaneously

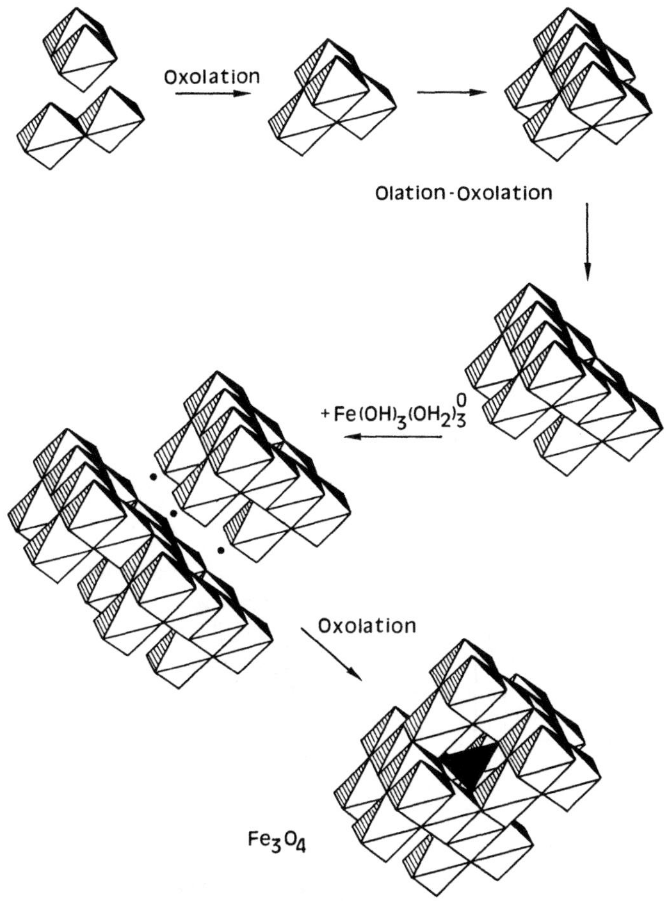

Fig. 17. Possible condensation pathway for the nucleation and growth of the spinel structure (Fe_3O_4) structure from aqueous solutions. The key point is the condensation of two mixed valence edge-sharing dimers to form a compact tetramer quite similar to that described in Fig. 12. Association of such tetramers through olation and oxolation then leads to octahedral chains and planes. During growth Fe(III) ions in excess become trapped between such planes adopting a four-fold coordination mainly for steric hindrance reasons (compare Fig. 8)

present various counterions are incorporated within the growing network to insure charge compensation. This favors tunnel structures which may be conserved upon leaching leading to various MnO_2 forms [93, 97]:

αMnO_2 has tunnels going through a framework of corner-sharing octahedral double chains and is isostructural with βFeOOH (akaganeite). This phase is obtained when large ions (Ba^{2+} or Pb^{2+}) are in the solution.

δMnO_2 is made of layers of edge-sharing MnO_6 octahedra (CdI_2 type). It is obtained in presence of Na^+ or K^+ ions.

λMnO_2 has a spinel framework. It is obtained with Li^+ as a counterion.

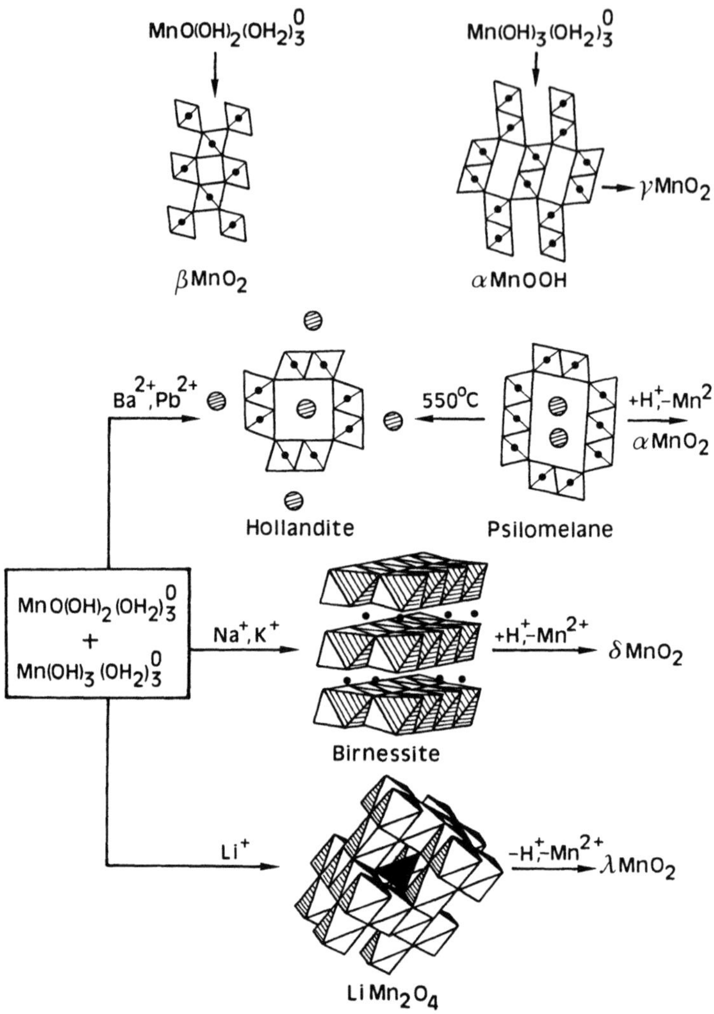

Fig. 18. The structural chemistry of manganese oxides according to Refs. [95] and [97]. Mn(IV) aqueous precursors condense in the same fashion as Ti(IV) (cf Fig. 15) while Mn(III) ones condense like Fe(III) (cf Fig. 11). Owing to the lack of electronic delocalization between Mn(IV) and Mn(III), cations must be incorporated within the network, to insure charge compensation, when both Mn(IV) and Mn(III) precursors are present in aqueous solutions (compare with Fig. 17 where electronic delocalization occur between Fe(II) and Fe(III)). Depending on the counter-cation, specific structures are promoted which after leaching of Mn^{2+} ions under acidic conditions leads to a wide variety of manganese oxides MnO_2

7 Complexation by Anionic Species

Chemical species other than water, protons or hydroxyls can play an important role in the formation of condensed phases from aqueous solutions. Anions are often added in order to control the shape and size of colloidal particles [5]. The crystalline structure and even the chemical composition of these particles also depend on the presence of anions even in small amounts [5, 98].

Anions such as acetates OAc^- can form complexes with metal cations in aqueous solutions:

$$[M(OH_2)_N]^{z+} + x\,[OAc]^- \leftrightarrow [M(OH_2)_{N-\alpha x}(OAc)_x]^{(z-x)+} + x\,H_2O \quad (43)$$

(α is the coordination mode of the anion)

Acetates only give outer sphere complexes with the solvated cation when $\alpha = 0$, while they give inner sphere complexes when $\alpha > 0$. They behave as mono or polydentate ligands depending on the value of $\alpha \geqslant 1$.

However, other chemical reactions can take place in aqueous solutions. Protonation of acetate ions occurs at low pH.

$$OAc^- + H_3O^+ \leftrightarrow AcOH + H_2O \quad (44)$$

while hydrolysis of the metal cation occurs at high pH.

$$[M(OH_2)_N]^{z+} + h\,H_2O \leftrightarrow [M(OH)_h(OH_2)_{N-h}]^{(z-h)+} + h\,H_3O^+ \quad (45)$$

As a result of these side reactions, acetate complexes remain stable over a rather narrow range of pH only. Apparent equilibrium constants can be deduced from literature for mononuclear complexes [99]. However, very little is known about condensed species. Therefore it would be useful to introduce some criteria in order to be able to predict the complexing ability of anions towards aqueous precursors.

7.1 *Theoretical Analysis*

Let us discuss the case of a monovalent and monodentate ($\alpha = 1$) anion X^-. The first complexation reaction can be written as follows:

$$[M(OH_2)_N]^{z+} + X^- \leftrightarrow [M(OH_2)_{N-1}(X)]^{(z-1)} + H_2O \quad (46)$$

It can be described as the nucleophilic substitution of one water molecule H_2O by X^-. Let us introduce side reactions such as:

$$[M(OH_2)_N]^{z+} + H_2O \leftrightarrow [M(OH)(OH_2)_{N-1}]^{(z-1)+} + H_3O^+ \quad (47)$$

$$H_3O^+ + X^- \leftrightarrow HX + H_2O \quad (48)$$

---

i.e. $$[M(OH_2)_N]^{z+} + X^- \leftrightarrow [M(OH)(OH_2)_{N-1}]^{(z-1)+} + HX \quad (49)$$

Therefore, complexation by the protonated form HX of the anion X^- has also to be taken into account:

$$[M(OH)(OH_2)_{N-1}]^{(z-1)+} + HX \leftrightarrow [M(OH)(OH_2)_{N-1}(HX)]^{(z-1)+} + H_2O \quad (50)$$

This reaction can also be considered as a nucleophilic substitution of one water molecule H_2O by HX. Therefore both species X^- and HX have to compete with water molecules in order to be able to complex metal cations:

$$\begin{array}{ccc} [M(OH_2)_N]^{z+} + X^- & \rightleftharpoons & [M(OH)(OH_2)_{N-1}]^{(z-1)+} + HX \\ +H_2O \upharpoonleft\downharpoonright -H_2O & & +H_2O \upharpoonleft\downharpoonright -H_2O \\ [M(OH_2)_{N-1}(X)]^{(z-1)+} & \rightleftharpoons & [M(OH)(HX)(OH_2)_{N-1}]^{(z-1)+} \end{array} \quad (51)$$

The complexing ability of a given anion X^- toward a metal cation M^{z+} then depends on the nature of charge transfers within the M–X bond i.e. on the electronegativities of the complexed precursor (χ_p), the anion $X^-(\chi_X)$ and its protonated form HX (χ_{HX}). Equation (51) is displaced upwards when the resulting M–X bond is highly ionic that is to say if $\chi_X > \chi_p$ (i.e. $\delta_X < -1$). It is also displaced upwards if the HX form cannot stay in the neighborhood of the hydrolyzed cation that is to say if $\chi_{HX} < \chi_p$ (i.e. $\delta_{HX} > 0$). Two limiting cases can then be defined from this analysis. They correspond respectively to $\delta_X = -1$ and $\delta_{HX} = 0$ within the complex. Equalization of the mean electronegativity χ_p of the metal precursor with the electronegativities χ_X and χ_{HX} of the basic and acid forms of the anion in an aqueous solution leads to an electronegativity range $\chi_X \leq \chi_p \leq \chi_{HX}$ where complexation is expected to occur. Outside this electronegativity range complexed [M–X] species are not stable.

The above analysis can be easily generalized to multivalent species H_nX. Complexation occurs if $\chi_X \leq \chi_p \leq \chi_{HnX}$, while outside this range complexes cannot be stable. Within this complexation range, two domains may be distinguished:

i) $\chi_{Hn-1X} \leq \chi_p \leq \chi_{HnX}$: In this range anions can act as network formers. They are tightly bonded in the coordination sphere of the metal. Upon precipitation of a solid phase, a basic salt structure is bound to result.

ii) $\chi_X \leq \chi_p \leq \chi_{Hn-1X}$: In this range, the anion is still bonded to the metal but its elimination is theoretically possible through ionic dissociation of a partly protonated form $H_qX^{(n-q)-}$ ($0 < q < n$). Therefore, despite complexation of solute precursors, anions can be removed during growth leading to an oxide network free of anions.

This analysis shows that multivalent anions can be used to monitor the size, shape, and morphology of oxide particles [5, 98]. In contrast monovalent anions cannot play such a role. They lead to the nucleation on growth of basic salts as soon as they show some complexing behavior. The following examples show how to handle the basic criteria $\chi_X \leq \chi_p \leq \chi_{HnX}$ in order to draw detailed conclusions concerning the behavior of aqueous precursors towards complexation.

7.2 Complexation at a given Hydrolysis Ratio

Such an analysis may be performed for zirconium precursors $[Zr(OH)_2(OH_2)_6]^{2+}$ corresponding to an hydrolysis ratio h = 2:

$$[Zr(OH)_2(OH_2)_6]^{2+} + m\,X^{n-} \leftrightarrow [Zr(OH)_2X_m(OH_2)_{6-m\alpha}]^{(2-mn)+} + m\alpha H_2O \tag{52}$$

Applying the criteria $\delta_X = -n$ and $\delta_{HnX} = 0$ within the complexed species leads to (see Appendix B for details):

$$\chi_D = \chi\{[Zr(OH)_2(OH_2)_{6-m\alpha}]^{2+}\} \tag{53}$$

$$\chi_H = \chi\{[Zr(OH)_{2+mn}(OH_2)_{6-m(\alpha+n)}]^{(2-mn)+}\} \tag{54}$$

Therefore, all ligands such as $\chi_X \leq \chi_D$ and $\chi_{HnX} \geq \chi_H$ are theoretically able to complex h = 2 aqueous precursors of Zr(IV). With monovalent anions (n = 1 and m = 2) acting as monodentate ligands ($\alpha = 1$) or with bivalent anions (n = 2 and m = 1) acting as bidentate ligands ($\alpha = 2$), Eqs. (53) and (54) lead to $\chi_D = 2.69$ and $\chi_H = 2.49$, respectively. On the other hand, with monovalent anions (n = 1 and m = 2) acting as bidentate ligands ($\alpha = 2$) the same relations lead to $\chi_D = 2.80$ and $\chi_H = 2.49$. Anions may therefore be classified according to the electronegativity of their protonated forms as shown in Fig. 19. From such a diagram, the following conclusions may be drawn [3, 99]:

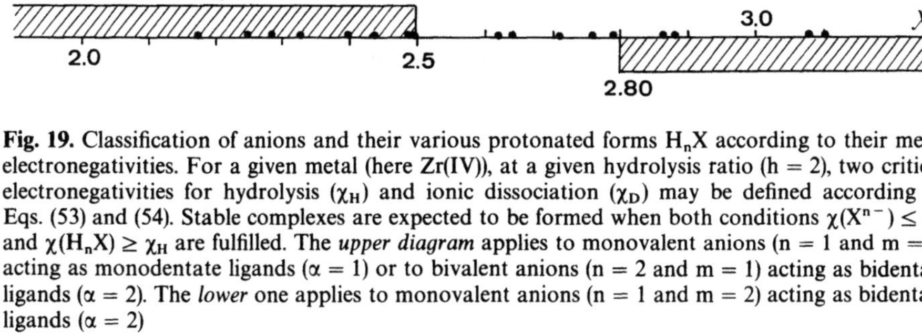

Fig. 19. Classification of anions and their various protonated forms H_nX according to their mean electronegativities. For a given metal (here Zr(IV)), at a given hydrolysis ratio (h = 2), two critical electronegativities for hydrolysis (χ_H) and ionic dissociation (χ_D) may be defined according to Eqs. (53) and (54). Stable complexes are expected to be formed when both conditions $\chi(X^{n-}) \leq \chi_D$ and $\chi(H_nX) \geq \chi_H$ are fulfilled. The *upper diagram* applies to monovalent anions (n = 1 and m = 2) acting as monodentate ligands ($\alpha = 1$) or to bivalent anions (n = 2 and m = 1) acting as bidentate ligands ($\alpha = 2$). The *lower* one applies to monovalent anions (n = 1 and m = 2) acting as bidentate ligands ($\alpha = 2$)

i) Anions such as ClO_4^- and Cl^- cannot complex h = 2 precursors whatever the coordination mode. With ClO_4^- ionic dissociation occurs ($\chi(ClO_4^-) = 2.86 > \chi_D = 2.69$ or 2.80), while with Cl^- dissociation is due to hydrolysis ($\chi(HCl) = 2.44 < \chi_H = 2.49$). Such anions are thus not involved in the formation of condensed species and behave as mere counterions. This may explain why in presence of such anions a cyclic tetramer $[Zr_4(OH)_8(OH_2)_{16}]^{8+}$ is formed via olation reactions. The molecular structure of this oligomer (Fig. 20a) was resolved both in the solid state [100, 101] and in solution [102] by X-ray diffraction.

ii) Anions such as NO_3^- are able to complex h = 2 precursors only if they act as bidentate ligands ($\chi(NO_3^-) = 2.76 < \chi_D = 2.80$). Accordingly, X-ray diffraction performed on the basic salt $ZrO(NO_3)_2 \cdot 5H_2O$, shows that nitrates substitute two water molecules giving rise to chain polymers $[Zr(OH)_2(OH_2)_2(NO_3)]_n^{n+}$ with terminal $(NO_3)^-$ groups (Fig. 20b) [102]. Moreover, ^{17}O-NMR measurements recently confirmed the complexing behavior of nitrate ions towards Zr(IV) in aqueous solutions [103].

iii) Anions such as CH_3COO^-, HSO_4^-, $H_2PO_4^-$, and $HC_2O_4^-$ are able to complex h = 2 precursors whatever their coordination mode. X-ray diffraction performed on $Zr(OH)_2SO_4$ single crystals show that sulfate ions behave as bridging ligands. They are able to substitute all water molecules giving rise to a tridimensional network in which sulfate ions link $[ZrO_4(OH)_2]_n$ infinite

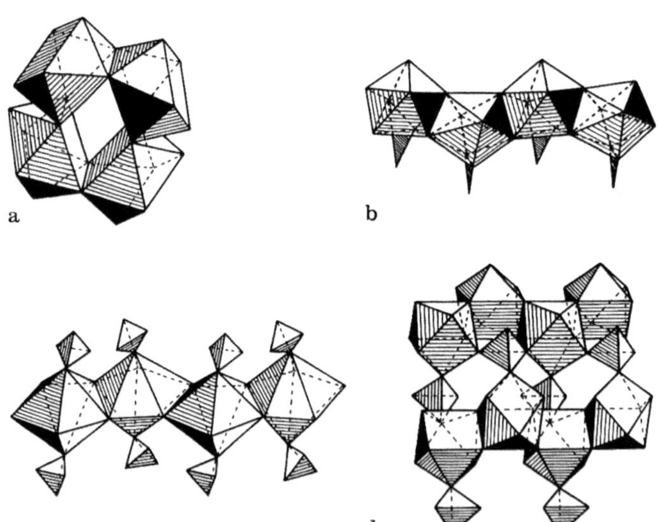

a b c d

Fig. 20a–d. The structural inorganic chemistry of h = 2-based basic salts of Zr(IV) according to Refs. [3] and [103] as a function of the counter-anion X^{n-} involved in solid phase formation. (*a*) The square antiprismatic cyclic tetramer obtained when $X = Cl^-$ or Br^-. (*b*) The dodecahedral chain polymer obtained when $X = NO_3^-$. (*c*) The hydrated pentagonal bipyramidal chain polymer obtained when $X = SO_4^{2-}$. (*d*) The anhydrous square antiprismatic chain polymer obtained when $X = SO_4^{2-}$

chains [102, 104] (Fig. 20d). In the hydrated form $Zr(OH)_2SO_4 \cdot H_2O$, sulfate ions also behave as bridging ligands linking $[ZrO_2(OH)_2(OH_2)_2]_n$ infinite chains [105] (Fig. 20c). Zirconyl acetate, oxalate, or phosphate are generally obtained by mixing aqueous solutions of $ZrOCl_2$ with acetic [106, 107], oxalic [108] or orthophosphoric [109] acids. This means that in all cases complexation between the corresponding anion and Zr(IV) occurs. Unfortunately, accurate structural data concerning such compounds are still lacking.

As shown in Appendix B, Eqs. (53) and (54) can be easily extended to any cation and any hydrolysis ratio h:

$$[M(OH)_h(OH_2)_{N-h}]^{(z-h)+} + m\,X^{n-} \leftrightarrow [M(OH)_hX_m(OH_2)_{N-h-m\alpha}]^{(z-h-mn)+} + m\alpha H_2O \quad (55)$$

$$\delta_X = -n \Rightarrow \chi_D = \chi\{[M(OH)_h(OH_2)_{N-h-m\alpha}]^{(z-h)+}\} \quad (56)$$

$$\delta_{HnX} = 0 \Rightarrow \chi_H = \chi\{[M(OH)_{h+mn}(OH_2)_{N-h-m(\alpha+n)}]^{(z-h-mn)+}\} \quad (57)$$

The complexing ability can thus be rationalized using seven readily available chemical parameters:

— oxidation state z of the metal,
— hydrolysis ratio h of the metal,
— Pauling electronegativity χ^0_M of the metal,
— coordination number N of the metal,
— valency n of the anion,
— stoichiometry m of the complex,
— coordination mode α of the anion.

7.3 The Electronegativity–pH Diagram

Section 7.2 shows that it is possible to predict the complexing ability of an anion at a given hydrolysis ratio h. In fact, we know from Sect. 3 that h increases with pH according to Eq. (15). This means that the electronegativity range where complexation is expected to occur must depend on the pH of the aqueous solution. In order to compute such pH ranges for complexation we just have to equalize χ_p, the mean electronegativity of the complexed precursor, with $\chi_q = \chi[H_qX^{(n-q)-}]$, mean electronegativity of the *q*-protonated form $(0 \le q \le n)$ of the X^{n-} anion. Applying charge conservation then allows to define critical hydrolysis ratio h^*_q which can be converted into critical pH^*_q through Eq. (15) [110]. Such a calculation is presented in Appendix B. It provides a quantitative guide to predict whether complexation occurs as a function of pH:

$$2.732 - 0.035\,pH^*_q = \frac{\chi_q(\Delta + 5.732\alpha + 2.064q) - 4.071(3.507\alpha + 2.064q)}{\Delta + (1.408\alpha + 0.507q)\chi_q - (3.507\alpha + 2.064q)} \quad (58)$$

where $$\Delta = z - 2.225N - (4.071 - \chi_M^0)/(1.36\sqrt{\chi_M^0}) \tag{59}$$

Let us consider the case of Al^{3+} cations for which we compute with z = 3, N = 6 and $\chi_M^0 = 1.47$: $\Delta = -11.923$. Inner-sphere complexation by bidentate anions ($\alpha = 2$) leads to:

$$(pH_q^*)_{in} = 28.6\left\{2.732 - \frac{(2.064q - 0.459)\chi_q - 4.071(7.014 + 2.064q)}{(2.816 + 0.507q)\chi_q - 18.937 - 2.064q}\right\} \tag{60}$$

while outer-sphere complexation ($\alpha = 0$) leads to:

$$(pH_q^*)_{out} = 28.6\left\{2.732 - \frac{(2.064q - 11.923)\chi_q - 8.403q}{0.507q\chi_q - 11.923 - 2.064q}\right\} \tag{61}$$

The results of Eq. (60) are displayed in Fig. 21 where χ_q is plotted as a function of pH_q^* for *q* values ranging from 0 up to 4. This "electronegativity–pH" diagram defines three domains:

i) Ionic dissociation occurs if $pH > pH_0^*$. Anions cannot stay within the coordination sphere of the metal as its aqueous precursors are not electronegative enough to polarize their electronic clouds.

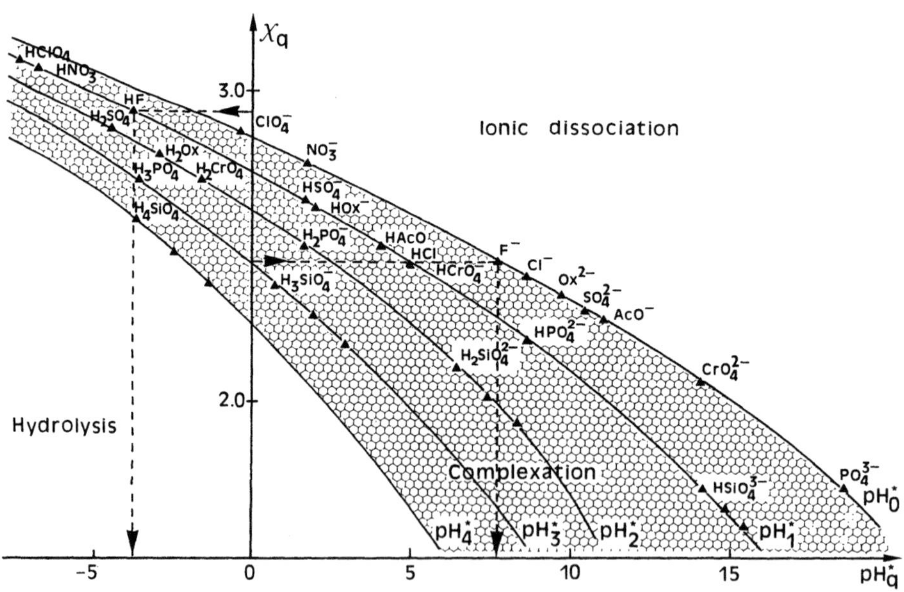

Fig. 21. The electronegativity-pH diagram for complexation. For a given metal cation (here Al^{3+}), n + 1 curves $\chi_q = f(pH_q^*)$ with $0 \leq q \leq n$ can be drawn according to Eqs. (58) and (59). Three domains are then defined corresponding to hydrolysis (*left*), ionic dissociation (*right*) or stability (*tint middle*) of the complexes. Knowing $\chi(H_nX)$ and $\chi(X^{n-})$ (HF and F^- ions in the diagram), a typical pH-range for complexation is easily obtained (see text for details)

ii) Hydrolysis occurs if $pH < pH_n^*$ as the aqueous metal precursors are now too electronegative. Protons are therefore preferentially fixed on the anion rather than on hydroxo groups, leading to the hydrolysis of these latter.

iii) Complexation is expected if $pH_n^* < pH < pH_0^*$. This range depends on the valency n of the anion. It becomes larger as n increases. Within such a range an optimum pH for complexation pH_m can be defined according to:

$$pH_m = (pH_n^* + pH_0^*)/2 \tag{62}$$

In the following we consider some typical anions covering a wide range of electronegativities and valencies in order to show the usefulness of such an "electronegativity–pH" diagram. For example, knowing the mean electronegativity of aqueous fluoride ions (χ_0) and aqueous hydrogen fluoride HF (χ_1) two horizontal lines can be drawn which intercept respectively the $q = 0$ and $q = 1$ curves. Drawing now vertical lines from these points leads to two critical pH_0^* and pH_1^* defining the pH complexation range for complexation of the metal by fluoride anion. Alternatively, one may use analytical Eqs. (60) and (61) to get the same values.

— ClO_4^-: According to the diagram inner-sphere complexation would be possible only if $-7.4 \leq pH \leq -0.3\,(pH_m - 3.9)$. For outer-sphere complexation this range becomes $-11.6 \leq pH \leq -3.6\,(pH_m - 7.6)$ a still more negative one. It follows that under standard conditions, perchlorate ions mainly behave as counterions. This is in agreement with the well-known behavior of such anions towards metallic cations in aqueous solutions.

— NO_3^-: Inner- and outer-sphere complexation leads to $-6.8 \leq pH \leq 1.7\,(pH_m - 2.6)$ and $-11.0 \leq pH \leq -0.9\,(pH_m - 6.0)$ respectively. Again under standard conditions and for outer-sphere complexation, nitrates are then expected to behave also as counterions. Accordingly, strong specific association of NO_3^- ions with Al^{3+} cations was evidenced in concentrated solutions only of $Al(NO_3)_3$ by infrared and Raman spectroscopies [111]. Moreover, ^1H NMR shows that complexation occurs mainly via the second coordination sphere of the Al^{3+} cation [111].

— Cl_{aq}^-: In order to compute the mean electronegativity of chloride ions in aqueous solutions, we have to include its first coordination sphere defining a solvated complex $[Cl(H_2O)_N]^-$. Recent results [112] show that $N = 6$ leading to $\chi_0 = \chi(Cl_{aq}^-) = 2.395$. Upon protonation this ion lead to the highly acidic HCl molecule which can be crystallized as an ion pair H_3O^+, Cl^- leading to $\chi_1 = \chi(HCl_{aq}) = 2.470$. These two values leads to $4.4 \leq pH \leq 8.6\,(pH_m = 6.5)$ for inner-sphere complexation and $4.6 \leq pH \leq 9.6\,(pH_m = 7.1)$ for outer-sphere complexation. This means that complexation by Cl^- anions occurs only under quasi-neutral conditions. Accordingly, anion exchange studies have shown that Cl^- does not complex Al^{3+} cations up to 12 M HCl [113]. However, upon precipitation of aluminum hydroxides from aluminum chloride solutions, it is found that chloride ions are retained within the precipitates by rather weak interactions [114]. This was not the case using aluminum perchlorate or aluminum nitrate in agreement with our analysis.

— CH_3COO^-: The two ranges are $4.0 \leq pH \leq 11.0$ (pH_m 7.5) and $4.0 \leq pH \leq 13.9$ (pH_m 9.0) for inner- and outer-sphere complexation, respectively. Complexation of aluminum by acetate anions is thus expected under weakly acidic or basic conditions. ^1H, ^{13}C, and ^{27}Al NMR confirm such a prediction. When Al^{3+} cations are in excess (acid conditions), protonation of the acetate anion occurs [115]. This prevents complexation through hydrolysis and induces dimerization of the resulting $h = 1$ precursors of Al^{3+}. When enough acetate ions are added ($CH_3COO^-/Al^{3+} \geq 1.5$), the pH increases and an inner-sphere complex is formed [115]. Such a complex can act as a precursor for precipitation of a basic salt having the $Al_2(OH)_2(CH_3COO)_4$ stoichiometry. Multinuclear NMR data are then consistent with the presence of dimeric $(H_2O)_3Al(OH)Al(OH_2)_3$ units with bridging acetate groups [116].

— F^-: Owing to the smaller size of F^- relative to Cl^-, the coordination number of fluoride ions in aqueous solution will be taken as $N = 4$ leading to $\chi_0 = \chi(F^-_{aq}) = 2.422$. Moreover, as HF is a weak acid we do not include any water molecule to compute the mean electronegativity of HF in aqueous solution leading to $\chi_1 = \chi(HF_{aq}) = 2.934$. This leads to $-3.8 \leq pH \leq 8.4$ (pH_m 2.2) and $-7.3 \leq pH \leq 8.9$ (pH_m 0.8) for inner and outer-sphere complexation respectively. These large values suggest that fluoride ions should interact strongly with Al^{3+} cations whatever the coordination mode. Accordingly, it was shown that fluoride ions form rather strong complexes $AlF_n^{(3-n)+}$ ($n = 1$–6) [117] which can be characterized by ^{27}Al and ^{19}F NMR [118].

— $C_2O_4^{2-}$: Strong complexation is also expected for oxalate ions as $-3.0 \leq pH \leq 9.7$ (pH_m 3.4) for inner-sphere complexation and $-5.3 \leq pH \leq 11.5$ (pH_m 3.1) for outer-sphere complexation. Potentiometry and ^{27}Al NMR confirm such a prediction as a series $[Al(Ox)_n]^{(3-2n)+}$ ($n = 1$–3) can be characterized for $7.2 \leq pH \leq 0.2$ [119]. NMR data also indicate the formation of an $[Al(HOx)]^{2+}$ complex in strongly acidic solutions. Polynuclear mixed hydroxo complexes $Al_3(OH)_3(Ox)_3$ and $[Al_2(OH)_2(Ox)_4]^{4-}$ were also reported [119].

— SO_4^{2-}: This anion is expected to show a behavior quite similar to that of oxalate ion: $-4.5 \leq pH \leq 10.4$ (pH_m 3.0) and $-7.3 \leq pH \leq 12.7$ (pH_m 2.7) for inner- and outer-sphere complexation, respectively. Such sulfate complexes have been known for quite a long time [117, 120] and were characterized by ^{27}Al-NMR spectroscopy [121]. Recent measurements [122] show that both inner- and outer-sphere complexation occur leading to a quite complex behavior in aqueous solutions.

— PO_4^{3-}: Complexation over the whole range of pH is expected for phosphate ions as $-3.6 \leq pH \leq 18.6$ (pH_m 7.5) for inner-sphere complexation. This means that all forms of phosphoric acid could act as ligands towards Al^{3+}. Accordingly, PO_4^{3-} ions react with Al^{3+} ions to form $AlPO_4$ precipitates [123]. Soluble complexes are formed with the HPO_4^{2-} and $H_2PO_4^-$ forms [117] which can be characterized by ^{27}Al NMR [118]. This technique also evidences the formation of a complex between the neutral phosphoric acid molecule and the Al^{3+} cation under highly acidic conditions [118].

It has to be pointed out that as n increases, complexation occurs over a wider range of pH. A direct consequence of this variation is that the complexing power of anions could be directly related to their valency n. Consequently, if complexation by phosphate ions (n = 3) occurs over the whole range of pH as shown above, a stronger complexation is expected for silicate ions (n = 4). Accordingly, for such species it is found from our diagram that $-3.7 \leq pH \leq 25$ (pH_m 10.7), meaning that strong chemical interactions are always possible between inorganic precursors of aluminum and silicon. This could explain why silico–aluminates are so widespread within the earth crust and in natural environment. Moreover, owing to their smaller z, aqueous precursors of silicon are able to undergo polycondensation reactions in contrast to phosphorus which remains monomeric. Consequently, alumino-silicates are expected to display a wide range of possible structures, ranging from zeolithes to clays and clays minerals, based on condensed ions. Such is not the case of alumino-phosphates which are mostly based on monomeric orthophosphates ions.

8 Conclusions

This paper was concerned with the aqueous chemistry of metal cations. Chemical reactions such as hydrolysis, condensation and complexation have been discussed in the frame of a simple theoretical model based on the old concept of electronegativity. It was shown that aqueous chemistry could be described in a three-dimensional "chemical space". The first dimension is provided by the oxidation state z of the cation, the second one by the pH of the solution, and the third one by the Pauling electronegativity of the chemical species (metal or ligands). Within such a triple basis (z, pH, χ), each chemical process could be described by a specific diagram:

i) By plotting z as a function of pH, a "charge–pH" diagram is obtained accounting for the hydrolysis of cations.

ii) By plotting z against χ^0_M, a "charge–electronegativity" diagram can be built accounting for condensation of cations.

iii) By plotting χ_q as a function of pH, an "electronegativity–pH" diagram is found accounting for complexation of cations.

When these three diagrams are used together, they provide a very useful and powerful tool to forecast the chemical behavior of an element in aqueous solution. Such a rationalization is of the utmost importance for the sol–gel chemist who has to deal with so many possible reactions and precursors by just dissolving a metallic salt in an aqueous solution at a given pH. The situation becomes quite hopeless when other salts are introduced. The (z, pH, χ) rationalization has the main advantage to provide some choice among so many reaction paths. Having determined the most reasonable chemical species which could interact in the solution, detailed structural pathways could be put forward to account for observable experimental facts. Such structural schemes are only indicative and tentative but afford great help to improve the knowledge of the system.

Such an analysis within the (z, pH, χ) frame is unfortunately at the present time only semi-quantitative. Other important chemical parameters such as temperature, pressure, ionic strength, or metal concentration are not yet explicitly involved in the model. Similarly, redox processes are not explicitly accounted for, as the oxidation state z of the cation must be assumed in order to apply the model. Consequently, a great amount of both theoretical and experimental work remains to be done in order to achieve a more quantitative model. Theoretical improvements must focus on the electronegativity concept itself. By extending present scales of electronegativity and hardness as a function of temperature and pressure, more detailed conclusions could be drawn. Experimental improvements have on the other hand been already brought about through the use of multinuclear NMR techniques. The amount of collected data remains however quite small with respect to the potentiality and power of the method. It is hoped that in the future this field of inorganic chemistry will grow faster and faster. This renewed interest for these cheap precursors of high-tech materials could provide a breakthrough for industrial applications, providing that the complex underlying solution chemistry is quantitatively understood and modelized.

9 Appendix A: The Partial Charge Model

The Partial Charge Model (PCM) is based on simple concepts such as electronegativity χ, hardness η, softness σ and partial charges δ. The starting point is the quantum-mechanical definition of electronegativity [8]:

$$\begin{aligned}\chi &= -\mu = -(\partial E/\partial N)_v = -\delta E_v[\rho(\mathbf{r})]/\delta\rho(\mathbf{r}) \\ &= -\{v(\mathbf{r}) + \delta F[\rho(\mathbf{r})]/\delta\rho(\mathbf{r})\} \qquad (63)\end{aligned}$$

In these relationships, μ is the electronic chemical potential, E the total molecular ground state energy, N the total number of electrons, $\rho(\mathbf{r})$ and $v(\mathbf{r})$ the electronic density and external potential at point $\mathbf{r}$, and F a quantum potential including electronic repulsions as well as exchange between indiscernable electrons. Equation (63) also tells us that at any space-point $\mathbf{r}$ the electronegativity of the system can always be partitioned between a purely quantum contribution $\delta F[\rho(\mathbf{r})]/\delta\rho(\mathbf{r})$ and a purely electrostatic one $v(\mathbf{r})$ in such a way that the sum of both terms is constant and equal to a real number χ over the whole space (Hellman-Feynman theorem). This is just the electronegativity equalization principle of R. T. Sanderson [6] which defines the mean electronegativity χ of the system. The chemical meaning of such a decomposition can be understood as follows:

i) When the external potential is zero everywhere, that is to say when each atom within the system has the same number of electrons than the free atom, integration over an atomic volume i shows that:

$$\chi_i^0 = -\langle \delta F[\rho(\mathbf{r})]/\delta\rho(\mathbf{r})\rangle_i \tag{64}$$

This is just the definition of the Pauling electronegativity of an atom within a molecule.

ii) According to Eq. (64) the χ_i^0 value depends on the volume over which integration is performed. This means that electronegativity varies from point to point within the system leading to a non-equilibrium situation. For example, if $\chi_A^0 < \chi_B^0$, electrons tend to move from atom A to atom B as electronegativity measures the power of atoms to attract electrons. Such charge transfers lead to a positive partial charge on A and a negative one on B, creating an external potential v(**r**). At equilibrium electronegativities become equalized according to Eq. (63), and the resulting potential may be approximated by integration over the various atomic volumes in chemical interaction [124]:

$$\langle v(\mathbf{r})\rangle_i \approx \delta_i/r_i + \Sigma_j\,\delta_j/R_{ij} \qquad i = 1, \ldots, n \tag{65}$$

In such a relation the first term arises from the presence around atom i of a partial charge δ_i distributed over a sphere whose radius is approximated by the covalent radius r_i of atom i. This covalent radius may be related to the hardness of atom i through the relation $\eta_i \approx 1/r_i$ [125]. The summation is the perturbative electrostatic potential, coming from the presence at distances R_{ij} of $n - 1$ other atoms j carrying a partial charge δ_j. When Eqs. (64) and (65) are brought together, they define the electronegativity χ_i of atom i within the system which must all equalize to the same mean value χ:

$$\chi = \chi_i = \chi_i^0 + \eta_i\,\delta_i + \Sigma_j\,\delta_j/R_{ij} \qquad i = 1, \ldots, n \tag{66}$$

Owing to electrical charge conservation, partial charge distributions are constrained through the relation:

$$\Sigma_i\,\delta_i = z \tag{67}$$

if z is the total electrical charge of the system. Equations (66) and (67) allows to compute structure-dependent partial charge distributions but requires the resolution of a linear set of equations. This could be avoided by making the drastic approximation $\Sigma_j\,\delta_j/R_{ij} \ll \chi_i^0$. This means that the electrostatic perturbation is supposed to be small compared to the covalent perturbation involving atom i. Making such an approximation leads to structure independent partial charges such as those used in this work. Moreover, if the Allred-Rochow scale such as those given in Table 5 is used to measure χ_i^0 values, Eqs. (66) and (67) with $\eta_i = k\sqrt{\chi_i^0}$ leads to the basic Eqs. (4) and (5).

The value of the constant k to be used with Table 5 may be found by considering HX molecules with X = F, Cl, Br and I. Knowing their dipolar

Table 5. Typical Allred–Rochow electronegativity scale used in this work compiled from Refs. [126–128]

H 2.10																	**He** 3.20
Li 0.97	**Be** 1.57											**B** 2.02	**C** 2.50	**N** 3.07	**O** 3.50	**F** 4.10	**Ne** 5.10
Na 1.01	**Mg** 1.29											**Al** 1.47	**Si** 1.74	**P** 2.11	**S** 2.48	**Cl** 2.83	**Ar** 3.50
K 0.91	**Ca** 1.04	**Sc** 1.23	**Ti** 1.32	**V** 1.56	**Cr** 1.59	**Mn** 1.63	**Fe** 1.72	**Co** 1.75	**Ni** 1.80	**Cu** 1.75	**Zn** 1.66	**Ga** 1.82	**Ge** 2.00	**As** 2.20	**Se** 2.50	**Br** 2.69	**Kr** 3.10
Rb 0.89	**Sr** 0.99	**Y** 1.19	**Zr** 1.29	**Nb** 1.45	**Mo** 1.56	**Tc** 1.67	**Ru** 1.78	**Rh** 1.84	**Pd** 1.85	**Ag** 1.68	**Cd** 1.60	**In** 1.49	**Sn** 1.89	**Sb** 1.98	**Te** 2.15	**I** 2.33	**Xe** 2.60
Cs 0.87	**Ba** 0.89	*	**Hf** 1.36	**Ta** 1.50	**W** 1.59	**Re** 1.88	**Os** 1.99	**Ir** 2.05	**Pt** 2.00	**Au** 2.02	**Hg** 1.80	**Tl** 1.60	**Pb** 1.92	**Bi** 2.03	**Po** 2.12	**At** 2.28	**Rn** 2.30
Fr 0.86	**Ra** 0.95	**															

*	**La** 1.18	**Ce** 1.17	**Pr** 1.18	**Nd** 1.19	**Pm** 1.20	**Sm** 1.20	**Eu** 1.13	**Gd** 1.27	**Tb** 1.24	**Dy** 1.26	**Ho** 1.28	**Er** 1.30	**Tm** 1.30	**Yb** 1.24	**Lu** 1.36
**	**Ac** 1.12	**Th** 1.24	**Pa** 1.22	**U** 1.24	**Np** 1.22	**Pu** 1.24	**Am** 1.25	**Cm** 1.20	**Bk** 1.20	**Cf** 1.20	**Es** 1.20	**Fm** 1.20	**Md** 1.20	**No** 1.20	**Lw** 1.20

moments μ (1.83, 1.08, 0.82 and 0.45 D respectively) together with the corresponding interatomic distances d (0.0917, 0.1275, 0.141 and 0.16 nm) allows to compute associated δ_X values through the classic relation $\mu(D) = 48d\,(nm)\delta$. On the other hand, for neutral diatomic molecules ($z = 0$), (4) leads to:

$$\chi = (\sqrt{\chi_H^0})(\sqrt{\chi_X^0}) \quad \text{and} \quad \delta_X = (\sqrt{\chi_H^0} - \sqrt{\chi_X^0})/k \tag{68}$$

Consequently, by making a linear least square analysis $\sqrt{\chi_X^0} = f(\delta_X)$, a straight line of slope k and ordinate $\sqrt{\chi_H^0}$ must be obtained. Using Table 5 and known δ_X values (-0.42, -0.18, -0.12, and -0.06, respectively) leads to $k = 1.36$ and $\chi_H^0 = 2.1$ with a regression coefficient $r = 0.997$.

10 Appendix B: Criteria for Hydrolysis, Condensation, and Complexation Reactions

10.1 Hydrolysis

Let us consider a solvated cation $[M(OH_2)_N]^{z+}$ which can exchange protons with the aqueous solution according to:

$$[M(OH_2)_N]^{z+} \leftrightarrow [MO_NH_{2N-h}]^{(z-h)+} + h\,H^+_{aq} \tag{69}$$

By writing charge conservation within the hydrolyzed precursor it comes:

$$\delta(\chi,M) + N\delta(\chi,O) + (2N-h)\,\delta(\chi,H) = z - h \tag{70}$$

$$\text{with} \quad \delta(\chi,X) = (\chi - \chi_X^0)/(1.36\sqrt{\chi_X^0}) \tag{71}$$

Summation of partial charges according to:

$$\delta[\chi,M(OH_2)_N] = \delta(\chi,M) + N\delta(\chi,O) + 2N\,\delta(\chi,H) \tag{72}$$

leads to:

$$h = \{z - \delta[\chi,M(OH_2)_N]\}/[1 - \delta(\chi,H)] = \Delta(\chi,z,N,M)/\Delta(\chi,1,0,H) \tag{73}$$

The quantity $\Delta(\chi,z,N,M) = z - \delta[\chi,M(OH_2)_N]$ is introduced in Eq. 32 (73). It characterizes an element M in a medium of electronegativity χ. If χ is given by Eq. (14), Eq. (73) may be used to compute h as a function of the pH as described in Sect. 3.

10.2 Condensation

Let us consider a neutral (i.e. $h = z$) inorganic precursor $[MO_NH_{2N-z}]^0$ (or $[MO_{z/2}(OH_2)_{N-z/2}]^0$) of mean electronegativity χ_0. In order to be able to condense, this precursor must fulfill at least three requirements:

i) The OH group must be negative at every stage of condensation to insure nucleophilic attack.
ii) Acid ionization ($MO–H \rightarrow MO^- + H^+$) must not occur.
iii) Basic ionization ($M–OH \rightarrow M^+ + OH^-$) must also not occur.

When these three criteria are met, condensation leads to hydroxides (olation) or oxides (oxolation) depending on the partial charges beared by water molecules (cf Sect. 6.1).

The first requirement is always met if $\chi(MO_{z/2}) < \chi(OH)$ because in the reverse case, OH groups would become positively charged before elimination of the $(N - z/2)$ water molecules. The condition $\chi(MO_{z/2}) = \chi(OH)$ then defines a criterion ruling the sign of the OH group. Applying charge conservation to the oxide phase $MO_{z/2}$ leads to:

$$\delta(\chi_{OH}, M) = (-z/2)\,\delta(\chi_{OH}, O) \quad \text{or} \quad 2.711 - \chi_M = 2(0.105)\sqrt{\chi_M^0} \tag{74}$$

as $\delta(2.711, O) = -0.31$. Solving this second-order equation for χ_M^0 defines the critical electronegativity χ_{OH}^* which varies with z as:

$$\sqrt{\chi_{OH}^*} = -0.105z + \{(0.105z)^2 + 2.711\}^{1/2} \tag{75}$$

The second requirement needs to consider the transfer of a proton from the precursor towards the aqueous solution. A criterion for such an acid ionization may be found by considering one proton shared between two species: $[MO_NH_{2N-z-1}]^- \ldots\ldots H^+ \ldots\ldots\ldots [H_2O]_n$. If proton jumps to the left the $[MO_NH_{2N-z}]^0$ precursor is formed, while if it jumps to the right a solvated proton $[H(H_2O)_n]^+$ is formed. A very simple means to know which way the proton goes is to look at partial charges on oxygen atoms within the species of electronegativity $\chi^- = \chi([MO_NH_{2N-z-1}]^-)$ and $\chi_S = \chi(H_2O) = 2.491$. If $\delta(\chi^-, O) > \delta(\chi_S, O)$ (i.e. if $\chi^- > \chi_S$) protons jump towards water molecules whereas in the reverse case they jump the other way. The criterion $\chi_S = \chi^-$ thus defines a critical point for acid ionization of the MO–H bond in aqueous solution.

For the third requirement, the transfer of an hydroxide ion towards the aqueous solution has to be considered giving rise to the following situation: $[MO_{N-1}H_{2N-z-1}]^+ \ldots\ldots OH^- \ldots\ldots [H_2O]_n$. We look now at the partial charges upon hydrogen atoms within the species of electronegativity $\chi^+ = \chi([MO_{N-1}H_{2N-z-1}]^+)$ and $\chi_S = \chi(H_2O) = 2.491$. As soon as $\delta(\chi^+, H) < \delta(\chi_S, H)$ (i.e. when $\chi^+ < \chi_S$), hydroxides are better solvated by the protons of water molecules while they prefer those of the solvated cation if $\chi^+ > \chi_S$. This leads to the criterion $\chi_S = \chi^+$ to characterize basic ionization in aqueous solution.

Finally, the last two criteria may be joined to a third one $\chi_S = \chi^0$ ruling the stability of ol bridges towards oxolation [3]. Applying charge conservation within the $[MO_NH_{2N-z-1}]^-$, $[MO_{N-1}H_{2N-z-1}]^+$, and $[MO_NH_{2N-z}]^0$ forms leads to the following equation (as $\delta(2.491, H) = +0.20$):

$$\delta(\chi_S,M) = (z + \omega)\delta(\chi_S H) - \omega$$

$$\text{or } 2.491 - \chi_M^0 = 2(0.136)(z - 4\omega)\sqrt{\chi_M^0} \quad (76)$$

with $\omega = +1$ or -1 for acid or base ionization, and $\omega = 0$ for olation/oxolation. Solving this second-order equation for χ_M^0 allows to define three critical electronegativities χ_ω^* for each process varying with z according to:

$$\sqrt{\chi_\omega^*} = -0.136(z - 4\omega) + \{[0.136(z - 4\omega)]^2 + 2.491\}^{1/2} \quad (77)$$

Equations (75) and (77) allows to draw the "charge–electronegativity" plot discussed in Sect. 4.

10.3 Complexation

The starting point is Eq. (55) involving X^{n-} anionic species:

$$[M(OH)_h(OH_2)_{N-h}]^{(z-h)+} + m\,X^{n-}$$
$$\leftrightarrow [M(OH)_h X_m (OH_2)_{N-h-m\alpha}]^{(z-h-mn)+} + m\alpha H_2O \quad (78)$$

If we make appear the *q*-protonated form of the $X^{n-}(H_qX)$ through intramolecular prototropic transfers, the right complex may be written $[M(OH)_{h+mq}(H_qX)_m(OH_2)_{N-h-m(\alpha+q)}]^{(z-h-mn)+}$. Now by applying both conditions for complexation to occur, it comes:

$$q = 0 \quad \delta_x = -n \Rightarrow \chi_D = \chi\{[M(OH)_h(OH_2)_{N-h-m\alpha}]^{(z-h)+}\} \quad (79)$$

$$q = n \quad \delta_{HnX} = 0 \Rightarrow \chi_H = \chi\{[M(OH)_{h+mn}(OH_2)_{N-h-m(\alpha+n)}]^{(z-h-mn)+}\} \quad (80)$$

We may also transform these critical electronegativities into critical hydrolysis ratio by writing charge conservation:

$$\delta[\chi, M(OH_2)_N] - m\alpha\delta(\chi, OH_2) - (h + mq)\delta(\chi, H)$$
$$+ m\delta(\chi, H_qX) = z - h - mn \quad (81)$$

making $\chi = \chi[H_qX^{(n-q)-}] = \chi_q$ (i.e. $\delta(\chi, H_qX) = q - n$) in Eq. (81) leads to:

$$\delta[\chi_q, M(OH_2)_N] - m\alpha\delta(\chi_q, OH_2) - (h + mq)\delta(\chi_q, H) = z - h - mq \quad (82)$$

Solving Eq. (82) for *h* then leads to the wanted hydrolysis ratio:

$$h_q^* = \frac{z - \delta[\chi_q, M(OH_2)_N] + m\alpha\delta(\chi_q, OH_2)}{1 - \delta(\chi_q, H)} - mq \quad (83)$$

$$\text{or} \quad h_q^* = \{[\Delta(\chi_q, z, N, M) + m\alpha\delta(\chi_q, OH_2)]/\Delta(\chi_q, 1, 0, H)\} - mq \quad (84)$$

using the Δ formalism of Sect. 10.1. We have now just to equalize Eq. (73) with $\chi = \chi_w = 2.732 - 0.035\,pH$ with Eq. (84) to work out the variation of pH as a function of χ_q. This is best done by noticing that for two electronegativities χ_1 and χ_2, the following relation always holds:

$$\Delta(\chi_1,z,N,M) = \Delta(\chi_2,z,N,M) + \sigma(N,M)(\chi_2 - \chi_1) \tag{85}$$

where $$\sigma(N,M) = [N\sigma(H_2O) + \sigma_M] = [1.408N + 1/(1.36\sqrt{\chi_M^0})] \tag{86}$$

is the softness of the $[M(OH_2)_N]$ complex. If we take the electronegativity of the bare proton $\chi(H^+) = 4.071$ as a common reference state to compute the Δs', it becomes:

$$\Delta(\chi_w,1,0,H) = 0.507(4.071 - \chi_w) \tag{87}$$

$$\Delta(\chi_q,1,0,H) = 0.507(4.071 - \chi_q) \tag{88}$$

$$\Delta(\chi_w,z,N,M) = \Delta(4.071, z,N,M) + \sigma(N,M)(4.071 - \chi_w) \tag{89}$$

$$\Delta(\chi_q,z,N,M) = \Delta(4.071, z,N,M) + \sigma(N,M)(4.071 - \chi_q) \tag{90}$$

With $\Delta = \Delta(4.071,z,N,M) = z - 2.225N - (4.071 - \chi_M^0)/(1.36\sqrt{\chi_M^0})$ and $\delta(\chi_q,OH_2) = 1.408\chi_q - 3.507$, combination of Eqs. (73), (84), and (86–90) leads to:

$$\begin{aligned}&[\Delta + \sigma(N,M)(4.071 - \chi_q) + m\alpha(1.408\chi_q - 3.507) - 0.507\,mq(4.071\\&\quad - \chi_q)](4.071 - \chi_w) = (4.071 - \chi_q)[\Delta + \sigma(N,M)(4.071 - \chi_w)]\end{aligned} \tag{91}$$

or after factorization of χ_w and elimination of $\sigma(N,M)$:

$$\begin{aligned}&\chi_w[\Delta + (1.408m\alpha + 0.507\,mq)\chi_q - (3.507\,m\alpha + 2.064\,mq)]\\&\quad = (\Delta + 5.732m\alpha + 2.064\,mq)\chi_q - 4.071(3.507\,m\alpha + 2.064\,mq)\end{aligned} \tag{92}$$

Making $m = 1$ and $\chi_w = 2.732 - 0.035\,pH_q^*$ in Eq. (93) leads to Eq. (58).

11 Symbols and Abbreviations

PCM: Partial Charge Model
χ: Electronegativity
μ: Chemical Potential
δ: Partial charge
σ: Chemical softness ($= 1/\eta$)
η: Chemical hardness
χ_i, χ_X or $\chi(X)$: Electronegativity of an atom or of a group of atoms
$\delta(X)$ or δ_x: Partial charge beared by an atom or a group of atoms
r: Covalent radius
Z_{eff}: Nuclear effective charge
k: Constant linking hardness to electronegativity (k = 1.36 in this work)
z: Oxidation state (valency) of a cation ($1 \leq z \leq 8$)

χ^0: Pauling electronegativity
N: Coordination number of a cation ($N \leq 12$)
h: Number of protons removed from an aquo-ion $[M(OH_2)_N]^{z+}$
χ_w: Mean electronegativity of an aqueous solution given by Eq. (13)
β: $= RTLn\ 10$ with R perfect-gas constant and T temperature
c_H: Proton concentration
λ: Constant linking electronegativity to pH(= 0.035 according to Eq. (12))
χ^*_+: Critical electronegativity for acid ionization given by Eq. (77) with $\omega = 1$
χ^*_-: Critical electronegativity for basic ionization given by Eq. (77) with $\omega = -1$
χ^*_0: Critical electronegativity for olation given by Eq. (77) with $\omega = 0$
χ^*_{OH}: Critical electronegativity for oxolation given by Eq. (75)
PZC: Point of Zero Charge of a solid phase
AcO^-: Acetate ion CH_3COO^-
α: Number of water molecules removed by a ligand upon coordination
χ_p: Mean electronegativity of a complexed precursor
n: Valency of an anion X^{n-}
m: Stoichiometric coefficient
q: Number of protons beared by an anion $H_qX^{(n-q)-}$
χ_D: Critical electronegativity for ionic dissociation given by Eq. (79)
χ_H: Critical electronegativity for hydrolysis given by Eq. (80)
h^*_q: Critical hydrolysis ratio associated to a $H_qX^{(n-q)-}$ anion given by Eq. (83)
pH^*_q: Critical pH value associated to h^*_q through Eq. (58)
χ_q: Mean electronegativity of an anion $H_qX^{(n-q)-}$
Δ: Charge differential characteristic of an element given by Eq. (59)
pH_m: Optimum pH for complexation by an anion X^{n-} given by Eq. (62)
ox: Oxalate anion $C_2O_4^{2-}$
$\delta(\chi,X)$: Partial charge upon the X group when mean electronegativity is χ
χ_S: Mean electronegativity of the aquo ligand (2.491 using Table 5)
ω: Index for Eqs. (76) and (77)
$\sigma(N,M)$: Softness of the $[M(OH_2)_N]$ complex given by Eq. (86)

12 References

1. Hench LL, West JK (1990) Chem Rev 90: 33
2. Dislich H, Hinz P (1982) J Non Cryst Solids 48: 11
3. Livage J, Henry M, Sanchez C (1988) Prog Solid State Chem 18: 259
4. Baes CF, Mesmer RE (1976) Hydrolysis of cations. Wiley, New-York
5. Matijevic E (1985) Ann Rev Mater Sci 15: 483
6. Sanderson RT (1951) Science 114: 670
7. Pauling L (1932) J Am Chem Soc 54: 3570
8. Parr RG, Donnelly RA, Levy M, Palke WE (1978) J Chem Phys 68: 3801
9. Parr RG, Pearson RG (1983) J Am Chem Soc 105: 7512
10. Yang W, Parr RG (1985) Proc Natl Acad Sci USA 82: 6723

11. Allred AL, Rochow E (1958) J Inorg Nucl Chem 5: 264
12. Jørgensen CK (1963) Inorganic complexes. Academic Press, London
13. Bach S, Henry M, Baffier N, Livage J (1990) Fr Pat 90, 02565
14. Henry M (1988) Application of the electronegativity concept to hydrolysis and condensation phenomena in inorganic chemistry. Thesis, Université Pierre et Marie Curie, Paris; Livage J, Henry M (1988) in: Mackenzie JD, Ulrich DR (eds) Ultrastructure Processing of Advanced Ceramics, Wiley, New-York, p 183
15. Tytko RH, Glemser O (1976) Adv Inorg Chem Radiochem 19: 239
16. Pope MT (1983) Heteropoly and isopolyoxometallates. Springer, Berlin Heidelberg New-York
17. Brown PL, Shying ME, Sylva RN (1987) J Chem Soc Dalton Trans 2149
18. Cruywagen JJ, Van der Merwe IFJ (1987) J Chem Soc Dalton Trans 1701
19. Tytko RH (1986) Polyhedron 5: 497
20. Maksimovskaya RI, Burtseva KG (1985) Polyhedron 4: 1559
21. Masters AF, Gheller SF, Brownlee RTC, O'Connor MJ, Wedd AG (1980) Inorg Chem 19: 3866
22. Richardson E (1959) J Inorg Nucl Chem 25: 575
23. Doucet Y, Bugnon S (1957) J Chim Phys 54: 155
24. Chemseddine A, Henry M, Livage J (1984) Rev Chim Minérale 21: 487
25. Chemseddine A, Babonneau F, Livage J (1987) J Non-Cryst Solids 91: 271
26. Heath E, Howarth OW (1981) J Chem Soc Dalton Trans 1105
27. Gharbi N, Sanchez C, Livage J, Lemerle J, Nejem L, Lefebvre J (1982) Inorg Chem 21: 2758
28. Ardon M, Bino A (1987) Structure and Bonding 65: 1
29. Grant M, Jordan RB (1981) Inorg Chem 20: 55
30. Xu FC, Krouse HR, Saddle TW (1985) Inorg Chem 24: 267
31. Fiat D, Connick RE (1968) J Am Chem Soc 90: 608
32. Stünzi H, Spiccia L, Rotzinger FP, Marty W (1989) Inorg Chem 28: 66
33. Singh KK, Sarode PR, Ganguly P (1983) J Chem Soc Dalton Trans 1895
34. Segal DL (1984) J Chem Tech Biotechnol 34A: 25, 355
35. Khoe GH, Brown PL, Sylva RN, Robins RG (1986) J Chem Soc Dalton Trans 1901
36. Blesa MA, Matijevic E (1989) Adv Colloids Interface Sci 29: 173
37. Akitt JW, Elders JM (1988) J Chem Soc Dalton Trans 1347
38. Thompson AR, Kunwar AC, Gutowsky HS, Oldfield E (1987) J Chem Soc Dalton Trans 2317
39. Bottero JY, Axelos M, Tchoubar D, Cases JM, Fripiat JJ, Fiessinger F (1987) J Colloid Interface Sci 117: 47
40. Axelos M, Tchoubar D, Bottero JY, Fiessinger F (1985) J Physique 46: 1587
41. Giese Jr RF (1976) Acta Cryst B32: 1719
42. Spiccia L, Marty W (1986) Inorg Chem 25: 266
43. Flynn Jr CM (1984) Chem Rev 84: 31
44. Christensen AN (1976) Acta Chem Scand A30: 133
45. Surayambunathan V, Liao YX, Meisel D (1989) Langmuir 5: 1423
46. Ervin Jr G (1952) Acta Cryst 5: 103
47. Thompson M, Connick RE (1981) Inorg Chem 20: 2279
48. Rotzinger FP, Stünzi H, Marty W (1986) Inorg Chem 25: 489
49. Monsted L, Monsted O, Sprinborg J (1985) Inorg Chem 26: 474
50. Spiccia L, Stoeckli-Evans H, Marty W, Giovanoli R (1986) Inorg Chem 24: 3496
51. Finholt JE, Thompson ME, Connick RE (1981) Inorg Chem 20: 4151
52. Stünzi H, Marty W (1983) Inorg Chem 22: 2145
53. Stünzi H, Rotzinger FP, Marty W (1984) Inorg Chem 20: 2160
54. Brown PL, Sylva RN, Batley GE, Ellis J (1985) J Chem Soc Dalton Trans 1967
55. Johansson G (1962) Acta Chem Scand 16: 403
56. Johansson G (1963) Arkiv Kemi 20: 305, 320
57. Akitt JW, Elders JM, Fontaine XLR, Kundu AK (1989) J Chem Soc Dalton Trans 1889
58. Van der Woude JHA, De Bruyn PL (1983) Colloids and Surfaces 8: 55, 79
59. Combes JM, Manceau A, Callas G, Bottero JY (1989) Geochimica Chemica Acta 53: 583
60. Brinker CJ, Scherer GW (1989) Sol–gel Science. Academic Press, New-York
61. Akitt JW, Gessner W (1984) J Chem Soc Dalton Trans 147
62. Baker BR, Pearson RM (1974) J Catal 33: 265
63. Yoldas BE (1972) J Appl Chem Biotech 23: 803
64. Eggleton RA, Fitzpatrick RW (1968) Clays and Clays Minerals 36: 111
65. Magini M (1977) J Inorg Nucl Chem 39: 409

66. Van der Woude JHA, Rijnbout JB, De Bruyn PL (1984) Colloids and Surfaces 11: 391
67. Van der Woude JHA, De Bruyn PL, Pieters J (1984) Colloids and Surfaces 9: 173
68. Chenavas J, Joubert JC, Capponi JJ (1973) J Solid State Chem 6: 1
69. Schneider W (1984) Comments Inorg Chem 3: 205
70. Patrat G, De Bergevin F, Pernet M, Joubert JC (1983) Acta Cryst B39: 165
71. Saraswat IP, Vajpei AC (1984) J Mater Sci Lett 3: 515
72. Spiccia L, Marty W, Giovanoli R (1987) Helv Chim Acta 70: 1737
73. Spiccia L, Marty W, Giovanoli R (1988) Inorg Chem 27: 2660
74. Christensen AN (1966) Inorg Chem 5: 1452
75. Hamilton WC, Ibers JA (1963) Acta Cryst 16: 1209
76. Ohtaki H, Yamaguchi T, Maeda M (1976) Bull Chem Soc Jpn 49: 701
77. Kolski GB, Kildahl NW, Margerum DW (1969) Inorg Chem 8: 1211
78. Johansson G, Olin A (1968) Acta Chem Scand 22: 3197
79. Perrin DD (1960) J Chem Soc 3189
80. Néher-Neumann E (1984) Acta Chem Scand A38: 517
81. Sylva RN, Davidson MR (1979) J Chem Soc Dalton Trans 232
82. Jaggi H, Ostwald HR (1961) Acta Cryst 14: 1041
83. Weiser HB (1923) J Phys Chem 27: 501
84. Hazell IF, Irving RJ (1966) J Chem Soc (A) 669
85. Ellis JD, Sykes AG (1973) J Chem Soc Dalton Trans 537
86. Ellis JD, Thompson AK, Sykes AG (1976) Inorg Chem 15: 3172
87. Åberg M (1977) Acta Chem Scand B31: 171
88. Clearfield A (1964) Rev Pure Appl Chem 14: 91
89. Einaga H (1979) J Chem Soc Dalton Trans 1917
90. Einaga H, Komatsu Y (1981) J Inorg Nucl Chem 43: 2443
91. Bekkerman LI, Dobrovol'skii IP, Ivakin AA (1976) Russ J Inorg Chem 21: 223
92. Narita E, Takeuchi H, Horiguchi N, Okabe T (1984) Bull Chem Soc Jpn 57: 1388
93. Fryer JR, Hutchinson JL, Paterson R (1970) J Colloid Interface Sci 34: 238
94. Jolivet JP, Tronc E (1988) J Colloid Interface Sci 125: 688
95. Clearfield A (1988) Chem Rev 88: 125
96. Parida KM, Kanungo S, Sant BR (1984) Electrochim Acta 26: 435
97. Burns RG, Burns VM (1975) in: First MnO_2 Symposium, p 306 Cleveland
98. Matijevic E (1986) Langmuir 2: 12
99. Ringbom A (1963) Complexation in Analytical Chemistry, Wiley, New York; Henry M, Jolivet JP, Livage J (1989) in: Uhlman DR, Ulrich DR (eds) 4th Int Conf on Ultrastructure Processing of Ceramics, Glasses and Composites, 19–24 Feb 1989. Tucson Arizona, USA
100. Clearfield A, Vaughan PA (1956) Acta Cryst 9: 555
101. Mak TCW (1968) Can J Chem 46: 3491
102. McWhan DB, Lundgren G (1963) Acta Cryst. 16: A36
103. Coddington JM, Howe RF, Taylor MJ (1989) Inorg Chim Acta 166: 13
104. Brahimi M, Durand J, Cot L (1988) Eur J Solid State Inorg Chem 25: 185
105. Hansson M (1973) Acta Chem Scand 27: 2614
106. Prozorovskaya ZN, Petrov KI, Komissarova LN (1968) Russ J Inorg Chem 13: 505
107. Paul RC, Baidya OB, Kumar RC, Kapoor R (1976) Aust J Chem 29: 1605
108. Zaitsev LM (1964) Russ J Inorg Chem 9: 1279
109. Baestlé L, Pelsmaekers J (1961) J Inorg Nucl Chem 21: 124
110. Livage J, Henry M, Jolivet JP, Sanchez C (1990) Mat Res Soc Bull 15: 18
111. Matwiyoff N, Darley PE, Movius WG (1968) Inorg Chem 7: 2173
112. Hunt JP, Friedman HL (1983) Prog Inorg Chem 30: 359
113. Kraus KA, Nelson F, Smith GW (1954) J Phys Chem 58: 11
114. Scott WB, Matijevic E (1978) J Colloid Interface Sci 66: 447
115. Akitt JW, Milic NB (1984) J Chem Soc Dalton Trans 981
116. Akitt JW, Elders JM, Fontaine XLR, Kundu AK (1989) J Chem Soc Dalton Trans 1897
117. Nordstrom DK, May HM (1989) in: Sposito G (ed) The environmental chemistry of aluminum, CRC Press, Boca Raton
118. Akitt JW, Greenwood NN, Lester GD (1971) J Chem Soc (A) 2450
119. Sjöberg S, Öhman LO (1985) J Chem Soc Dalton Trans 2665
120. Nishide T, Tsuchiya R (1965) Bull Chem Soc Jpn 38: 1398
121. Akitt JW, Greenwood NN, Lester GD (1969) J Chem Soc Dalton Trans 803

122. Akitt JW, Farnsworth JA, Letellier P (1985) J Chem Soc Faraday Trans I 81: 193
123. Vogel RF, Marcelin G (1983) J Catal 80: 492
124. Mortier WJ, Ghosh SK, Shankar S (1986) J Am Chem Soc 105: 4315
125. Mortier WJ, Van Genechten K, Gasteiger J (1985) J Am Chem Soc 107: 829
126. Little EJ, Jones MM (1960) J Chem Educ 37: 231
127. Batsanov SS (1968) Russ J Struct Chem 37: 332
128. Zhang Y (1982) Inorg Chem 21: 3886

Optical Properties of Colorants or Luminescent Species in Sol–Gel Glasses

Renata Reisfeld[1] **and Christian K. Jørgensen**[2]

[1] Enrique Berman Professor of Solar Energy; Department of Inorganic and Analytical Chemistry, Hebrew University of Jerusalem, Jerusalem 91904, Israel
[2] Département de Chimie Minérale, Analytique et Appliquée, Université de Genève, CH 1211 Geneva 4, Switzerland

Inorganic ions and organic molecules or ions having characteristic absorption and/or luminescence spectra can be incorporated into glasses formed by the sol–gel process from appropriate precursors. The typical spectra of the dopants change during the sol–gel transformation and may serve as optical probes for the process. A variety of new materials can be designed from the doped glass bulks or films. Some examples include nonlinear materials, tunable lasers, luminescent solar concentrators, and optical sensors. Typical glasses are doped by Co(II), Eu(III), Tb(III), RuBPY, CdS, fluorescein and its derivatives, oxazine, 2,2′-bipyridyl-3,3′-diol, methyl orange, methyl red, acridine orange, and acridine yellow. Composite materials (either formed from porous inorganic glasses impregnated with a dissolved monomer, subsequently polymerized in the pores; or ormocers formed by simultaneous treatment of alkylsilicon alkoxides, or a precursor containing reactive glycidyloxopropyl substituents) can dissolve (otherwise almost insoluble) perylimide dyes, which are exceedingly photostable, and used in tunable lasers in the visible.

Structure and Bonding 77

1 Introduction

The sol–gel technique offers a low-temperature method for synthesizing amorphous materials which are either totally inorganic in nature or composed of inorganics and organics. The process is based on hydrolysis and condensation reactions of organometallic compounds in alcoholic solutions. The most widely investigated system involves silica-based glasses which are prepared by polymerization of a silicon alkoxide, $Si(OR)_4$ [1, 2, 26].

The ability to synthesize inorganic polymers using sol–gel processing with little or no heating makes it possible to dope these gels with a variety of organic and organometallic molecules [3–11] especially dyes having desired luminescent properties. The emission properties of the molecules have been used to optically probe sol–gel chemistry [3–8] and structure [3, 10–12]. Another active area for organic doped sol–gel glasses is to use dopants to induce selected optical properties and to synthesize new optical materials [13]. Among the properties reported to date are solar light concentrators [5–7], photochromism [14], nonlinear optical effects [5–7], and tunable laser action [15–18]. The latter was achieved by incorporating organic laser dyes in a sol–gel matrix and thus represents a potentially important direction for solid-state laser material in the visible range of the spectrum.

Organic laser dyes have been widely reported as having desirable properties for optical gain and laser applications. Finding suitable host materials for these organics presents a challenge and, in the case of laser dyes, devices have been primarily restricted to liquid-state applications. There are a number of devices, however, where solid-state gain media would be quite advantageous. Attempts have been made to fabricate solid-state organic dye materials using polymeric hosts such as polymethylmethacrylate, polycarbonate, polystyrene, and polyvinyl alcohol [19–21]. These hosts, however, have been shown to be inherently lacking in mechanical and thermal properties, photostability, and refractive index uniformity. Inorganic glasses, on the other hand, do possess extremely good optical, thermal, and chemical stability; however, the processing temperatures of conventional glasses promote the rapid decomposition of most organic species. In view of the number of dyes and dopants available, the prospect of achieving a wide range of sol–gel-based tunable solid-state lasers is of substantial interest to the optical materials community.

As mentioned above, the sol–gel process opens many possibilities for the synthesis of optical materials [22]. The process is based on inorganic polymerization reactions, starts from molecular precursors, and a macromolecular oxide network is obtained via hydroxylation–condensation reactions [23, 24] which can be controlled by the chemical design of molecular precursors. The viscosity properties of sols can be adjusted allowing easy deposition of transparent coatings onto glass, ceramic, or polymeric substrates [25, 26]. Sol–gel chemistry is performed in solution at lower temperatures than conventional chemical methods. Homogeneous doping by mixing components at a molecular level, synthesis of metastable or amorphous phases allowing larger concentrations of chromophores, and synthesis of mixed organic–inorganic materials can then be performed.

Silica [3, 4, 17, 28] or alumina–silica [16, 29] have been studied extensively. Sanchez [60] has studied the incorporation of organic luminophores (such as Rhodamine 6G, Rhodamine 640, and Coumarin 4) in titanium(IV) oxide gels prepared, for instance, by hydrolysis and polycondensation of $(n\text{-}C_4H_9O)_4Ti$ in a mixture of *n*-butanol and aqueous acetic acid. As seen below, Eu(III) was also studied. The analogous zirconium(IV) gels can also be prepared [60]. Genet and Brandel [61] have prepared almost stoichiometric, limpid $Th_3(PO_4)_4$ by appropriate treatment of thorium(IV) solutions, and incorporated Nd(III), Eu(III), Er(III), and Cr(III) in this material, as well as Rhodamine 6G. This phosphate should not be confused with the α-modification of $Zr(HPO_4)_2$, H_2O known to be an ion exchanger with most of the H^+ able to be replaced with alkali ions A^+. Also [62] hafnium(IV) forms ion-exchanging $Hf(HPO_4)_2$ of which the saturated Na^+ adduct can be dried at 180 °C to $Na_2Hf(PO_4)_2$. The thorium glass does not depend on hydrogen bonds, but has a relation to unreactive mixed oxides intended to confine fission products and transuranium elements (though this can only be obtained by a gel glass after strong heating).

In the case of organosilicate glasses (organically modified ceramics or ormocers), the silicate network may be modified by organic substituents such as alkyl (e.g. methyl) groups or other functional groups (e.g. 3-glycidyloxypropyl) [30] which may form organic copolymers that penetrate the silicate structure. In principle, Si may be substituted by Al, B, Ti, or Zr to yield ceramics of variable mechanical properties [31]. At present there is no consistent picture about the microstructure of these materials. In the investigation of glassy systems, several analytical methods [31, 32] have been used among which are neutron and light scattering.

The coating material based on *ormocers* [33] can be synthesized from alkoxides of Si, Zr, or Ti and from silanes with functional groups (e.g. 3-glycidyloxypropyltrimethoxysilane; short name Glymo) by sol–gel processing. During the time of the process a low degree of hydrolysis and condensation was achieved by the moisture of air. The coating solution was then poured on PMMA plates and dried for 3 h in air. The coatings were finally densified for one hour at 80–100 °C. A coating thickness of 10 to 20 μm is necessary to obtain a sufficient extinction. For a good adhesion of the coating to PMMA it was necessary to pretreat the substrates with $SOCl_2$ (5 min at room temperature).

To improve the long-term stability of the fluorescence dyes a new type of dye (A2) was developed by BASF containing $Si(OEt)_3$ groups to be chemically bonded to the ormocer backbone by cocondensation. The coating material is of high optical transparency. No scattering losses can be detected in layers of 100 μm [34].

Through the impregnation of porous silica gels with organic monomers, such as methylmethacrylate, followed by in situ polymerization, transparent silica gel-PMMA composites can be prepared [35, 36]. Such transparent composites, prepared at relatively low temperatures, have potential applications as optical hosts for organic dye molecules [13].

The main reason that most polymer–glass composites were opaque until now was due to light scattering. Glass fiber-reinforced epoxy, for example, typically uses fibers of 5 to 20 microns in diameter and, often, many meters in length. These large "phase dimensions" result in light scattering, despite the fact that both glass and epoxy are quite transparent in bulk form. In the new composites, however, the phase dimensions are in the order of 10 nm, much smaller than the wavelengths of visible light (390 to 780 nm). Hence, this new class of glass–polymer composites is quite transparent, which opens up a wide variety of applications for which composite materials had been previously excluded.

The chemical synthesis of gels is performed in organic solutions around room temperature. Organic molecules or dyes can therefore be easily incorporated into an oxide gel matrix. These mixed organic–inorganic compounds provide new opportunities as optical coatings for applications such as lasers, nonlinear optics, or luminescent materials [7, 37, 38].

2 Optical Probes for Glass Transformation Processes

2.1 *Europium(III) and Terbium(III)*

Ions or molecules whose optical properties (absorption and fluorescence) change during the sol–gel transition, are a sensitive probe for studying this transition. Trivalent europium is an excellent ion for this purpose. The total fluorescence of Eu(III) from the $^5D_0$ electronic state exhibits a gradual increase and a relative increase of intensity of the hypersensitive transition $^5D_0 \rightarrow {}^7F_2$ as a function of time and temperature of dehydration of the gel prepared by hydrolysis of tetramethoxysilane TMOS. This behavior is explained by formation of a chemical bond between Eu(III) and the oxygens of the silicate group whereby Eu(III) acquires a symmetry similar to that in a regular glass obtained by melting the appropriate oxides. The degeneracy of the $^7F_1$ and $^7F_2$ states is removed in low-symmetry sites in the glass as reflected by the half-width of the emission bands.

While in glasses prepared from TMOS the magnetic $^5D_0 \rightarrow {}^7F_1$ transition is higher than the electrical dipole transition $^5D_0 \rightarrow {}^7F_2$ followed by reversal of the intensities of these peaks with dehydration and glass formation [6], in glasses prepared from tetraethoxysilane TEOS the electrical dipole transition is higher already in glasses prepared at 40 °C [39]; see Fig. 1 showing emission spectra of europium in sol–gel glasses and Fig. 2 of terbium in sol–gel glasses (with addition of aluminum). The ratio of TMOS:H_2O:MeOH in the starting solution was 1:4:5 and in TEOS 1:4:4.

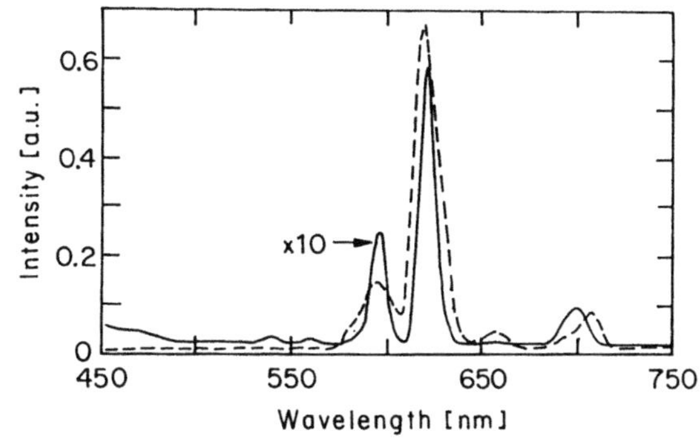

Fig. 1. Emission spectra of Eu(III) in sol–gel glasses, excitation wavelength 388 nm; ——— dried at 40 °C, density, 1.6 g/cm³ ------- heated and dried at 500 °C, density 2 g/cm³

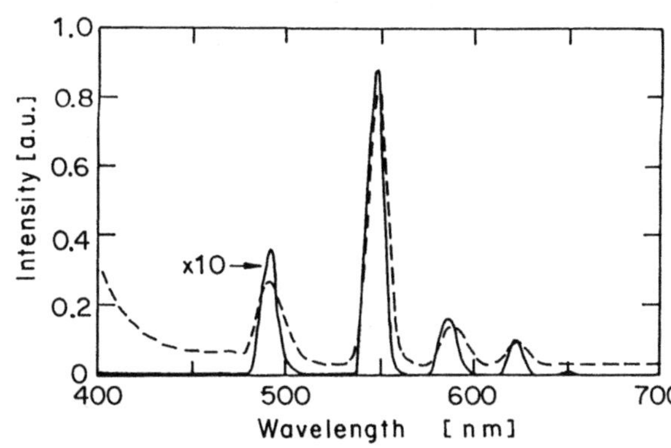

Fig. 2. Emission spectra of Tb(III) in sol–gel glasses, excitation wavelength 373 nm; ——— dried at 60 °C, density 1.9 g/cm³ --------heated and dried at 400 °C

The conclusion from these experiments is that glasses prepared from TEOS are more dense and the probe ion is surrounded sooner by the silica groups which replace water.

Sanchez [60] provided evidence from the relative intensities of luminescence of $^5D_0$ to the five lowest ^7F_J levels that a part of the Eu(III) forms intimate connections with the polycondensated Ti_nO_m network already at room temperature, whereas silica needs some 200 °C for such a change.

So far we know, neodymium(III) does not luminesce perceptibly in the near infrared; however, sufficient concentrations can be dissolved in various sol–gel glasses [25] to show the sharp absorption bands strongly. Besides nonradiative quenching due to OH vibrations [62], cross-relaxation between adjacent Nd pairs may play a role.

2.2 *Pyrene and its Dimers*

The planar pyrene $C_{16}H_{10}$ shows in dilute solution finely resolved vibronic components of fluorescence in the violet, but tend in higher concentration to give broad-band blue emission due to one or several dimeric adducts.

The influence of the disappearance of pyrene excimers was used in order to compare the quality of the glasses obtained from TMOS and TEOS [40]. The relative concentrations of pyrene monomer to excimer indicate the velocity of gel formation which is much higher in methoxysilane than in its ethoxy homolog. This is concluded from a study of time-dependent absorption and emission behavior.

Figure 3 shows the absorption spectra of pyrene (Py) in (A) TEOS glass and (B) TMOS glass (the full line represents 7 days after preparation, the dashed line 9 days, and the dashed-dotted line 11 days). The final absorption maxima obtained at 334, 319, 305, 273, and 262 nm are in agreement within 1 nm with the absorption spectra of Py in solvents [41]. The absorption and emission spectra reveal dramatic time-dependent changes which stop when the gelation is final.

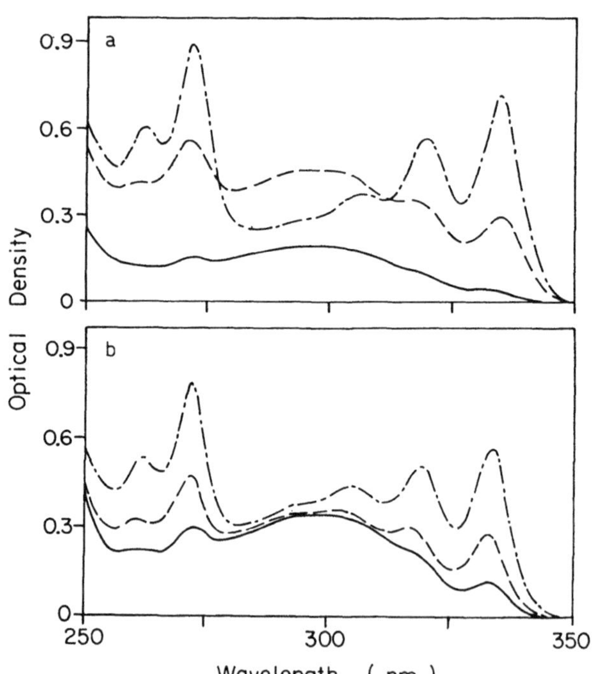

Fig. 3. Absorption spectra of pyrene in (**A**) TEOS glass and (**B**) TMOS glass; ——— 7 days after preparation -------- 9 days after preparation - - - - - - - - 11 days after preparation

A remarkable increase in the Py optical density (despite the decrease of glass thickness during the dessication stage) suggests that in the first alkoxysilane hydrolysis stage, when a slight amount of water is present, Py molecules tend to aggregate [42]. Therefore, at the beginning of the gelation a weak absorption spectrum is obtained. As soon as the polymerization terminates, the Py becomes more "dissolved". As individual molecules are dispersed into the sol–gel glass, the optical density increases and due to the isolation of the monomer molecules a more defined absorption spectrum is observed. The spectrum does not change further in the final glass at longer times.

The fluorescence is notably altered during the sol–gel process: the intensity of the excimer band becomes weaker and disappears, and simultaneously an enhancement of the monomer band becomes with its typical vibronic structure is observed. These results strengthen the theory of Py aggregate formation at the start of gelation. When the Py aggregates are being formed the distance between the Py molecules decreases and allows diffusion-controlled excimer formation between adjacent Py molecules. Following the development of gelation, a continuous increase in cross-linking traps and isolates Py molecules into narrow pores causing an enhancement in the absorption and emission monomer spectra and the disappearance of the excimer emission band. This phenomenon is slower in the TEOS glass than in the TMOS glass because of its slower gelation rate [43]. On the other hand, it is more homogeneous than in TMOS glass, where different parts of the same sample give different absorption spectra. For this reason measurements were always taken at the same spot in the samples.

The usual definition of an excimer is an adduct formed between one system in an excited state and one system in its groundstate. However, conceptually, two species in their groundstate may be so close together that they show both absorption and emission at lower energy than the isolated monomer. This behavior may be studied in sol–gel glasses of high viscosity and very scarce inter-species collisions. Actually, pyrene in sol–gel silica glass [63] from $Si(OC_2H_5)_4$ shows picosecond time-resolved fluorescence spectra similar to those previously found for pyrene in Langmuir-Blodgett films, and suggest that the longer-lived excimer E_2 (35 ns) is formed from the ground-state dimer, emitting the fluorescence band D, whereas the excimer band E_1 (0.2 ns) originates from a weak molecular interaction (providing a conformation different from E_2). By the same token [64] naphthalene distributed in a sol–gel glass shows similar time-resolved picosecond scale fluorescence.

8-Hydroxo-pyrene-1,3,6 tris(sulfonate) [also called pyranine] shows a variety of emission bands [29] in aluminosilicate gels evolving as a function of polymerization, aging, and drying of the gel. The green emission peak at 515 nm (due to the excited state of the deprotonated pyranine^{-4}, whereas the blue emission at 430 nm is due to the hydroxo compound pyranine^{-3}) is strongly decreased by higher water concentration in propanol mixtures, related to a much more acidic excited state (pK = 1) of pyranine^{-3} than pK = 7.5 of the ground state. Perhaps related to the high anionic charge, no evidence for excimer luminescence was mentioned.

2.3 Cobalt(II) Absorption Spectra as Internal Probe

The rather detailed transitions within the $3d^7$ shell of cobalt(II) has been used as an indicator of transformation of the gels into a final glass during heating [12]. Like in the mixed methanol–water–$Si(OCH_3)_4$ precursor solution, Co(II) is mainly octahedral (but shows intensities considerably higher than of $Co(OH_2)_6^{+2}$), also by drying the gel at 60 °C. At temperatures between 200 and 600 °C, Co(II) adapts a lower coordination number, with three strong absorption bands.

Sol–gel glasses were prepared from $Si(OCH_3)_4$. Precursor 0.1 molar solutions of $Co(NO_3)_2 \cdot 6H_2O$ or $Co(CH_3CO_2)_2 \cdot 4H_2O$ were formed by adding 3 ml water containing the appropriate amount of cobalt(II) nitrate or acetate to 17 ml p.a. methanol and 6.5 ml TMOS (Aldrich) corresponding to a molar ratio of 1 TMOS:9.5 CH_3OH:4.4 H_2O:0.06 Co (taking into account also the salt hydration) to which drops of 12 molar nitric acid (0.05 per TMOS) were subsequently added with rapid mixing. The supporting flat-plate iron-free soda-lime glass parallelepiped (a few cm^2 area; 3 mm thickness) was dipped into this solution at reproducible rate, dried at 60 °C for 30 min and heated up to e.g. 500 °C at the rate of 10 °C per min, and then cooled for one hour in the oven now turned off.

In the following Sect. 3, the appropriate identifications of the absorption bands of samples, having been dehydrated at various T, are discussed, partly using AOM (angular overlap model) arguments, and partly induction from Co(II) in chromophores of known symmetry.

3 Tetrahedral vs. Octahedral Cobalt(II)

A prominent factor of the characterization of cobalt as a new element (around 1730) was the blue coloration of Delft faience (as previously observed empirically in China). A major blue pigment is the cubic spinel $CoAl_2O_4$ invented by Thenard in 1805. Weyl [44] reviewed the subsequent work on silicate, borosilicate, and phosphate glasses, and concluded that the pink (weakly colored) cobalt(II) is essentially octahedral (like the aqua ion) and the intensely blue chromophore close to a regular tetrahedron CoO_4. Any absorption spectrum of intermediate type was ascribed to a mixture of CoO_6 and CoO_4 as previously proposed by Brode [45] in transmission spectra of lithium, sodium, potassium, and cesium borate glasses (high A^+ concentration per borate, and large ionic radius of A^+ both favoring the blue form). Though the blue form showing bands at 650, 605, and 550 nm is not really compatible with CoO_4 in the point-group T_d (as discussed below) the conclusion by Weyl [44] was later modernized to "in a given glass, there is one pink (O_h) and one dark blue (T_d) species".

When ligand-field theory was applied to $3d^7$ since 1954, this sharp alternative remained unchallenged. One argument is available for an abrupt change in stereochemistry. When large amounts of $H_3O^+Cl^-$ or Li^+Cl^- are added to aqueous $Co(OH_2)_6^{+2}$, a distinct (100 times more light-absorbing) $CoCl_4^{-2}$ is formed, and the only intermediate step perceived is [46] a raspberry-red octahedral species likely to be $ClCo(OH_2)_5^+$ [though $Cl_2Co(OH_2)_4$ is known from the crystal structure of $CoCl_2 \cdot 6H_2O$]. It should be emphasized that this change does not involve modified spin S of the ground state [as when apple-green $Ni(OH_2)_6^{+2}$ with triplet ($S = 1$) ground state forms diamagnetic yellow $Ni(CN)_4^{-2}$ and orange, tetragonal-pyramidal $Ni(CN)_5^{-3}$ in cyanide solution] although dark-green $Co(CN)_5^{-3}$ with doublet ground state is formed under similar circumstances.

$CoCl_4^{-2}$ shows a large number of rather sharp absorption bands in the visible [47, 48]. The band at 531 nm (18800 cm^{-1}) has the molar extinction coefficient $\varepsilon = 9$ (still twice the ε of the broad transition between the two $^4T_1$ of the octahedral aqua ion) and, like the 8 bands (with ε1 to 2) at higher energy, due to numerous doublet states, as can be rationalized [48] by the Tanabe-Sugano diagram for point-group T_d. However, the two strong transitions in the near infrared ($\varepsilon = 80$) and in the red (highest $\varepsilon = 600$ to 700 in various solvents) ascribed to the two highest quartet levels 1^4T_1 and 2^4T_1 (the ground state being the Jahn-Teller and Tanabe-Kamimura stable [49] $^4A_2$) imply persistent problems. It is a fact [47] that 2^4T_1 has at least three (if not four) maxima. The 3 prominent bands are roughly equidistant; the difference from the lowest-energy (and most intense) to the middle maximum is close to 630 cm^{-1} in 7 situations, and from the lowest-energy to the third component 1470 cm^{-1}. $CoBr_4^{-2}$ has also been measured [48] in 7 solvents, and the mean values of these two band distances are 500 cm^{-1} (with slightly larger scattering) and 1840 cm^{-1}. The near-infrared transition to 1^4T_1 is more difficult to measure (being superposed absorption bands due to solvent overtones) but it is smeared out to a plateau, 1800 cm^{-1} wide in $CoCl_4^{-2}$ and 1200 cm^{-1} in $CoBr_4^{-2}$.

There are two appealing justifications for such triple maxima: *first-order spin-orbit coupling*, and intermediate coupling with adjacent *doublet states* such as ^2E and $^2T_1$. The $^4T_1$ sets should split like ^4P in three double-group Γ_j with distances in the ratio 5:3. Unfortunately, the highest quartet 2^4T_1 is expected [49] to split by only $2/3\zeta_{3d}$ in the asymptotic case of definite sub-shell configuration $e^3(t_2)^4$ discussed below, and (in the same order) by $4/3\zeta_{3d}$ in the spherical limit ^4P known to be mixed with squared amplitudes 80 percent $e^3(t_2)^4$ and 20 percent $e^2(t_2)^5$. There is general agreement that the total splitting of $^4T_1$ involves a Landé parameter ζ_{3d} not much below the value 540 cm^{-1} in gaseous Co^{+2}. In such d^7 systems, magnetochemists tend to write $\zeta_{3d} = -3\lambda$ and values close to 500 cm^{-1} are reported [50, 51]. An overall [52] calculated splitting 400 to 600 cm^{-1} is too low (by a factor 2 to 3) in the near infrared 1^4T_1 and even more insufficient in the red 2^4T_1. In close analogy to the almost coincidence [49] of the first spin-allowed transition from the $^4A_2$ ground state to $^4T_2$ in many octahedral $3d^3$ chromium(III) systems with the two spin-forbidden transitions to ^2E and $^2T_1$ it can be argued that all 11 Kramers doublets of $^4T_1$, ^2E, and $^2T_1$ play a role for the spin-orbit coupling (and this cannot be disproved against a clever choice of numerical parameters) but it is surprising [50] that both 1^4T_1 and 2^4T_1 of $Co(N_3)_4^{-2}$ and $Co(NCO)_4^{-2}$ show triple structures 1200 to 1700 cm^{-1} wide (in spite of strongly bent CoNX angle) although [51] $Co(NCS)_4^{-2}$ in solution and the solid $Co[NCS]_4Hg$ show much less structure. The two latter spectra are distinctly shifted toward higher wave-numbers than in $CoCl_4^{-2}$.

An early AOM application [65] to $CoCl_4^{-2}$ with local symmetry close to T_d and representing the point-group D_{2d} gave qualitative agreement with the structure of 2^4T_1 but was not able to rationalize splittings larger than 708 cm^{-1}.

In addition to CoX_4^{-2} complexes, it is interesting to compare with reflection spectra [53] of cubic spinels such as $MgGa_2O_4$, $MgAl_2O_4$, Mg_2SnO_4, and Mg_2TiO_4 containing a few percent cobalt(II). These four crystals are mentioned in order of the expected Co–O distance increasing: 1.85, 1.92, 1.94, and 2.00 Å. As one expects, 2^4T_1 shifts, on the whole, 1100 cm^{-1} toward the red along this series, but it invariantly shows three maxima separated typically by 1100 to 1400 cm^{-1} between two adjacent components. $Co_{0.1}Mg_{1.9}TiO_4$ shows a sharp spin-forbidden transition at 19500 cm^{-1} analogous to the first isolated distinct doublet state of $CoCl_4^{-2}$ or of $Co_{0.1}Zn_{0.9}Al_2O_4$ at 20800 cm^{-1}.

Perhaps the most alarming discrepancy from the Tanabe-Sugano diagram (for T_d) is a shoulder [47, 48] of $CoCl_4^{-2}$ at 16200 cm^{-1} far too intense to be due to an octahedral impurity (though a low-symmetry species cannot be absolutely excluded) and rather too intense for intermediate coupling with $^2T_1$. Most spinels [53] also show weaker shoulders at higher energy than the third maximum of 2^4T_1. The thiourea [54] complex $Co(tu)_4^{+2}$ and the corresponding ethylene-thiourea complex [54] show a broad, quite intense shoulder at 16000 cm^{-1}. The dithiophosphate $Co(S_2P(OC_2H_5)_2)_2$ dissolved in CCl_4 has a similar shoulder on a sharp maximum [56] at 15100 cm^{-1}.

3.1 *Four-Coordinated Cobalt(II) Chromophores CoA_2B_2*

It seems unlikely that the spectra (Fig. 4) of Co(II) in sol–gel glasses [12] having been heated (e.g. to 200 or 600 °C) are due to a mixture of a pink, octahedral, and one tetrahedral blue species. Though the band positions in the reactant solution at room temperature before heating clearly indicate octahedral coordination and ε a few times that of the aqua ion, the three bands observed at 640, 585, and 520 nm (acetate heated to 200 °C), 635, 590, and 525 nm (nitrate heated to 200 °C) and 645, 595, and 525 nm (quite similar for both samples heated to 500 °C) show *slightly* different relative intensities (excluding presence of only one definite species). Rather than an intensity ratio of order 100 in $CoCl_4^{-2}$ vs. $Co(OH_2)_6^{+2}$, a ratio of order 10 would be entirely plausible in the sol–gel glass. Nevertheless, this would mean less than 20 percent transformation of the red to the blue species (in spite of the 525-nm band growing higher, and not weaker)

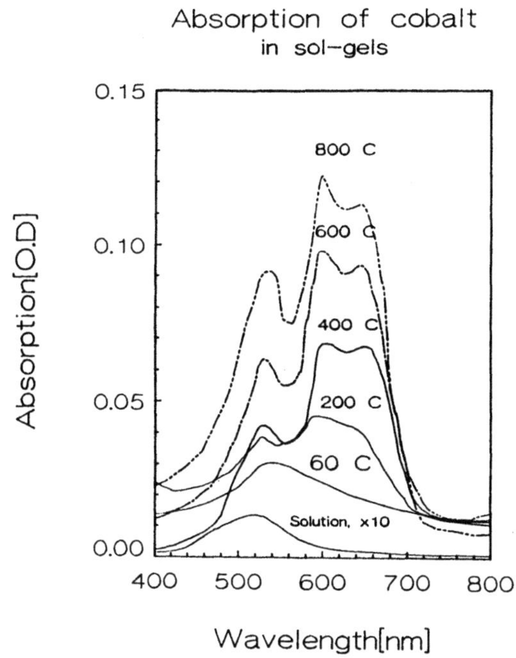

Fig. 4. Absorption spectra, measured at 25 °C, of a thin sol–gel film doped by cobalt(II) having been heated to 60 °C, 200 °C, 400 °C, 600 °C and 800 °C and of precursor solution at room temperature.

and also meaning a blue species having two strong bands separated by 1300 cm^{-1}.

Actually, a similar solution spectrum is known [57], that of $CoCl_2$ in acetone with a maximum at 674 nm (14800 cm^{-1}, $\varepsilon = 306$), a pronounced shoulder at 630 nm (15900 cm^{-1}, $\varepsilon = 225$) and a maximum at 575 nm (17400 cm^{-1}, $\varepsilon = 150$) taking into account that the complex almost certainly involves $Cl_2Co(OC(CH_3)_2)_2$ with Cl_2CoO_2 chromophore, whereas four (more or less non-equivalent) oxygen atoms (from polyatomic anions) are expected to give higher wave-numbers, as known [51] for N-bound complexes of thiocyanate. The advantage of the low-di-electric solvent acetone is that the anion complexes (X = Cl, Br, I) CoX_2, CoX_3^-, and CoX_4^{-2} can be studied almost without re-arrangement [57] to other complexes (here disregarding acetone solvation) in contrast to aqueous solution [46, 58] where a given MX_n usually rearranges to considerable proportions of MX_{n-1} and MX_{n+1} (X also being neutral ligands like ammonia).

A clear-cut L_2CoX_2 complex [51, 59] has L = triphenylphosphine $P(C_6H_5)_3$ and X = Cl with three bands (in methyl-ethyl ketone solution) at 682 nm (14700 cm^{-1}, $\varepsilon = 310$), 638 nm (15700 cm^{-1}, $\varepsilon = 290$) and 581 nm (17200 cm^{-1}, $\varepsilon = 235$) strikingly similar in shape to Fig. 4. The corresponding complex with X = Br shows all three bands (in dichloromethane) shifted 1600 cm^{-1} toward lower energy, the quite sharp bands at 755 and 665 nm having the much higher $\varepsilon = 610$ and 970. These two peaks [51] have ε close to 1000 for X being S-bound SCN^-.

3.2 Plausible Structures of Cobalt(II) in Sol–Gel Glasses

It is not absolutely evident that the blue species [Fig. 4] in sol–gel glasses is anywhere close to regular tetrahedral, in spite of the close similarity to the three bands of $CoCl_2$ in acetone [57] at 674 nm ($\varepsilon = 306$), 630 and 575 nm (especially when allowing for a shift toward lower wavelengths [53] in the glass co-ordinating with four non-equivalent oxygens) and to the blue cobalt(II) species in potassium and cesium borate glass [44, 45] (where it seems ε is well above 50). The situation reminds one of 10^{-3} molar $Co(tu)_4^{+2}$ in acetone (720 and 686 nm, ε close to 650) which is only obtained in the presence of 0.05 molar free thiourea [54], but at lower concentrations of thiourea exhibits three broad bands with ε between 120 and 200; or even of the blue dithiophosphate in ethanol [56] for which the formula $[(C_2H_5O)_2PS_2]_2Co(C_2H_5OH)_2$ was suggested.

Though the blue films studied here were shown by interference to have roughly uniform thickness, we cannot report more precise ε values than typically 200 to 250 at 640 nm, and 150 to 200 at 530 nm (corrected for the weak background due to scattering seen on Fig. 4).

The blue species may perhaps be a distorted five-coordinated complex. Taken at face value, even the splittings of 1^4T_1 and 2^4T_1 in CoX_4^{-2} and in spinels might be explained by a very moderate distortion of the ground state $^4A_2$ away from T_d, in spite of $^4A_2$ being Tanabe-Kamimura stable [49]. Projecting this minor distortion up on the multi-dimensional potential surfaces of $^4T_1$ (each surface divides in six sheets belonging to Kramers doublets) may have perceptible effects on the absorption band shape, much like the (only π-induced) Jahn-Teller instability observed in $^5T_2$ iron(II) and cobalt(III) spectra [69] and in $3d^1$ titanium(III) [70], another perplexing example for textbooks, since $(3z^2 - r^2)$ and $(x^2 - y^2)$ do not seem to coincide as excited states.

The description of tetrahedral cobalt(II) as $3d^7$ may not be [12] as satisfactory as $3d^8$ for octahedral nickel(II), or the Pauli-related $3d^3$ chromium(III). The rather unique presence of three σ-anti-bonding electrons may induce mixing and deformation of some of the d-like orbitals due to slight p-character (not necessarily LCAO involvement of $4p$, but rather p-like orbitals, with quite contracted radial functions, but still keeping orthogonal on cobalt $2p$ and $3p$ inner shells). Since the average radii of $3d$ are not much greater than that of $3p$, the deformation orbitals might simply be the $3p$ shell with an ionization energy about 67 eV.

Several Co(II) complexes have already antecedents of having been presumed tetrahedral, such as $Co(NO_3)_4^{-2}$ with a band at 538 nm ($\varepsilon = 107$). Its present status (crystal structure) is eight-coordinated $Co(O_2NO)_4^{-2}$ but not with a simple bidentate nitrate ligand (known today in many compounds) but [71] the planar groups CoO_2NO have one Co–O distance much shorter (2.04 Å) than the other (2.4 to 2.6 Å), and very similar to $Co(O_2CCF_3)_4^{-2}$. Cases are also known of sky-blue Co(II) being octahedral (such as anhydrous $CoCl_2$) and it is needed to scrutinize the spectra carefully, and not assume *a priori* that approximately tetrahedral symmetry is the only other alternative. It is also a fascinating question whether the numerous blue species with three broad bands separated by about 2500 cm^{-1} (with intensity ratios close to 4:3:2) are CoA_2B_2 chromophores having a similar stereochemistry. The angular overlap treatment [66–68] of the observed cobalt(II) spectra has been reviewed [12] and several (slightly paradoxical) conclusions drawn.

4 Ruthenium Bipyridyl Complexes in Sol–Gel and Related Glasses

Organometallic dyes, like transition metal complexes, can be used as photosensitizers in photochemical conversion of solar energy [72], and their incorporation into solid matrices may be advantageous. In particular, the photophysical behavior of the ruthenium tris[2, 2′-bipyridyl] (+ 2) cation [in this review RuBPY] at low temperatures in solids, and in solutions at room temperature, has been the subject of extensive research [73–79]. The spectroscopic properties of RuBPY in sol–gel films and in porous Vycor glass had been studied previously [4, 80]. The luminescent, first excited state of RuBPY is attributed to triplet ($S = 1$) metal to ligand charge transfer (MLCT) states having one of the six 4*d*-electrons in the ground state transferred to the lowest empty orbital of a single bipyridyl (recently also called 2,2′-bipyridine) ligand [73]. Weak inter-ligand interactions seem to permit the migration of the excited electron from one ligand to another on a nanosecond timescale [74]. The lifetime reveals the quantum efficiency and nonradiative relaxation of the excited state due to triplet–triplet quenching by atmospheric oxygen, and temperature dependent interactions with the solvent (or solid surroundings). The latter can be explained by a four-level model of ruthenium complexes proposed by van Houten and Watts [75] and by Demas and Crosby [76]. The results and explanations outlined in these papers indicate that at low temperatures the lifetimes are dependent on the Boltzmann population of three adjacent triplet levels with double-group quantum numbers A_1, E, and A_2 of which transitions to the singlet ground state are spin-forbidden, each characterized by different temperature independent radiative and nonradiative decay constants [76, 77]. At higher temperatures (above 77 K) these levels can be already characterized by average decay rates, radiative, $\langle k_r \rangle$, and nonradiative, $\langle k_{1q} \rangle$. The latter is sensitive to the nature of the surroundings [75]. An additional manifold of *d*-parentage is believed to be placed at about 3600 cm^{-1} above the first manifold and has a nonradiative decay rate, k_{2q}, of the order of 10^{13}/s. At

77 K its contribution to the total decay is totally negligible but at room temperature it already contributes significantly to the decay rate. This situation is similar to the delayed fluorescence in fluorescein [80–82]. The radiative lifetime of the d-π^* triplet state is almost constant in different media of the same refractive index as reported in numerous works and equals about 14–15 μs when thermal equilibrium between the three components of the triplet state exists. The quantum efficiency from the d-π^* states at temperature T is the ratio of the thermalized radiative decay rate $\langle k_r \rangle$ and the sum $\langle k_r \rangle + \langle k_{1q} \rangle + k_{2q}^{\ddagger} \exp(1 - E/kT)$ where E is the average energy difference between the $^3(d$-$\pi^*)$ state and the $^3(d$-$d)$ states [75]. Thus comparing the different hosts at the same temperature where all parameters are equal with the exception of $\langle k_{1q} \rangle$, we may estimate the role of triplet quenching by the atmospheric oxygen and by the surroundings. [The influence of refractive index n on radiative lifetimes is easily estimated by using the approximate relation, $\tau_{1r}/\tau_{2r} \sim (n_2/n_1)^2$.] Thus we may expect that part of the quenching mechanisms present in solutions will be eliminated in solids and the quantum yield should increase. Therefore it is a matter of interest to test the theory for solids at room temperature as well as the practical interest of the increase of quantum yield of the RuBPY complex for applications such as LSC [83] and phase conjugation in saturable absorbers [81, 82].

Williams [84] pointed out in 1955 that conjugated, but colorless, organic ligands such as 2,2′-bipyridine; 1,10-phenanthroline; and even pyridine [85] may form strongly colored complexes of fairly reducing central atoms having a partly filled [e.g. iron(II)] or full [copper(I)] d shell, and showing *inverted electron transfer* bands, one d-like electron jumping to an empty, delocalized orbital of the (usually heterocyclic) conjugated system. It is common to speak [72] about MLCT (metal to ligand charge transfer) excited states (the transfer from reducing ligands to M being LMCT in a desire for a symmetric notation). For whatever reason, the inverted electron transfer bands (i.e. MLCT) are only known for organic ligands (including cyanide) and are far closer related to the conventional organic colorants [87] than to the familiar orange to red or brown colors of most iron(III) compounds, here excepting practically colorless FeF_6^{-3} and very pale violet $Fe(OH_2)_6^{+3}$. Some conjugated ligands, such as acetylacetonate [88] show both normal and inverted electron transfer bands, varying as one would expect with the oxidizing or the reducing character of M in the oxidation state occurring.

There exist a large number of recent reviews [72, 86, 89, 140–145] treating at length, or even exclusively, the tomato-red, diamagnetic $4d^6$ ruthenium(II) tris(2,2′-bipyridine) complex. The abbreviations of this ligand *bipy*; *bpy*; *dipy*; and *dip* occur in the literature; it was often [85] called 2,2′-dipyridyl. We took above the step to call the cation $Ru(bpy)_3^{+2}$ for RuBPY. There is no doubt that the intense absorption band observed very close to 452 nm (molar extinction coefficient $\varepsilon = 14300$) in 28 different solvents [89] is due to a singlet MLCT state. However, the band does not vanish in any clear-cut way toward lower energy and allows one to suspect shoulders in the region 500 to 600 nm. One explanation might be internal $4d^6$ transitions well known from the isoelectronic rhodium(III) complexes [90, 91]. The two weak bands ($\varepsilon = 11.5$ and 13.5) at 535 and 393 nm of $Ru(OH_2)_6^{+2}$ indicate [92] the sub-shell energy difference $\Delta = 20050\ cm^{-1}$ when identified as transitions to $^1T_1$ and $^1T_2$. A comparison with the nickel(II) aqua and bpy complexes [90, 93] suggests Δ close to $26700\ cm^{-1}$ in RuBPY, providing $^1T_1$ not far from 400 nm. The regularities [90, 94] in the distance from $^1T_1$ to the lowest $4d^6$ triplet $^3T_1$ should put it so close to the intense MLCT maximum at 452 nm as to be undetectable, and with certainty below 500 nm.

Gafney [86] reviewed spectroscopy, photophysics, and photochemistry of RuBPY in porous Vycor glass. The limpid samples of this glass have cavity sizes between 2 and 10 nm; larger cavities induce opalescence. The cations are bound to deprotonated silanol sites (p$K \sim 9$). Gases such as O_2 quench fluorescence of the excited state *RuBPY by energy transfer producing the lowest singlet state of O_2; and NNO and SO_2 by electron transfer forming Ru(III) rapidly. If iron(III) or copper(II) is present in the Vycor glass, an electron is also transferred from *RuBPY, much like the strongly colored double ferrocyanides of Fe(III) and Cu(II) without direct collision, but as an electron transport.

Figure 5 shows the absorption and emission spectrum (at 300 K) of RuBPY in a composite glass (cf. Sects. 6 and 7). We have already reported [95] on the spectroscopic behavior of RuBPY in sol–gel glasses, heavy fluoride glasses, polyvinylalcohol (PVA) films, boric acid (molten at 230 K, and largely dehydrated) and compared the results with RuBPY in aqueous solution. In all these cases (cf. Fig. 5) the absorption spectra show a maximum around 450 nm assigned to metal-to-ligand charge transfer (^{1}MLCT or $d\pi^*$) [73]. The fluorescence spectra of all the samples peaking at 600 nm are assigned to the Stokes-shifted emission of the ^{3}MLCT excited state to the ground state. These results demonstrate that essentially the mechanisms of absorption, internal conversion, and intersystem crossing are similar for all solids studied here. This conclusion is supported by the fact that the surroundings have little effect on the shape and on the molar absorptivity of the CT band.

The behavior of the luminescence decay profiles is similar for different types of glass, showing a double exponential dependence (except for the complex

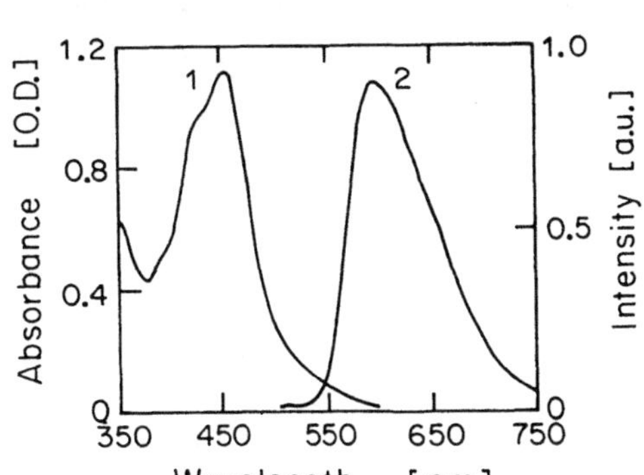

Fig. 5. Absorption (*curve 1*) and emission (*curve 2*) spectra of $[Ru(bpy)_3]^{2+}$ in composite glass at room temperature. The optical density $lg_{10}(I_0/I)$ scale is plotted on the *left*, and the normalized luminescence scale on the *right*

hosted in boric acid, where the decay shows a multiexponential decay). The average lifetimes measured at room temperature were: 1.34 ± 0.05 μs in sol–gel, 1.74 ± 0.03 μs in PVA, 2.00 ± 0.02 μs in boric glass and 1.98 ± 0.02 μs in lead–tin fluorophosphate glass. All the lifetimes are longer than in water solution (0.46 ± 0.04 μs at concentration of 10^{-4} M/L). This increase is the result of the fixation of the ruthenium complex into a rigid solid matrix where a majority of the mechanisms that can quench the excited states are removed. In addition, concentration quenching can be neglected due to the isolation of the complex molecules in the matrix, for example in the fluorophosphate glass a small dependence of the lifetime on the complex concentration has been observed, 1.98 μs for $\sim 10^{-4}$ Molar and 1.89 μs for 0.0025 Molar.

These results are consistent with the scheme of relaxation mechanism proposed by van Houten and Watts [75] in their study of temperature dependence of the luminescence lifetimes of RuBPY in water and with low temperature measurements of temperature dependence of the luminescence in ethanol/methanol glass [76] and in polyvinyl alcohol (PVA) [74]. Our results indicate that the increased lifetime of RuBPY in solids is due to decreased k_{1q}, i.e. decreased interaction of the lowest triplet manifold of the complex with the solid surroundings as compared to water and alcohol at room temperature. At constant room temperature no change of the nonradiative decay constant from the *d-d* manifold is expected relative to liquids [75].

Quantum yield of RuBPY in water is about 0.04 [73] at room temperature and its intrinsic lifetime is 14.4 μs independent of temperature [75]. This should be compared with the lifetimes of the compound at 77 K in PVA, 5.25 μs and in ethanol/methanol, 5.2 μs [74] and the same lifetimes given by Demas and Crosby [76] where also quantum efficiency in the alcohol and intrinsic lifetime are given, 0.377 and 13.9 μs, respectively.

Lifetimes and quantum yields of RuBPY in solids can be estimated as follows; since it is well established that the molar absorption coefficient of RuBPY does not depend on its surroundings, we may assume that the radiative lifetime of the compound in a solid is directly proportional to the radiative lifetime in alcohol times the square of the ratio of refractive indices in alcohol and in the solid, respectively. Independent measurements of quantum yield by comparison of emission of the sample of heavy glass doped by RuBPY with the emission of sample of heavy glass doped by Rh6G reveals an estimation of the quantum yield of 30% ± 10%. The dramatic increase of the quantum yield in solids as compared with water can be rationalized by two main factors. The first is the absence of molecular oxygen in the solids and the second is the limited freedom of motion of the complex in the solid matrices. Both of these should be attributed to the decrease of the decay constant k_{1q}. This thesis can be supported by quite similar lifetimes of the compound in frozen alcohol and in PVA at 77 K [74], where the only common features of the matrices are the absence of oxygen and high viscosity, their structure apparently being irrelevant in respect to the spectroscopic behavior of the complex. The remaining factor responsible for quenching of RuBPY in the solid matrices is probably the fourth level of the *d* parentage discussed in Ref. [75], which, in the opinion of the authors is not luminescent. This level influences the lifetime of RuBPY through the relation:

$$\tau = (k_{1nr} + k_{1r} + k^{*}_{2nr}\exp(-E/kT)) \qquad (1)$$

Using this relation we can estimate also the trend of decreasing k_{1nr} in the solid samples by an additional assumption that k_{1nr} is much more sensitive to the surroundings than k_{2nr}. First we extract the total nonradiative decay rates from the calculated and observed lifetimes. Then, by assumption of constant k_{2nr}, the values of k_{1nr} are extracted and compared with those of RuBPY in water which are taken from ref. [75]. It turns out that the nonradiative decay rate for the lowest triplet level of RuBPY is quite dependent on the nature of the material, where there is an indication that the more dense the material the lower the nonradiative rates. The sol–gel sample studied here has a density of 1.2 g/cm^3 and refractive index of 1.34. Thus more than 50% of its volume is empty pores [96]. We can expect that denser samples of sol–gel will show lower nonradiative rates than those presented in this work.

Lifetimes and quantum yields of RuBPY in PVA, boric acid, tin-lead fluoride glass, and sol–gel glass indicate that in dense matrices part of the quenching mechanisms, such as oxygen quenching and freedom of motion, are absent. This conforms to the theory of four-levels used in Ref. [75] in order to explain the high temperature behavior of RuBPY in liquids. The slightly lower lifetime of the complex in sol–gel glass indicates that the glass is still porous, which permits both some measure of oxygen quenching and freedom of rotation. The fact that the estimated quantum yields are still much lower than unity can be rationalized by intersystem crossing to a quenching level, governed by Boltzmann population of this level. We expect that at lowered temperatures the quantum yield should increase significantly.

Besides more common processes of nonradiative relaxation, the Ru(II) complexes in many cases undergo various photochemical reactions. In principle, four molecules retaining an optical excitation energy even as low as 2.1 eV (the lowest triplet MLCT state of RuBPY) would be able to split water [97] in $2H_2$ and one O_2 and the standard oxidation potential $E^\circ = +1.26$ V of the ground state is decreased (by the triplet excitation energy) to -0.80 V in the fluorescent state. By the same token, $E^\circ = -1.3$ V of $Ru(bpy)_3^+$ (known as a deep purple transient by flash illumination) *increases* by the same amount, to $+0.8$ V, meaning that the excited *RuBPY is not only quite reducing (and would evolve H_2 if had time) but at the *same moment also* oxidizing to a stronger extent than ferricyanide (cf. the reviews [86, 140]).

Slama-Schwok, Avnir, and Ottolenghi [146] studied in sol–gel-produced silica glass a yellow isomer (the same stoichiometry) as the tris (bidentate) yellow $Ir(bpy)_3^{+3}$. The former isomer was known since 1977 to contain one unidentate bpy ligand, but for several years, the sixth ligator was assumed to be a water molecule, which could be reversibly deprotonated ($pK = 3$) to OH^-. However, a crystal structure [147] showed that in the perchlorate, the ring (having difficult, but not impossible, rotation around the central C–C bridge) was turned, so its 2′-nitrogen was not coordinated (and presumably can be protonated with $pK = 3$) and the bidentate character was due to 2-N and, in the other ring, what would be the deprotonated 6′-carbon in usual organic nomenclature [but called C3 by [147] the crystallographers]. Such a bidentate carbanion (like 2′-deprotonated 2-phenylpyridine, which is strictly isoelectronic with 2,2′-bipyridine) is a case of the numerous orthometalations [148, 149] known of Pd(II), Pt(II), Ir(III), . . presenting first a basic nitrogen to the d^q ion, and then losing a proton from an adjacent CH group. The special conditions in nearly dehydrated gels may favor such orthometalations. Anyhow, the purpose of the IrN_5C complex [146] was to study photochemical electron transfer from 1,4-dimethoxybenzene (able to form a cation, like substituted 1,4-diaminobenzenes).

The solvent shifts of MLCT absorption bands [150] of (necessarily *cis*-) $Ru(bpy)_2(CN)_2$ are spectacular, and probably represent a superposition of CN ··· H hydrogen bond effects; electric dipole solvatochromism [151]; and the rigidochromic effect [10] mentioned in Sect. 2.2 of strong molecular distortion in the fluorescent state, e.g. $ReCl(CO)_3(bpy)$ [being octahedral d^6 like Ru(II) and Ir(III)].

It seems that ruthenium(II) luminescence continues to provide many surprises. Specific optical techniques [143] allow distinction between MLCT emission, e.g. in Ru(bpy) $(pyridine)_4^{+2}$ and Ru(bpy) $(CN)_4^{-2}$, and triplets from the "aromatic" excitation of the conjugated ligand. Thus, 2,2′-isobiquinoline has lower energy of its own triplet states [78] in $Ru(i\text{-}biq)_3^{+2}$ than the MLCT states. The opposite extreme of $4d^6\ {}^3T_1$ emission occurs in some rhodium(III) complexes [91] and has been extensively studied [49] in $Co(CN)_6^{-3}$.

5 Nonlinear Materials Based on Sol–Gel Technology

Organic materials in polymers for nonlinear optics have been of great interest in recent years [98, 99]. The second-order nonlinearity connected with the second-order susceptibility χ^2 can be utilized in the generation of second-order harmonics, e.g. creating the green light from the Nd laser. Degenerate four-wave mixing, on the other hand, is based on the third-order nonlinearity connected to χ^3 of the electric susceptibility tensor.

There is a great need for nonlinear optical materials that can be used with low intensity light sources for applications such as phase-conjugation, image processing and optical switching.

The phase conjugated waves are able to restore distorted optical beams to their original unaberrated state under reflection from a nonlinear medium [100].

The origin of the large nonlinear susceptibility at room temperature of organic molecules having good singlet-triplet transfer incorporated in a glass is illustrated in Fig. 6. Following optical excitation from the ground single state of

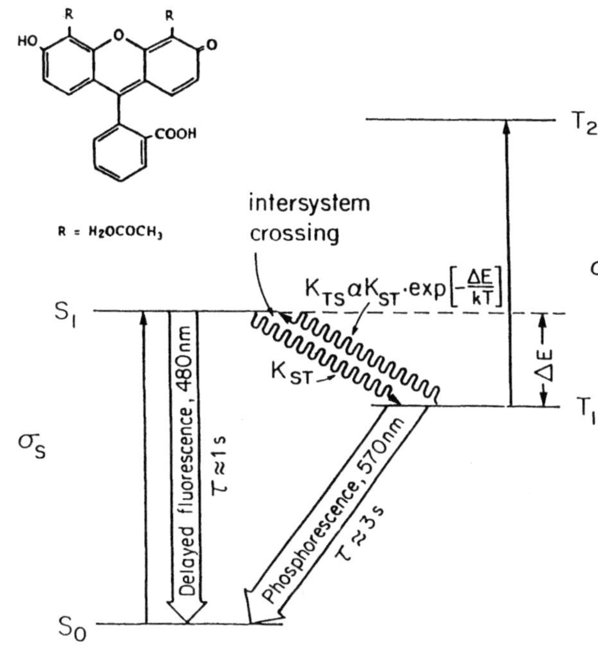

Fig. 6. Scheme of energy levels and saturation mechanism for diacetoxy-mercury fluorescein

the molecule S_0 an intersystem crossing transfers population to the lowest lying triplet state T_1. Provided ΔE, the energy difference between the first excited singlet S_1 and the triplet state T_1 is of appropriate value, delayed fluorescence from the excited singlet state occurs after a back intersystem crossing that returns the population to the excited singlet state from vibrational excited states of the lowest lying triplet manifold. Many of the relaxation mechanisms that can quench the triplet state do not exist when the dye is held rigidly in a solid matrix, and hence the lowest lying triplet state has a very long lifetime. Because of the longevity of the triplet state, significant population can be trapped in that state even by weak optical fields. The saturation intensity I_s is therefore quite small, less than 100 mW cm^{-2}, and since the nonlinear susceptibility χ^3 varies inversely with I_s, the nonlinear χ^3 susceptibility is extremely large, ~ 0.1 esu as compared with $\sim 10^{-12}$ for the liquid CS_2. This treatment has, for instance, been applied to diacetoxymercury fluorescein [81].

The dynamic grating formed by the interaction of the three beams in the four-wave mixing experiment in such material has a lifetime of tens to hundreds milliseconds.

The third-order susceptibility of such a system may be described by [101]:

$$\chi^3(\omega) = (n^2c^2\alpha/24\pi^2 I_s)(\delta + i) \quad (2)$$

where n is the unsaturated refractive index of the material, α is the unsaturated small field attenuation coefficient, I_s is the saturation intensity, defined as the light intensity required to bring half of the population of the saturable absorber to the metastable state, ω is the frequency, and δ is the normalized detuning from the absorption line center. The phase conjugated reflectivity is proportional to the third-order susceptibility, which is in turn proportional to the detuning factor and inversely proportional to the saturation intensity. One must note that the trends of α and δ are opposite, but their frequency dependences differ, therefore the optimum magnitude of χ^3 can be found at some place off the absorption center.

The value of the saturation intensity can be calculated by fitting the differential equation for intensity, $dI/dz = -(\alpha + \alpha_e)I$, to the experimental saturation curve (transmittance versus applied optical field), where α_e is the absorption of the excited state and α, the absorption coefficient for linearly polarized light, is given [101] by:

$$\alpha = N\sigma/S[(1 - 1/\sqrt{3S}\tan^{-1}\sqrt{3S})] \tag{3}$$

where σ is the unsaturated absorption cross-secton and S is the saturation parameter, I/I_s.

For the preliminary estimation of applicability of a saturable dye for phase conjugation, the saturation intensity can be calculated without actually performing the saturation experiment. This can be done by using the well-known expression derived for solutions of saturable absorbers:

$$I_s = h\omega/\sigma\tau \quad [\mathrm{mW/cm^2}] \tag{4}$$

where τ is the lifetime of the metastable level, which is determined from its decay curve after pulsed laser excitation. This formula ignores both the absorption of the excited state and the fact that in solids the saturable absorber excited by a polarized light is saturated more easily than in solution [101]. Hence the formula overestimates the saturation intensity by a factor of two to three, i.e. one may expect that the actual measurement will reveal saturation intensity lower than predicted by Eq. (4). Proceeding further, we can also, by using the predicted value of the saturation intensity and measured values of absorption coefficients, calculate the third-order susceptibility from Eq. (2). As a figure of merit we can use only the real part of the formula which, apart from the constants, may be expressed as:

$$\chi^3 \propto N\sigma^2(\omega\tau/h\omega^2)\delta \tag{5}$$

where N is the concentration of the saturable absorber (molecules/cm^3) and σ is the cross-section of absorption.

In order to make these materials applicable for technological uses the dyes must be incorporated in stable rigid matrices and then, their absorption coefficients and lifetime measured. The well-known technology of dyes in polymer matrices is not sufficient since both plastics and dyes undergo decomposition under prolonged irradiation and under harsh external conditions. On the other hand, one cannot use the current glass technology for incorporation of dyes because of the high temperature of the process at which the dyes decompose.

First, fluorescein [102] and its derivatives [81, 82] have been studied as materials for nonlinear optics. Fluorescein appears in different forms which depend on the pH of the surroundings. The absorption spectra reveal the various species. At pH 0 the cationic form is obtained, at pH 3 the neutral form, at pH 5.5 the mono-anion, and at pH 11 or 12 the di-anion. Addition of a heavy ion of mercury in the form of mercury acetate [81, 82], see Fig. 6, increases the singlet $S_1 \rightarrow T$ transition rate followed by the decrease of saturation intensity I_s.

Figure 7 presents the absorption spectra of fluorescein in sol–gel glass (curve 1), in lead–tin-fluoride glass (heavy glass [102, 103]) (curve 2) and in boric acid (curve 3) and composite glass (curve 4). The absorption spectrum of fluorescein in the heavy glass is red shifted in respect to the spectrum in boric acid and to the spectrum in sol -gel. The emission spectra are also given of fluorescein in impregnated sol–gel glass. The maximum of absorption and emission spectra are strongly shifted to blue as compared to the spectra in sol–gel and water [104]. In this respect the spectra consist partly of that of the acidic form of fluorescein, easily observed in boric acid and in acidic solution in water [105]. It is well established that the acidic (cationic) form of fluorescein is

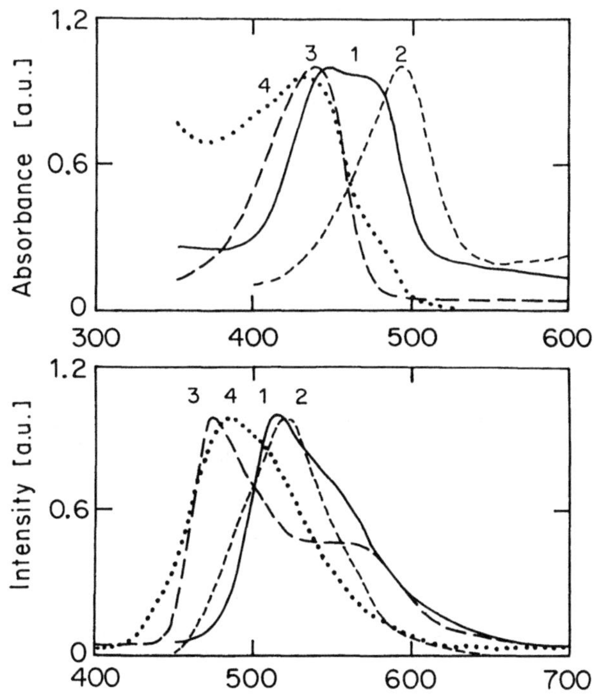

Fig. 7. Upper figure: absorption spectra of fluorescein in sol–gel glass (*curve 1*); in lead–tin fluoride glass (*curve 2*); boric acid glass (*curve 3*); and in composite glass (*curve 4*)

Lower figure: emission spectra of fluorescein in sol–gel glass (*curve 1*); in lead–tin fluoride glass (*curve 2*); boric acid glass (*curve 3*); and in composite glass (*curve 4*)

distinguished by exceptionally long lifetime [82, 103, 105]. From the decay curve of the glass a lifetime of 175 ms was observed. This is less than in boric acid, but more than in heavy glass. This lifetime, together with the benefits of the composite sol–gel glass makes the fluorescein doped composite one of most promising materials for phase conjugation.

Bathochromism of fluorescein in liquids was investigated in the past and it was found that in neutral or basic surroundings (above pH 7) fluorescein exists in its most stable anionic form and its absorption spectrum peaks at about 480 nm, while at acidity of around 0 or less the spectrum shifts to about 435 nm, which is known to be due to the creation of a cationic fluorescein [104]. Various forms of fluorescein are shown in Fig. 8. In solids we observe the same phenomena, however interpretation of these spectra must be in more general terms of Lewis acids. In a boric acid glass the cationic form exists because of the action of boron as a strong Lewis acid [106]. The behavior of fluorescein in sol–gel glass suggests a slightly basic character of the glass [81]. In the heavy glass the absorption spectrum is predominantly anionic and the additional red shift can be ascribed to the high refractive index of the glass. In this glass we expect also a strong singlet–triplet coupling because of the presence of heavy metals. While in boric acid the emission peaks at about 480 nm, in the heavy glass and the sol–gel glass the emissions are characteristic of a stable anionic form of fluorescein peaking at about 515 nm. These observations are consistent with the strong acidic character of the dehydrated boric acid and with the basic character of the heavy and sol–gel glasses.

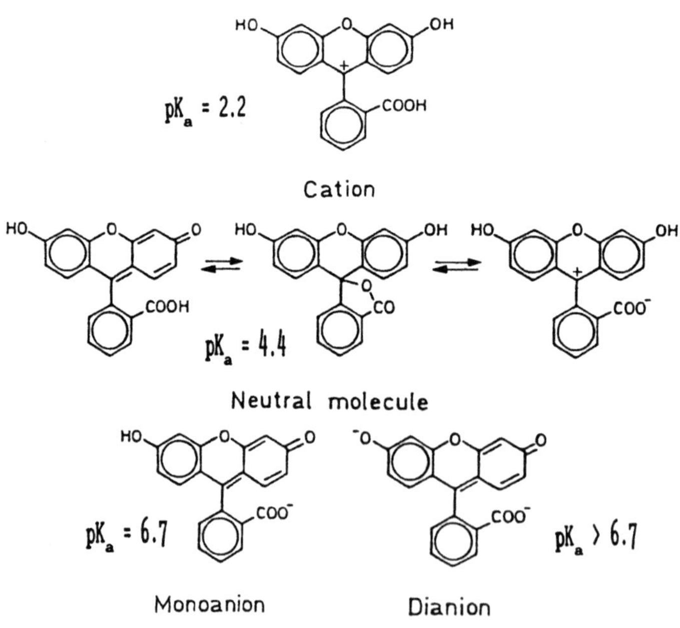

Fig. 8. Various forms of fluorescein as a function of pH

From the semi-log plots of emission intensities of fluorescein in boric acid (curve 1) and in heavy glass (curve 2) monitored at 580 nm and excited by a nitrogen laser. Lifetimes calculated from these plots are 1.15 ± 0.1 s and 25 ± 2 ms, respectively. The former is in fair agreement with literature values [105, 106]. In sol–gel glass the approximate lifetime of the emission at 580 nm is about 200 μs, much too short to be presented on this figure. The lifetime of the excited singlet was found to be 3.5 ns. The exceptionally long lifetime of fluorescein in boric acid can be attributed to the blue shift of the first singlet and to a weak triplet quenching in this material. The first increases the energy gap between the singlet and triplet. This process decreases the rate of back-transfer of the energy from triplet to singlet which is determined by the product of the singlet–triplet rate and the Boltzmann population of upper levels of the triplet. On the other hand, due to a weak triplet quenching, the triplet population decays mainly via the back-transfer channel. These conditions are rarely met in other dyes; either the singlet–triplet coupling is too strong, which results in a strong coupling to the ground state and relatively short lifetime, or the energy gap between singlet and triplet is too small, which increases the rate of back-transfer, the result of which is again decreased lifetime of delayed luminescence. This is best illustrated by the lifetime of fluorescein in the heavy glass, which, in our opinion is due to both increased singlet–triplet rate, because of the presence of heavy atoms, and to the smaller energy gap between singlet and triplet.

6 Composite Glasses

Composite sol–gel samples were prepared by mixing tetraethoxysilane (TEOS) with ethanol, water, and hydrofluoric acid at the ratio of 1:4:4:0.1, respectively, by mixing first the TEOS with ethanol during 1 h, after which water and HF were added and mixed together for another 30 min. The solution was allowed to gel and dry at 40 °C in sealed polypropylene containers for a week. After drying, the samples were heated to 600 °C at a rate of 100 °C/h.

MMA monomer was deinhibited by washing with NaOH and rinsing three times by distilled water. The bulk sol–gel were slowly cooled and then soaked in a warm (35 °C) solution of 2% benzoyl peroxide and organic dye (acridine orange or methyl orange) in the MMA. Then the samples were re-immersed in MMA catalysed by 0.5% of the peroxide in sealed containers and left there for 5 days at constant 60 °C. Finally, the samples were cleaned from excess polymer and polished.

The behavior of fluorescence serves as a model for saturable absorbers. Calculation of saturation intensity and third-order susceptibility are the same (except the input data) also for acridines and for pyrene.

Figure 9 presents the excitation and emission spectra of pyrene in sol–gel glass (upper) and in impregnated sol–gel glass (lower) (Absorption spectrum of

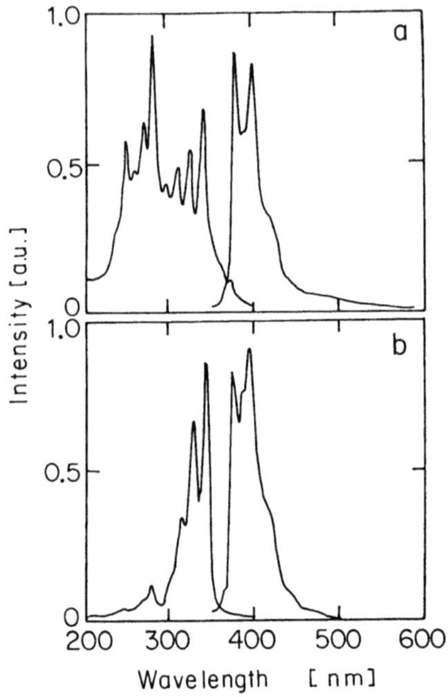

Fig. 9a, b. Excitation and emission spectra of pyrene in sol–gel glass (**a**) and in composite sol–gel glass (**b**). Difference in the structure of excitation and emission bands should be noted

pyrene appears on a strong background of absorption of the glass and does not provide information). The phosphorescence peak at about 650 nm, clearly seen in the emission spectrum of pyrene in heavy glass, is absent in both sol–gel glass and in the impregnated sol–gel glass. It is not surprising, therefore, that phosphorescence lifetime of pyrene in both materials is short. These spectra, however, show some interesting features; one is the disappearance of eximer fluorescence at about 500 nm in the impregnated sample. The second is the quite different structure of excitation and emission spectra in these samples, which in our opinion is due to different interactions of pyrene with the solid surroundings in these samples.

Absorption and emission spectra of acridine orange in a composite glass are presented in Fig. 10. The spectra of acridine yellow are similar. They are

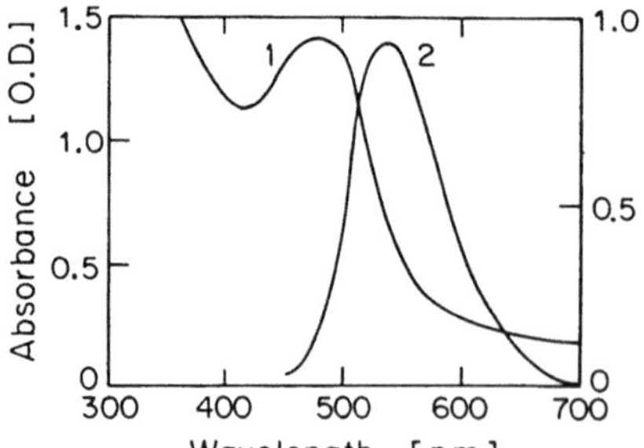

Fig. 10. Absorption (*curve 1*) and emission (*curve 2*) spectra of acridine orange in composite sol–gel glass

less red-shifted than in heavy glass, but their delayed luminescence lifetimes are longer, 65 ± 4 ms for acridine yellow and 52 ± 3 ms for acridine orange. Together with the better transparency of the impregnated glass it makes it a material more favorable for nonlinear applications.

Figure 5 presents the absorption and emission spectra of RuBPY in impregnated sol–gel glass. Lifetime of its luminescence is about 2 μs.

Table 1 summarizes the lifetimes, saturation intensities, and figures of merit for fluorescein, acridine yellow, acridine orange, and pyrene. The results for nonimpregnated doped sol–gel bulk glasses are not included, since all these materials show lifetimes shorter than 1 ms, and consequently their saturation intensities are orders of magnitude higher than in impregnated glasses.

The optical and nonlinear properties of dyes in sol–gel glasses can be improved by applying a new technique of composite sol–gel glass impregnated by an organic polymer which possess all the outlined properties and also the property of largely reduced oxygen quenching of the triplet states, due to filling up of the pores of the sol–gel glass, which provides lifetimes much longer than in ordinary sol–gel. By using this technique a variety of molecules were introduced.

Table 1. Lifetimes and saturation intensities of organic dyes in boric acid, heavy glass and in impregnated sol–gel bulks [mW/cm^2]

Compound	Matrix	Lifetime (ms)	Saturation intensity (mW/cm^2)	Estimated κ [$\chi^{(3)} = \kappa(i + \delta)$]
Acridine	Boric acid	46 ± 5	493	0.005
Yellow	Heavy glass	27 ± 2	200	0.06
	composite	65 ± 4	300	0.05
Acridine	Boric acid	25 ± 1	652	0.006
Orange	Heavy glass	13 ± 1	273	0.06
	composite	52 ± 3	310	0.04
	Boric acid	1115 ± 50	2.7	7.6
Fluorescein	Heavy glass	25 ± 2	91	0.3
	composite	175 ± 10	28	0.8
	Boric acid	160 ± 5	* ≃ 0.004 J/pulse	0.4
Pyrene	Heavy glass	34 ± 5	* ≃ 0.004 J/pulse	0.1
	composite	< 1	–	–

Remarks to Table 1:

1. Saturation intensities are calculated at the center of the absorption band. The formula ignores absorption of excited state and polarization of the exciting laser beam. It has been shown that polarized beam saturates the saturable absorber incorporated in a solid much faster than in solutions [103]. Thus calculations based on the formula may serve as an indicator only, correct within a factor of 2–3.
2. The third order susceptibilities (in units of esu) are estimated for the concentration of the saturable absorber of 10^{-3} M/L (about $6 \cdot 10^{17}/cm^3$) and at the peak absorption of the absorber. This situation requires that the normalized detuning, δ, equals zero. However, the figure of merit, determining the applicability of the saturable dye to phase conjugation is the real factor κ. Given the normalized detuning, δ, which for practical purposes is between 0 and 5, the third-order susceptibility is readily calculated by multiplication of suitable κ by (δ + i).
3. Saturation intensity for pyrene calculated for 3 ns nitrogen laser pulse (337 nm).

The saturable absorption of fluorescein and acridines arising from singlet–triplet is followed by long lifetimes of delayed luminescence. Methyl orange and methyl red change their absorption under irradiation due to change of conformation [107].

6.1 Low-Temperature Emission Spectra of Acridine Orange

In order to verify the assumed mechanism of saturable absorption of acridine orange [101] we measured the emission spectra of composite glass doped with acridine orange at 20 K, 146 K, and at room temperature. Figure 11 presents the result. The emission intensity is highest at the lowest temperature having two distinct emission peaks at about 625 nm and at about 540 nm. At room temperature the emission spectrum becomes weaker and the emission peak at 625 nm almost disappears. We measured also the lifetimes of these peaks and found that the lifetime of the peak at 625 nm is about 27 ms and at 540 nm is a few ns. These data lead to the conclusion that the 625 nm peak is due to the phosphorescence from the first triplet of acridine, while the short-lived emission at 540 nm is due to the first-excited singlet. At higher temperatures a thermal deactivation of the excited levels takes place and emission intensity decreases, both of triplet and of singlet levels. In addition, thermally controlled back-transfer from triplet to singlet also occurs, contributing to further decrease of intensity of the phosphorescence.

Due to large difference of lifetimes of the triplet and the singlet levels, excitation energy is trapped in the triplet level and the ground-state population decreases. Further excitation populates the triplet state, which can be excited to a higher triplet. A complicated situation arises which usually results in an absorption bleaching. However, if the absorption cross-section of the excited state is higher than the cross-section of the ground state, it is possible to obtain induced absorption as well. This is shown in the appendix.

The formula of acridine orange is presented in Fig. 12 (upper) and of azo dyes in Fig. 12 (lower). The pump and probe measurements as well as dynamic grating were described in Refs. [108–110]. Similar experiment done on methyl

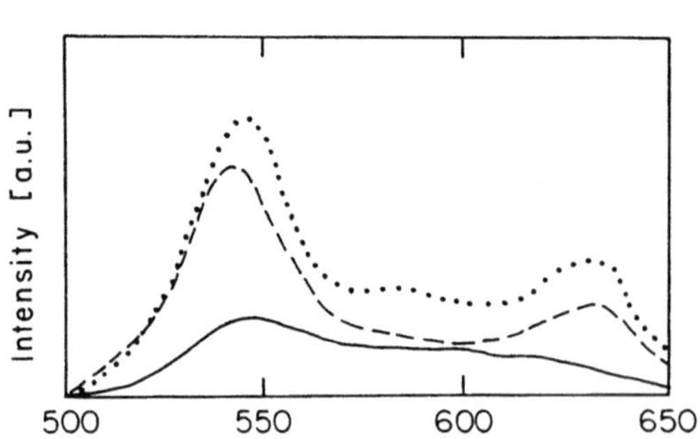

Fig. 11. Luminescence of acridine orange in composite glass as a function of temperature; .... 20 K – – – – – – 146 K ———— room temperature. The phosphorescence peaking at about 625 nm is distinct at low temperatures and disappears at room temperature

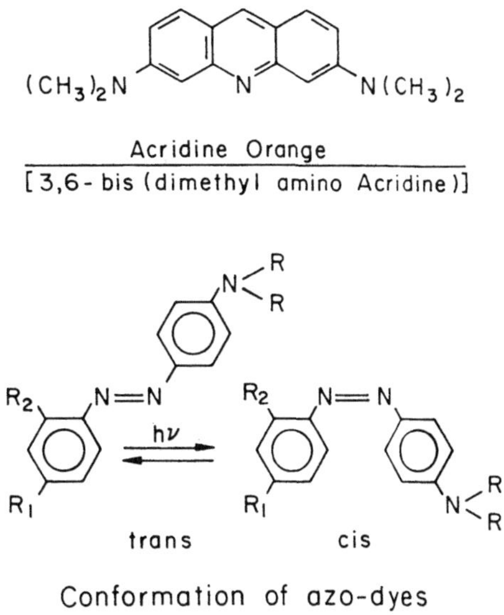

Fig. 12. *Upper*, chemical formula of acridine orange; *lower*, scheme of *cis-trans* transformation for azo-dyes. R: CH_3; R_1: H for methyl red; R_1: $NaSO_4$ for methyl orange. R_2: COOH for methyl red; R_2: H for methyl orange

orange are described in Ref. [109]. We observe induced absorption at both low and high excitation powers and a similar energy dependence of the induced absorption as in acridine. However, in contrast to the acridine, the relaxation of the absorption after switch-off of the laser is completed within less than a minute. At a continuous-wave (CW), low power excitation, methyl orange undergoes induced bleaching due to different absorption coefficients of its *cis* and *trans* forms [107].

6.2 Self-Diffraction Experiments

Self-diffraction was studied on the samples of composite glass doped by acridine orange or methyl orange. The samples were irradiated by two coherent chopped laser beams of Ar^+ laser at 514.5 nm (max. power about 600 mW, minimum power about 10 mW). The two beams were crossed on the sample and the first-order diffracted beams superimposed on the background zeroth-order beams were captured on a photodiode. The intensities of the beams were determined by the use of attenuators to within the ratio of 0.95/1 for acridine orange samples and 0.98/1 for methyl orange samples. Reason for this is that for equal intensities no first-order gratings could be observed due to the compensation of the first- and zeroth-order diffractions. We prefered to observe the signals at margins of the light spots to avoid the effects of saturation of the diode. The transient growth-and-decay curves were observed on a screen of a recording oscilloscope. Higher diffraction orders could be also observed at higher light intensities.

The recorded temporal profiles were then analyzed in the following manner: First, the background due to transmitted intensity of the laser beam was subtracted. Then, the grating efficiency was calculated by division of maximum signal intensity by the difference between the transmitted zeroth-order beams, using formulae evaluated for the "thin grating" case, the second inequality given in below Eq. (11).

According to theory [111], it is possible to arrive to a reasonable approximation of the grating efficiency in two extreme cases: one is the "thin grating" case, for which many diffracted orders can be observed; and the second is the "thick grating" case, for which, except of special conditions, only the first diffraction order can be observed, higher orders being quenched by interference. Thickness of our samples falls somewhere between these two cases and second diffraction order could be observed, albeit weakened by a strong interference. The fact that at such thickness we were still able to observe the higher diffraction orders indicates an efficient nonlinear medium.

An experiment was done on a sample of acridine orange in the composite glass. Highest efficiency of this grating was about 6.6% for sample thickness of 1.4 mm, grating constant of 35 μm and under the irradiation of 360 mW.

By using chopped pulses of different length, it was possible to observe also the decay of the grating, which, as expected, has a short component of about 100 μs and a long one of few ms. Figure 13 presents the response of the diffracted intensity to the applied chopped light. It resembles the shape of a bistable thermal system [112] behavior. This behavior is easily understood since a saturable absorber always responds faster to a rising part of the excitation pulse (excitation of the singlet level) than to the falling part (decay of the triplet level). Thus it possesses a built-in bistability.

The experiments on methyl orange reveal similar information. Figure 14 presents the transient behavior of the diffracted laser light. Here the decay curve of the grating lacks the long component present in acridine-doped grating.

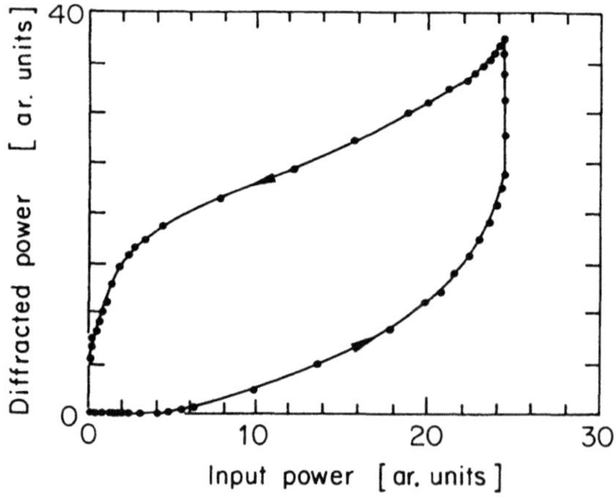

Fig. 13. Diffracted intensity vs input power in two beam experiment for acridine orange in composite glass

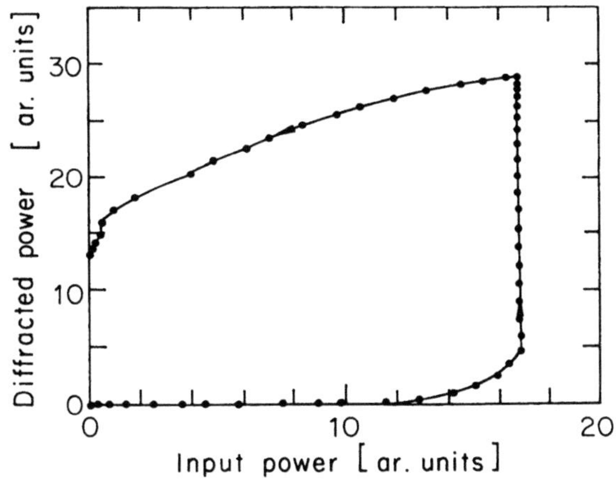

Fig. 14. Diffracted intensity vs input power in two beam experiment for methyl orange in composite glass

Highest grating efficiency observed for this material is about 1% at 1.8 mm thickness, 35 µm grating constant, and 400 mW power.

These efficiencies can be compared with the corresponding number found for the polarized holographic gratings experiment by Todorov et al. [107], where the grating efficiency for methyl orange in polyvinyl alcohol varies from 1% to 35%, depending on polarizations of writing and reading beams. In a three-beams experiment on thin platelets of CdS, 7% grating efficiency was found [113].

Figure 15 presents the response of the system to the applied chopped light. Here the thermal relaxation from the *cis* to *trans* form is slower than the *trans-cis*

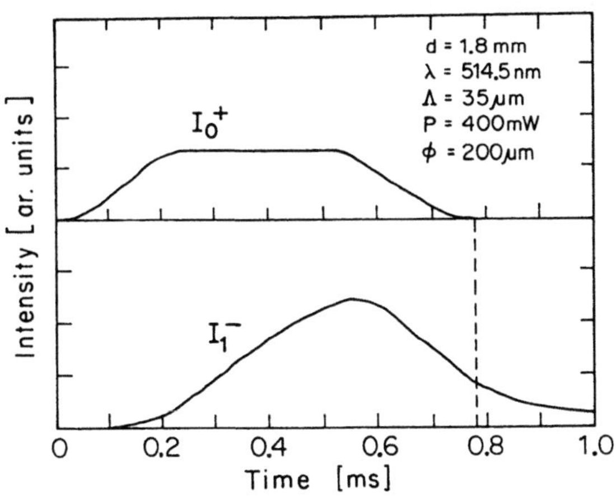

Fig. 15. Transmitted, I_0^+ and diffracted, I_1^- intensities in two-beam experiment on methyl orange in composite glass. The transmitted pulse, Io^- is longer than Io and not shown. The decay of the diffracted beam measures the lifetime of the induced grating.

transformation. Moreover, the response curve for methyl orange shows an interesting property: it is near zero for almost all of the rising part of the excitation pulse, then it rises almost vertically to the maximum diffracted power and then decays slowly with the falling portion of the excitation pulse. Such behavior may find application in digital switching devices.

The *power dependence* of the intensity of the diffracted beams was measured on the two-beams setup. The intensities of the crossed beam were varied by means of polarized light and measured prior to and after each experiment [109].

Figure 16, empty circles, presents the power dependence of the intensity of the diffracted beams in the sample doped by acridine orange, the dependence is $Is \propto I_0^{1.5}$, while the calculated power dependence for a saturable absorber at infinitely high saturation ratio is linear (see Eq. (29)). This result suggests that the laser-induced grating is of a mixed origin, i.e. both absorption coefficient change and refractive index change contribute to the grating.

Figure 16, full circles, presents the power dependence for methyl orange. This sample shows a power threshold and the diffracted intensity is proportional to the third power of the transmitted pump beam. This dependence is consistent with the dependence expected for a refractive index grating (vide infra).

The power dependence of diffracted intensity for the methyl orange is consistent with the assumption that the major contribution to the nonlinear coupling comes from refractive index change, which is the real part of nonlinear third-order susceptibility. Using Eq. (9) below and setting $\Delta\alpha$ to zero, we obtain the refractive index change, $\Delta n = 9.1 \times 10^{-6}$.

Using Eq. (19) below we get $\mathrm{Re}(\chi^3) = 3.1 \times 10^{-7}$ [esu]. This value is similar to the value for some semiconductors [111]. As stated in 6.22, the diffraction efficiency of methyl orange in composite glass can be increased to about 10%. This should increase the value of $\mathrm{Re}(\chi^3)$ to 10^{-6}.

In the case of acridine orange the power dependence indicates that in this material the grating is of a mixed origin. Thus by using Eq. (17) we find (for

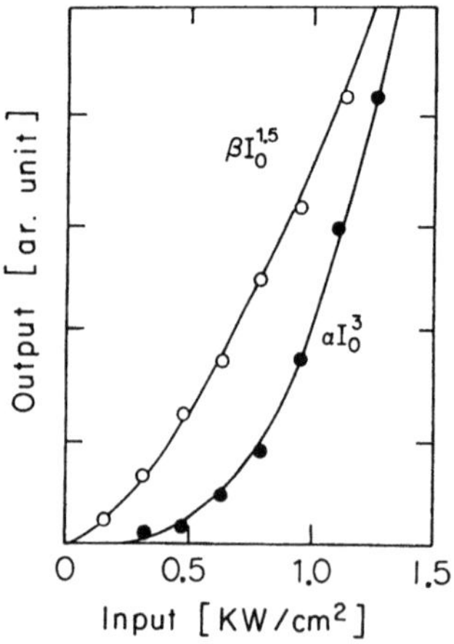

Fig. 16. *Open circles*: power dependence of diffracted intensity for acridine orange in composite glass. The line represents the function $I_1 = 1.5$. *Solid circles*: power dependence of diffracted intensity for methyl orange in composite glass. The line represents the function $I_1 = \alpha Io^3$

$\eta = 0.066$, d = 1.4 mm, and $\lambda = 514.5$ nm) $|\Delta X| \sim 10^{-4}$. This value is lower than the values calculated from the data of Kramer et al. [101] and from the data of Boothroyd et al. [114] (7.6×10^{-2} and 7.3×10^{-2}, respectively) and lower than the values calculated by us (cf. Table 1). We shall reveal the source of these discrepancies in further experiments.

Figure 17 presents schematically the diffraction induced by two coherent laser beams falling on a sample at an angle 2Θ.

The periodic modulation of absorption and refractive index in nonlinear medium induced by the laser beams can be described by the formula:

$$n = n + \Delta n \cos(2\pi x/\Lambda) \quad (6)$$

where Δn is the light-induced change of the complex refractive index. It is defined as,

$$\Delta n = \Delta n + i\Delta\alpha/4\pi \quad (7)$$

and Λ is a grating constant given by $\Lambda = \lambda/2\sin\Theta$.

The transmission function is therefore given by:

$$t(x) = \exp(i\Phi\cos(2\pi x/\Lambda)) \quad (8)$$

where Φ is a phase shift due to the change of the complex refractive index Δn.

For small phase shift the intensity of the first-order diffracted beam, I_1 is given by [110]:

$$I_1 = I_0|\Phi/2|^2 = I_0 d^2[(\pi\Delta n/\lambda)^2 + (\Delta\alpha/4)^2] \quad (9)$$

and for second-order diffraction this relation is:

$$I_2 = I_0|(\Phi/4)^2|^2 \quad (10)$$

where I_0 is the transmitted zeroth-order beam, d is the sample thickness, and $\Delta\alpha$ is the nonlinear change of absorption coefficient induced by the laser beams. The first term in the square brackets is responsible for the phase grating (induced by the change of refractive index) and the second term is responsible for absorption grating. The number of diffracted orders observed depends on thickness of the

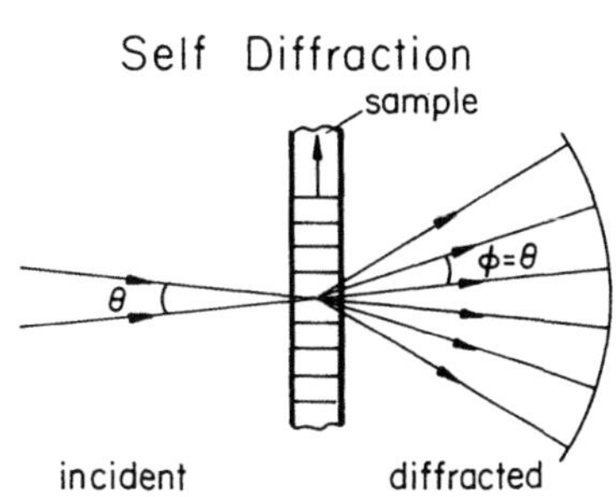

Fig. 17. Self-diffraction in two-beams experiment

sample. A wide-used criterion for classification of gratings is the comparison of the thickness, d, with the quantity $2\Lambda^2/\lambda$;

$$\begin{array}{ll} d \gg 2\Lambda^2/\lambda & \text{thick grating: only first order observed} \\ d \ll 2\Lambda^2/\lambda & \text{thin grating: higher orders observed} \end{array} \tag{11}$$

In our particular case, typical sample thickness is of 1–2 mm, excitation wavelength is 514.5 nm, and grating constants are 30–50 μm. Thus the samples thickness is of the same order of magnitude as the criterion quantity and the gratings can be classified as intermediate between thin and thick. In practice, only the first diffracted orders were measured.

6.2.1 *Third-Order Susceptibility*

The change of polarization of materials induced by electromagnetic fields is usually expressed as:

$$P = \varepsilon_0 \chi(E) E \tag{12}$$

where $\chi(E)$ is electric susceptibility and P and E are complex.

To account for effects induced by strong electromagnetic fields it is convenient to expand the susceptibility into a power series:

$$\chi = \chi^1 + \chi^2 E_1 + \chi^3 E_1 E_2 + \dots \tag{13}$$

where the second- and third-order *coefficients* are tensors.

In disordered media the second term is zero due to symmetry reasons. One can write formally:

$$\chi = \chi^1 + \Delta\chi \tag{14}$$

where the delta term collects all the effects induced by electromagnetic fields.

This can be expressed explicitly by using the relationship of the susceptibility with the complex refractive index. For $\alpha \ll 4\pi/\lambda$:

$$\chi = (n + i\alpha\lambda/4\pi)^2 - 1 \simeq n^2 - 1 + in\alpha\lambda/2\pi \tag{15}$$

Using this relation, we get:

$$\Delta\chi = 2n\Delta n + in\Delta\alpha\lambda/2\pi \tag{16}$$

Comparing equations (9) and (16) we get,

$$\eta = I_1/I_0 = (\pi d/2n\lambda)^2 |\Delta\chi|^2 \tag{17}$$

and the third-order susceptibility, χ^3 can be calculated approximately by:

$$\chi^3 = \Delta\chi/I_0 \tag{18}$$

or [111]:

$$\chi^3[\text{esu}] = 19 n^2 \Delta n / I_0 \quad [\text{cm}^2/\text{W}] \tag{19}$$

where Δn is the complex refractive index change as defined in Eqs. (6) and (7).

In our particular experiments, we could determine only the moduli values of the complex quantities $\Delta\chi$ and Δn. The power dependence of the diffracted intensity (see below) for methyl orange shows that the dominant part is the real part of the susceptibility (refractive index change). Thus, in this case we can approximate the value of χ^3 by neglecting its imaginary part. In the case of acridine the grating is of a mixed nature and only the value of $|\Delta\chi|$ can be calculated.

6.2.2 *Power Dependence of Diffracted Beam Intensity*

For *purely phase-shift, thermal gratings*:

$$\Delta n = (dn/dT)\Delta T \tag{20}$$

where ΔT is the spatial modulation of temperature induced by the two beams.

On the other hand, since [115]:

$$\Delta T \propto I\Lambda^2 \tag{21}$$

hence, in a narrow range of temperatures:

$$\Delta n \propto I\Lambda^2 \tag{22}$$

Therefore, from Eqs. (9), (10), and (22):

$$I_1 \propto I^3\Lambda^4 \tag{23}$$

Such kind of power dependence was observed for methyl orange in composite glass (Fig. 16). Although the observed grating efficiency was relatively low, still it can be increased considerably by increase of the grating constant, Λ. For example, increase of the grating constant to 55 μm, should lead to a grating efficiency of about 10%, some ten times higher than the observed for the grating constant of 35 μm.

For *absorption gratings*, the power-dependent absorption coefficient, α, of a saturable absorber may be expressed as a sum of two terms:

$$\alpha = (N\sigma_0/(1 + S)) + N_e\sigma_e \tag{24}$$

where N is the molecular concentration and N_e is the population of excited state, σ_0 is the absorption cross-section of the ground state and σ_e is the cross-section of the excited state. S is defined as $S = I/I_s$ where I is the actual light intensity and I_s is the saturation intensity (a power needed to transfer half of population to the excited state). The first term in Eq. (16) is a saturable term, since it decreases with increased light intensity. The second term is unsaturable in the sense that it does not depend directly on the applied intensity. It is, however, reasonable that also the second term may become saturable under high enough irradiation intensity. In first approximation, N_e is given by:

$$N_e = N(1 - \alpha_g/\alpha_0) \tag{25}$$

where α_0 is the absorption coefficient of the ground state under infinitesimally weak excitation, $\alpha_0 = N\sigma_0$ and $\alpha_g = \alpha_0/(1 + S)$.

Inserting Eq. (25) into Eq. (24) we get:

$$\alpha = N(\sigma_0 + S\sigma_e)/(1 + S) \tag{26}$$

In our power-dependence experiments, the intensities are by far higher than 100 W/cm^2, while saturation intensity I_s is of the order of tens of mW/cm^2 and thus $S \gg 1$. Using this fact we can approximate Eq. (19) by:

$$\alpha = N[\sigma_0/S + \sigma_e] \tag{27}$$

and:

$$\Delta\alpha = \alpha - \alpha_0 = N[\sigma_0/S + \sigma_e - \sigma_0] \sim N(\sigma_e - \sigma_0) \tag{28}$$

since, if $S \gg 1$, then $\sigma_0 - \sigma_0/S \sim \sigma_0$ for any value of S.

Applying Eq. (28) to Eq. (9) we get two proportionalities:

$$I_1 \propto I(\Delta\alpha)^2 \propto I \text{ (for invariant } \Delta\alpha) \tag{29}$$

Such linear form of power dependence can be fitted to the power dependence for acridine orange in the composite glass (Fig. 16), however, the best fit is of the form $I_1 \propto I_0^{3/2}$, suggesting that the grating is of a mixed character, i.e., also the change of the refractive contributes significantly to the grating.

Looking at the form of Eq. (28) one can conclude that under high excitation, $S \gg 1$, the absorption coefficient of a saturable absorber should be higher than the unsaturated absorption coefficient when $\sigma_e > \sigma_0$ and lower when $\sigma_e < \sigma_0$. In CW experiments done on saturable absorbers reported to date the absorption coefficient was observed to decrease in respect to the unsaturated absorption coefficient, i.e. $\sigma_e < \sigma_0$. However, it is a matter of interest to perform an analysis of a saturable absorber possessing $\sigma_e > \sigma_0$, such as some tetracene dyes [116].

7 Solid-State Tunable Lasers in the Visible

Stable tunable solid-state lasers existing today are based on transition metals [49], such as Cr(III) and Ti(III) ions which emit in the infrared part of the spectrum. Because of the importance of solid-state tunable lasers in the visible, a considerable effort has been extended towards the search for tunable lasers made of solid inorganic materials operating in the visible part of the spectrum [16–18, 28, 117, 118], since, as mentioned above, the sol–gel enables single or multiple component glasses to be prepared at temperatures low enough in order not to destroy the organic molecules [3]. The glass preparation allows adding of an organic molecule to the solution to be homogeneously incorporated in the final glass, provided the dye dissolves in the solvent from which the glass is prepared [3, 13, 15]. The resulting photostability of the dye is enhanced in the inorganic host matrix and surpasses that of organic media such as polymethyl methacrylate. However, despite this increased photostability, lasers prepared by the sol–gel method from conventional laser dyes can survive only a few pulses of exciting laser sources before degradation [16–17, 28, 117, 118].

7.1 *Perylimide Dyes*

Recently perylimide dyes were developed by Seybold and Wagenblast [132] with a structure (derived from perylene) now published [190]. They are characterized by their extreme photostability and negligible singlet–triplet transfer which is responsible for nonradiative relaxations. The absence of a long-lived triplet state providing enough time for photochemistry to take place allows the existence of stable dyes with high quantum efficiency of fluorescence. Optical characteristics of several perylene dyes are presented in Table 2.

The perylene dyes have limited solubility in conventional solvents from which the sol–gel glasses are prepared because of the planar structure of the molecules which facilitates the formation of crystals with high lattice energies. Fortunately perylene dyes can be introduced into composite glasses which are composed of mostly an inorganic matrix with an addition of organic polymer [13]. We were able to introduce a variety of perylene dyes into the composite glasses either in bulk glasses [15, 120] or into thin films [121]. An example of bulk glasses doped by the perylene dye BASF-241 is presented in Fig. 18 which gives the excitation and emission spectra of the dye in a composite glass. The laser action of this dye was tested by frequency doubled Nd:YAG. The measured threshold was less than 6 μJ and the slope efficiency was around 8%. This efficiency can be increased by improving the optics of the laser cavity. The laser could also be excited by a copper laser source. There was no change in the laser operation after several thousand excitation pulses. Laser action was also observed in the films prepared by the sol–gel method with dopants BASF-241 and Rot-300, the absorption and emission spectra of which is given in Fig. 19 [122]. A different dye based on naphthylimide derivative was recently prepared by Dr. Seybold of BASF Farbenlaboratorium, Ludwigshafen. The absorption and emission spectra of this dye in a sol–gel glass film is presented in Fig. 20. The quantum efficiency of this dye is 94%.

Table 2. Optical characteristics of perylene dyes

Dye	λ_{abs} max (nm)	λ_{em} max (nm)	Extinc. coeff.	Quantum effic. %
yellow	474	508	48000	91
orange	525	540	85000	100
red	578	613	44000	96
blue	610	685	63000	92

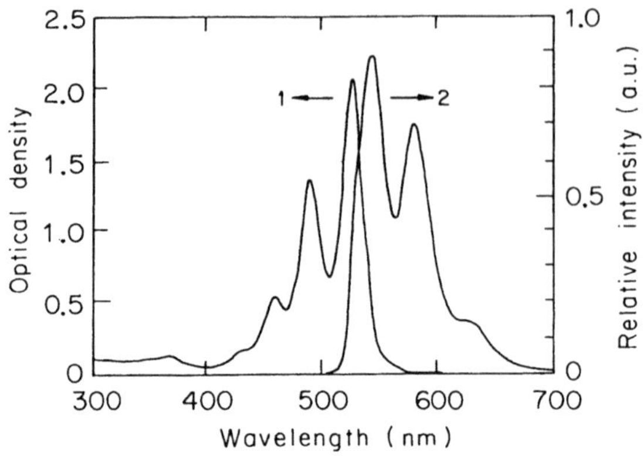

Fig. 18. Absorption and emission spectra of BASF-241 in composite glass. *1*, absorption; *2*, emission

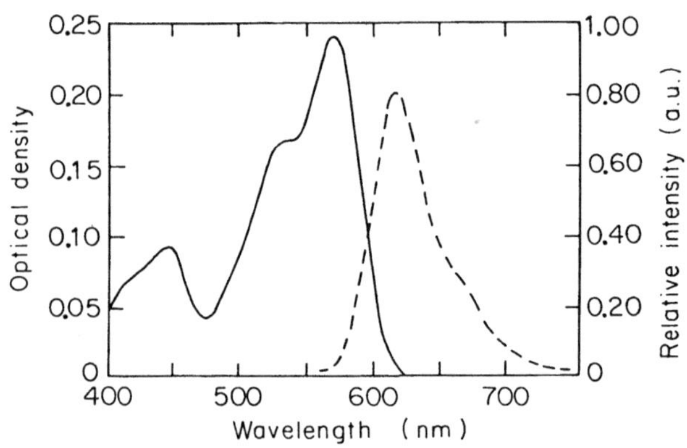

Fig. 19. Absorption and emission spectra of Rot-300 in composite glass

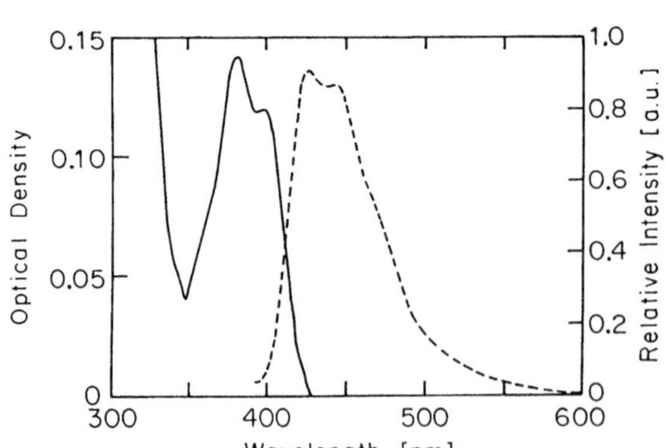

Fig. 20. Absorption and emission spectra of KF-42 in sol–gel film

Film lasers may have some advantage over bulk lasers in the future because of the good thermal dissipation and waveguide properties of the films. If the refractive index of the film is higher than that of the glass support, the laser radiation is trapped and guided by a total internal reflection in the thin film.

As mentioned above the class of the perylimide dyes are only very slightly soluble in polar organic solvents but dissolve well in apolar solvents such as chloroform. Introduction of a bulky tetrabutyl group into the perylimide ring [123] makes the dye soluble also in a number of polar solvents while retaining a high quantum efficiency of fluorescence. We have recently prepared glasses based on acid hydrolysis of TEOS doped by this dye [124], the absorption and emission spectra of which is presented in Fig. 21. Laser action was detected on excitation with the green line of a copper laser.

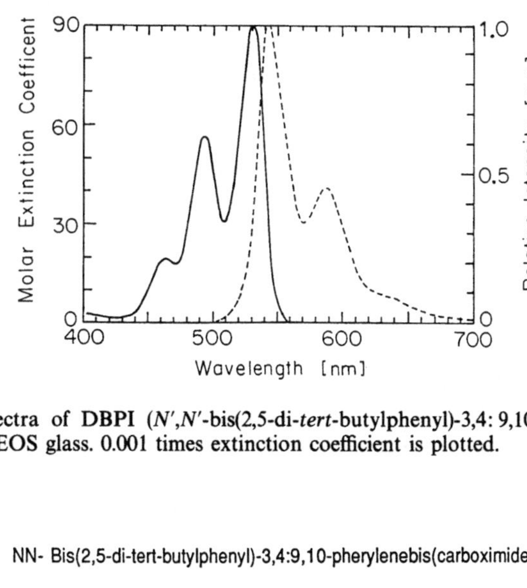

Fig. 21. Absorption and emission spectra of DBPI (N',N'-bis(2,5-di-*tert*-butylphenyl)-3,4: 9,10-perylene-bis(carboximide)) in a bulk TEOS glass. 0.001 times extinction coefficient is plotted.

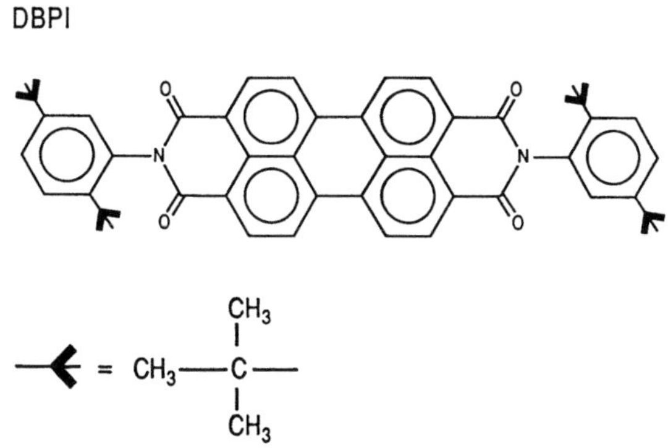

Fig. 21a. Formula of DBPI

7.2 Optical Properties of [2, 2'-Bipyridyl]-3,3'-Diol in Sol–Gel Glasses

The absorption and emission spectra of [2,2'-bipyridyl]-3,3'-diol ($BP(OH)_2$) in solution were studied by the Grabowski group [125–127]. This molecule undergoes an excited-state proton transfer reaction as seen from the formula in Fig. 22 resulting in a very strong Stokes shift. It has also been recommended as an interesting laser material. We introduced [128] this molecule (prepared by L. Kaczmarek in Warsaw) into glass prepared by hydrolysis of tetraethoxysilane (TEOS) and into glass bulks prepared (a) from Glymo [33] (glycidyloxypropyltrimethoxysilane) with a small addition of tetramethoxysilane (TMOS) and (b) by hydrolysis of PST (3-(trimethoxysilyl)propylmethacrylate) with a small addition of TMOS. The latter glasses (a) and (b) in their final form are a composite material mainly composed of an inorganic SiO_2 lattice with a residual organic backbone. The molecule was also introduced into PMMA (polymethylmethacrylate). An example of absorption and emission spectra of the molecule in a thin glass film, in ethanol solution, and in PMMA is presented in Fig. 23. The spectra of the molecule in bulk glasses resemble those of PMMA.

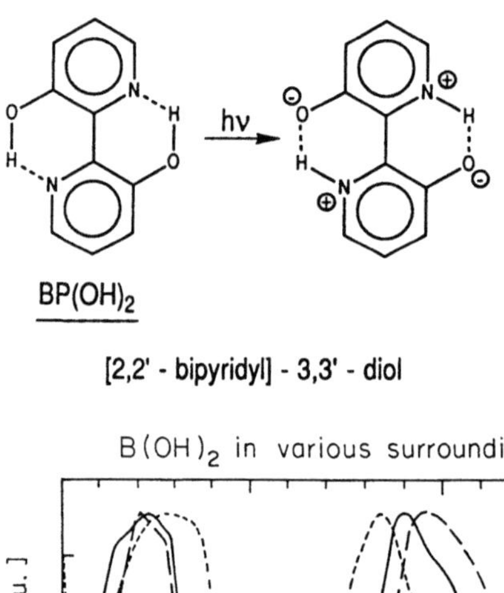

Fig. 22. Excited state proton transfer in $BP(OH)_2$

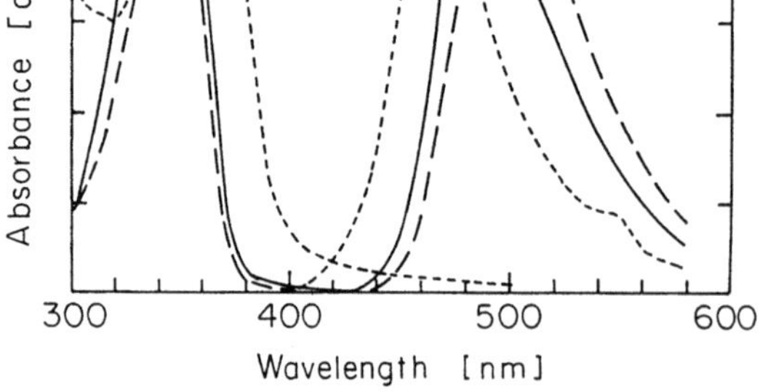

Fig. 23. Absorption and emission spectra of $BP(OH)_2$. ------ in sol–gel glass film. ———— in ethanol. ——— in PMMA

Table 3. Quantum yields, lifetimes and peak positions of $BP(OH)_2$ in various media

Material	Concentration (M/L)	Absorption λ_{nm}	Emission λ_{nm}	Lifetimes ns calculated	measured	Relative quantum yield
Chloroform	5×10^{-6}	341	485	7.95	3.3	0.40 ± 0.05
Ethanol	6.5×10^{-6}	342	480	4.9	2.2	0.45 ± 0.05
Water	7×10^{-7}	365	445	5.6	1.0	0.15 ± 0.05
PMMA*	2×10^{-5}	342	490	5.2	2.4	0.45 ± 0.05
Thin glass§ film	2×10^{-4}	350	470	5.0	0.7	0.15 ± 0.05
PST glass**	1.2×10^{-4}	345	485	8.0	4.2	0.50 ± 0.05
Glymo glass***	1.5×10^{-4}	345	480	7.9	4.0	0.50 ± 0.05

* polymethylmethacrylate. ** 3-(trimethoxysilyl)propylmethacrylate. *** glycidyloxypropyl-trimethoxysilane § see note added in proof

The absorption and emission maxima together with the radiative lifetimes and measured lifetimes and quantum efficiencies of luminescence obtained from the ratio of the measured and calculated lifetimes are presented in Table 3. The high quantum efficiencies of the molecule in PST and Glymo glass and a large Stokes shift which circumvents the self-absorption of the emitted radiation make these new materials promising for lasers in the 480–530-nm range provided that their photostability will be sufficiently high.

8 Luminescent Solar Concentrators

The operation of a luminescent solar concentrator is based on the absorption of solar radiation in a flat collector containing one or more fluorescent species in which the emission bands have little or no overlap with the absorption bands. The fluorescence emission is trapped by total internal reflection and concentrated at the edges of the collector which is usually a thin plate [83, 130–136]. LSCs have the following advantages over conventional solar concentrators: they collect both direct and diffuse light; there is good heat dissipation of non-utilized energy by the large area of the collector plate in contact with air so that essentially "cold light" reaches the photovoltaic cells; tracking the sun is unnecessary; the luminescent species can be chosen to allow matching of the concentrated light to the maximum sensitivity of the photovoltaic cell [83, 131–136].

The performance of LSCs which absorb light in a large plate area and convert it to luminescence which ultimately concentrates at the edges of the plate has been discussed in Refs. [83, 131, and 133] and is schematically presented in Fig. 1 of Ref. [6].

The optical efficiency of an LSC may be expressed by:

$$\eta_{opt} = (1 - R)\eta_{abs}\eta_{fl}\eta_{st}\eta_{tr}\eta_{par} \qquad (30)$$

where R is the Fresnel reflection coefficient, η_{abs} is the efficiency of the absorbed light, η_{fl} is the fluorescence efficiency, η_{st} is the Stokes efficiency, η_{tr} is the efficiency of the trapped fluorescence due to internal reflections, and η_{par} is the efficiency including parasitic losses. η_{abs}, η_{fl}, and η_{st} are parameters characteristic of a dye in a given medium. The trapping efficiency η_{tr} is dependent on the material of the collector. The parasitic losses η_{par} depend both on the dye (self-absorption) and the collector material (scattering of light).

While a large number of papers have been published about luminescent plates in which the dye is incorporated in the entire bulk of the plate [83, 131–136], the configuration in which the plate is covered by a thin film incorporating the colorant deposited in optical contact with the plate is relatively new [137–139]. The advantages of doped thin films having optical contact with the transparent plate is that the luminescence emitted from the thin film is trapped in the plate and that the parasitic losses due to self-absorption and scattering from impurities can be greatly reduced as compared to bulk doped plates [137–139].

As an example of such a system we may examine rhodamine 6G (Rh6G) incorporated in a sol–gel film. Based on the experimental data of absorption and emission of Rh6G in sol–gel glass, its quantum efficiency of 0.95, molar extinction coefficient of 82000, and the overlap of the absorption of Rh6G with the solar spectrum using a 50 micron thick film deposited on a plate having refractive index of 1.5, a Monte Carlo computer method was performed for an optical efficiency of a plate having dimensions of 1 m^2. Details of the computation procedures may be found in Ref. [138]. The optical efficiencies of such plates are approximately 15%.

While Rh6G is not stable enough to be used in a practical device for LSCs it can serve as a model for computation in sol–gel glasses. Much better results can be obtained with the photostable dye BASF-241, the spectrum of which is presented in Fig. 18, in combination with additional dyes, provided it can be introduced into the sol–gel system. In a recent work we have indeed been able to introduce this dye into a polymer/sol–gel system [15]. Also a combination of dyes increases the overlap of absorption with the solar spectrum followed by an increase in optical efficiency.

9 The Novel Technique of Incorporated Colorants in Sol–Gel Glasses

The incorporation of organic colorants into homogeneous inorganic or composite solids has many basic and applied consequences. The differing ratio between monomers and dimers of pyrene [40, 42, 43, 63] and of naphthalene [64] is a result of caging effects on the molecules in the glass pores of different size and distribution, which can be controlled during the preparation.

The spectroscopic consequences of incorporating molecules into glass are intimately connected with solvatochromism [151] to a large extent modifying the position, intensity, and shape of each absorption and each emission band. A bathochromic (red) shift or a hypsochromic (blue) shift may be the result of increasing host polarity and serve as an indicator for the interaction between the molecule and its host. Gibson [152] suggested that two main resonance structures: one quinoidal, essentially not polarized; and one zwitterionic, highly polarized (and fully aromatic) are responsible for the positions and intensities of the absorption bands. The relative contribution of these two forms to the resonance hybrid determines the latter quantities [152]. The asymmetric hydrogen bonds between the colored species and its surrounding medium inducing displacements or even tautomeric migration of protons have been studied by many authors [81] in fluorescein.

Seen from the point of view of quantum chemistry, a given system contains a definite number of electrons, and a set of N nuclei. When N is at least 3, these nuclei have $(3N - 6)$ mutually independent internuclear distances on an instantaneous picture [87, 153]. In modern crystallography (determining the average content of the unit cell, if differing, and the time average of the nuclear positions) the dispersion of these two (conceptually distinct) average distributions is evaluated as "ellipsoids of thermal motion". The temperature dependence of the three principal axes is asymptotically proportional to the square-root of T (for high T) but they remain somewhat above this expression by cooling, and at very low T, they achieve asymptotically constant, small values (like zero-point vibrations of a molecule). Correspondingly, well crystallized solids can show very sharp absorption bands at low T. In a disordered solid (or an exceedingly viscous liquid, such as a glass) the absorption spectra yield an impression of the high T prevailing, when the rearrangement of nuclear positions ceased on a practical time-scale, and a glass cannot lose by cooling its apparent high T suggested by absorption spectra. This is *a fortiori true* for luminescent spectra [153, 154] where the broad emission bands are determined by a transition from the minimum of a $(3N - 5)$ dimensional potential surface belonging to an excited electronic state, by Franck-Condon projection down on the $(3N - 5)$-dimensional ground-state surface. The study of band shapes [154] both of luminescence and of absorption from an excited electronic state to an even higher state can be quite informative.

Hence, one expects that sol–gel glasses show many aspects of apparently high effective T. The study of local heterogeneity may become very significant. Like in the related case of glass-ceramics [49] with microcrystallites between 10 and 50 nm diameter, sol–gel glasses having cavities sufficiently small to confine only one colored (or luminescent) molecule (or ion) avoid formation of excimers and oligomers, which frequently may induce nonradiative deactivation, quenching luminescence. Another, similar, mechanism of quenching is provided by collisions (so prominent in solutions of low viscosity) when the sol–gel glass can act as a "molecular sieve" connected with sufficiently open pores. If the pore size becomes comparable to the wavelength of visible light, opalescence occurs, like in glass-ceramics with 200 to 1000 nm microcrystallites.

As far as choices of sol–gel precursors and of incorporated colorants go, it seems that we are today just on the starting line of a rapidly extending field of great importance for both academic and applied research. In addition to the widely studied silicon alkoxides, mixtures involving alkoxides of the d group elements like Ti, V, Y, Zr, Nb, and W, the f element cerium, and p group B, Al, and P are subjects to current intense activity.

The transition elements have a lower electronegativity and, usually [87, 93], a much higher coordination number than silicon. As a result, transition metal, rare earth, and aluminium alkoxides are far more reactive than the corresponding silicon alkoxides [25]. Neodymium and europium(III) ions, as well as the colorants Rhodamine 6G and Coumarin 4 doped into these glasses [38, 155] show strong luminescence, even at quite high concentrations. It may be noted that efficient energy transfer (presumably without direct contacts via collisions) has been observed [60] betwen Coumarin 460 and both europium(III) and terbium(III) dispersed in thorium (IV) glass. Ormocers having the high refractive index 1.68 were recently prepared, involving TiO_2 and ZrO_2 [156, 157]. The option of gradually changing the refractive index is highly useful in devicing new molecular materials and wave-guiding instrumentation.

Although the historical evolution of the sol–gel technology has been reviewed by Dislich [26], Hench and West [158], and by the contributors to this volume, there is hardly any doubt that the field at present rapidly ramifies in many directions. One example is the important photomorphogenic effects of the relative illumination of various wavelength intervals in the near infrared and in the visible [159]. Not only the photosynthetic yield, but also the growth of the plant in a definite habitus (shape; extent of leaf area, etc.) can be modified. This photomorphogenic result can be achieved by filtering solar light by sol–gel glasses doped with various vanadium(IV) complexes containing the linear ion VO^{+2} with unusually short internuclear distance.

Sol–gel-doped glasses may also play an important role in developing *sensors* and monitoring devices for detection and computerized control of compounds of environmental, biological, and agricultural significance. Such an example was recently described [160]. Oxazine 170 can exist in four differently colored species: neutral, monoprotonated, diprotonated, and triprotonated, and was incorporated in sol–gel glasses, that could then be transformed into thin glass films adhering to colorless glass tubes trapping light. The different, reversibly colored pH indicators depend critically on even very low concentrations of NH_3 in the ambient air. This detection and monitoring of ammonia may be especially useful in avoiding the dispersion of useless excess amounts of fertilizers.

9.1 Semiconductor-Doped Glasses

Conventional (or sol–gel prepared) glasses doped with extremely finely dispersed semiconductors have attracted wide interest, since they show a large third-order nonlinear susceptibility χ^3 [161–163]. The importance of this property is described here in Sect. 5. Further on, very small crystallites of CdS or CdSe dispersed in glasses exhibit quantum-confinement effects [164] which are relevant for "quantum dots".

One of the fundamental questions behind the study of finely dispersed semiconductors in glass (or as colloidal solutions in solvents of low viscosity) is the title "Semiconductor crystallites: a class of large molecules" of a recent review [165]. The earlier evidence accumulated [166–168] indicates that diameters d from 2–5 nm provide distinctly differing properties (especially as far as luminescence goes) of CdS and ZnS, whereas d around 10 nm provide another set of characteristics rather different from the bulk compounds. Nevertheless, crystallinity as cubic blende [CuCl type] is clearly perceived at d = 10 nm. In one way, one would have expected interactions with next-nearest and third-nearest neighbor atoms to have induced ordered crystals [87] for lower d values, in view of the typical M–S distance being slightly below 0.3 nm (or in other words, d = 3 nm corresponds to some 500 atoms).

The theoretical treatment of such data and their "quantum effects" was thoroughly reviewed [169]. Since ZnS, CdS, and CdSe belong to the same class (though frequently hexagonal wurtzites) of adamantoid semiconductors as GaP, GaAs, InP, and elemental Si and Ge, the energy band description (which has been far more successful here than in so many other types of binary inorganic compounds [87]) rationalizes an increase of the band gap (in sufficiently small crystallites) with concomitant blue shift of both absorption edge and of the Stokes-shifted luminescence. However, this is not the whole truth; the next-neighbor internuclear distances agree much less perfectly with the unit cell of the bulk compound. A major consequence is then a red shift of the emission peak, typically achieving 650 nm around d = 10 nm for CdS (in the band model, ascribed to larger dispersion of the one-electron energies in the quite disordered structure) but, on the other hand, reaching 390 nm for much lower d-values. The asymmetrical emission band from bulk CdS peaks at 550 nm.

Lianos and Thomas [170] obtained CdS particles as small as d = 1 nm in inverted micelles in solution, having the emission peak around 470 to 490 nm, whereas d = 10 nm correspond to red emission (640 to 670 nm). Modes and Lianos [171] studied silica sol–gel glasses containing either CdS or $Ru(bpy)_3^{+2}$. Stramel, Nakamura, and Thomas [172] incorporated CdS in porous Vycor glass and in zeolitic solids such as faujasite X and sodalite. The Vycor samples emit a broad band at 590 nm, but very dilute (small d) CdS emit also at 480 nm (like in zeolites).

Rajh et al. [173] prepared a large number of finely divided sulfides, selenides, and iodides (CdS, CdSe, Bi_2S_3, PbS, HgSe, In_2Se_3, AgI, . .) typically with d close to 3 nm, dispersed in silica sol–gels obtained by hydrolysis by aqueous ammonia at the moderate pH 8. The luminescence intensity (yield) in sol–gel glasses is much higher than in aqueous colloidal solutions.

Another technique of preparing finely dispersed CdS in sol–gel glasses is to diffuse gaseous H_2S into the glass samples already doped by cadmium nitrate [174]. By the same token, $CdS_{1-x}Se_x$ can be formed from the precursor solution containing $Cd(NO_3)_2$ and K_2SeO_4 when treated with an aqueous solution of thioacetamide. The color varies smoothly from luminous yellow (x = 0) to deep orange-red shades for high x [175].

Henglein [176] reviewed "small-particle research". A typical example is 2×10^{-4} molar AgI sol fluorescing at 450 nm, to which addition of very dilute (10 to 30 micromolar) H_2S produces a new emission peak at 850 nm. This luminescence moves further out in the infrared at slightly higher H_2S concentrations. It is also possible to prepare long-lived Ag_n^- clusters in solution [176]. Several other adamantoid semiconductors have been studied, such as CuCl [164, 177] showing a fairly narrow fluorescence band at 390 nm. It is conceivable that appropriate incorporation in sol–gel glasses would allow red and/or infrared luminescence to be detected in GaP, GaAs, and CuI particles, and likely to vary with their size.

9.2 Strong Colors Due to Mixed Oxidation States and Electron Transfer

Recently, $3d^6$ iron(II) and $3d^1$ vanadium(IV) were incorporated in silica sol–gel glass [178], providing strong absorption in the near infrared. This observation has relations to the obtention of the photomorphogenic effect [159] mentioned above. There is hardly any doubt that the dark color [when Fe(II) and V(IV) are simultaneously present] is due to an electron transfer band (MM′CT in the notation used in Sect. 4) having as excited state the oxidation states [93, 179–181] $3d^5$ iron(III) and $3d^2$ vanadium(III). This type of sol–gel glasses, including those containing tungsten(VI) also able to accept electrons, have a strong potential use as electrochromic devices [182–184].

Above, the question was implicitly asked how small a semiconducting particle has to be for being considered a large molecule [165, 176]. One may also ask a question how large a fragment of a metallic element or alloy has to be in order to be a genuine metal. Most textbooks refer to the model of one-electron energy bands, and argue that a metal has vanishing energy gap. This opinion runs the risk of neglecting several problems connected with two-electron operators, which imply the usual rationalization of stabilizing high spin quantum numbers *S* in gaseous Pr^{+3}, Eu^{+2}, and Gd^{+3} or in the analogous Pr(III), Eu(II), and Gd(III) compounds, but nevertheless is subject to inherent difficulties [129, 185, 186]. There is no doubt that 100 atoms do not possess a continuous distribution of Schrödinger many-electron eigenvalues immediately above the ground state [169], and the energy band paradigm would agree with the more pragmatic approach that a metal has the *same* electron affinity as ionization energy (or, at least, that these two quantities do not differ by more than the thermal kT). This line of thought brings up the idea of the dielectric constant D (or, perhaps, rather the reciprocal value 1/D vanishing in metals, although, like conductivity of heat, it is only a necessary, but not sufficient, condition for being a metal, since $1/D = 0.013$ for liquid water). Sophisticated experimentation with external electromagnetic fields can decide whether a metal globule in an insulating glass is indeed metallic.

It would seem that there is no clear-cut limit between very small fragments of semiconductors and of metals, already because of the optical transparency. Very thin films of gold are translucent green (a minimum light absorption) and of most other metals blue (absorption increasing monotonically toward the infrared). The situation is entirely different for d = 5 to 20 nm silver particles in glass [187] having their absorption maximum shifting from 420 to 520 nm. The singularity in the electromagnetic field plays a role, for instance in the fluorescence of simultaneously present Rhodamine 6G in the sol–gel [187]. Faraday had discovered the colloidal gold solutions in water, with average d around 20 to 15 nm measured by Turkevich, Stevenson, and Hillier [188]. A very attractive red silicate (so-called "ruby") glass [44] contains such finely dispersed gold particles of comparable size. It may be noted that they contain 10^5 gold atoms, thousands of times more than clusters [87] such as Au_{55} covered with triphenylphosphine and chloride ligands. Wiesner and Wokaun [189] studied the absorption spectra of aqueous solutions, finding all to show a quite narrow band close to 525 nm, and a second band (not much broader on a wavenumber scale) moving smoothly from 1070 to 770 nm when the concentration of Au originally added as seeds for colloid growth is varied from 3.6 to 28 micromolar. The position of the infrared absorption, as well as the decay time of a transient birefringence, depend on both the volume *and* the eccentricity [deviation from spherical shape] of the gold particle [189]. From an analytical chemical point of view, it may be added that Cassius purple obtained from $AuCl_4^- + SnCl_2$ solutions must be colored by a similar colloid, whereas the deep-red solution from $PtCl_6^{-2} + SnCl_2$ contains [93] the trigonal–pyramidal complex $Pt(SnCl_3)_5^{-3}$.

In the last years, literally thousands of papers have been published on superconducting mixed oxides with high upper limits T_c of superconductivity of order 100 K. Actually, $YBa_2Cu_3O_{7-x}$ can be obtained [191] by sol–gel technique from the stoichiometric mixture of acetates, and heating the gel (in steps) to 950 °C under O_2. The T_c value 92 K is almost identical with T_c previously measured for bulk, fairly crystalline samples. The unit cell parameters indicate $x = 0.2$, that is an average oxidation number 2.2 for Cu. The sol–gel preparation of such ternary oxides may be more suitable than the rather tricky reaction between fine powders, needing rather strong and prolonged heating.

Acknowledgements. The writing of this review was rendered feasible by the grant 20.27938.89 from the Swiss National Science Foundation. The experimental studies cited in this review were to a large extent carried out as a common research project between the European Community and the Israeli Ministry of Science and Energy; and for other parts also subsidized by the Krupp Foundation and by the Volkswagen Foundation.

We are grateful to Dr. V. Chernyak, Dr. M. Eyal, and Dr. H. Minti, Hebrew University, Jerusalem for communication of important experimental results, and to Professor C. Klingshirn and his group at the University of Kaiserslautern, for cooperation on nonlinear optics.

One of us (R. R.) is very grateful to Professor A. Revcolevschi and Professor M. Genet, both at Université Paris-Sud for initiating her invitation to Orsay, at which occasion extremely useful discussions and exchange of information were enabled.

10 References

1. Mackenzie JD, Ulrich DR (eds) (1988) Proc Third International Conference on Ultrastructure Processing. Wiley, New York
2. Brinker CJ, Clark DE, Ulrich DR (eds) Better ceramics through chemistry II [MRS Symposium Vol 73 (1986)] and Better ceramics through chemistry III [MRS Symposium Vol 121 (1988)] Materials Research Soc, Pittsburgh
3. Avnir D, Levy D, Reisfeld R (1984) J Phys Chem 88: 5956
4. Avnir D, Kaufman VR, Reisfeld R (1985) J Non-Cryst Solids 74: 395
5. Reisfeld R (1989) Optical behaviour of molecules in glasses prepared by the sol–gel method. Proc Winter School on Glasses and Ceramics from Gels, SOL–GEL Science and Technology, Brazil, August 1989, Aegerter MA, Jafelici M, Souza DF, Zanotto ED (eds) World Scientific, Singapore, p 323
6. Reisfeld R (1990) J Non-Cryst Solids 121: 254
7. Reisfeld R (1990) Theory and applications of spectroscopically active glasses prepared by the sol–gel method, Sol–Gel Optics, SPIE Internat. Symposium on Optical and Optoelectronic Applied Science and Engineering, San Diego, California, July 8–13, 1990, SPIE Proc 1328: paper 1328-04
8. Reisfeld R, Zusman R, Eyal M (1988) Chem Phys Lett 147: 142
9. Kaufman VR, Avnir D, Pines-Rojanski D, Huppert J (1988) J Non-Cryst Solids 99: 379
10. McKiernan J, Pouxviel JC, Dunn B, Zink JI (1989) J Phys Chem 93: 2129
11. Makishima A, Tani T (1986) J Am Ceram Soc 69: 4
12. Reisfeld R, Chernyak V, Eyal M, Jørgensen CK (1989) Chem Phys Lett 164: 307
13. Pope EJ, Mackenzie JD (1987) MRS Bull 12: 29
14. Levy D, Einhorn S, Avnir D (1989) J Non-Cryst Solids 113: 137
15. Reisfeld R, Brusilovsky D, Eyal M, Miron E, Burstein Z, Ivri J (1989) Chem Phys Lett 160: 43
16. Kobayashi Y, Kurokawa Y, Imai Y (1988) J Non-Cryst Solids 105: 198
17. Dunn B, Knobbe E, McKiernan J, Pouxviel JC, Zink JI (1988) Proc MRS Symp 121: 331 [cf. ref. 2]
18. Salin F, LeSaux G, Georges P, Brun A, Bagnall C, Zarzycki J (1989) Opt Lett 14: 785
19. Gromov DA et al (1985) J Opt Soc Am B 2: 1028
20. Itoh U, Takakusa M, Moriya T, Saito S (1977) Jap J Appl Phys 16: 1059
21. O'Connel RM, Saito TT (1983) Opt Eng 22: 393
22. Zelinski BJJ, Uhlmann DR (1984) J Phys Chem Solids 45: 1069
23. Brinker CJ, Scherrer G (1989) Sol–gel science, the physics and chemistry of sol–gel processing, Academic Press, San Diego CA
24. Livage J, Henry M, Sanchez C (1988) Prog in Solid State Chem 18: 259
25. Sanchez C, Livage J (1990) New J Chem 14: 513
26. Dislich H (1988) Thin films from the sol–gel process, in: Klein LC (ed) Sol–gel technology for thin films, fibers, preforms, electronics and speciality shapes, pp 50–79, Noyes Publishers, Park Ridge NJ
27. Pouxviel JC, Parvaneh S, Knobbe ET, Dunn B (1989) Solid State Ionics 32: 646
28. Reisfeld R (1987) J Phys Colloq 48 (C 7): 423
29. Pouxviel JC, Dunn B, Zink JI (1989) J Phys Chem 93: 2134
30. Scholze H, Strehlow P (1986) In: Tagungsband zum dritten Otto Schott-Kolloquium, Jena
31. Fricke J (1986) Phys unserer Zeit 17: 151
32. Fricke J, Reichenauer G (1987) J Non-Cryst Solids 95: 1135
33. Schmidt H, Seiferling B, Philipp G, Deichmann KJ (1989) In: Ultrastructure Processing of Advanced Materials, p. 651, Mackenzie JD, Ulrich DR (eds) New York, Wiley

34. Haas KH, Schmidt H, Roggendorf H (1989) Incorporation of fluorescent dyes into ormocer coatings, Extended abstract, pp 6811–6814, Proc Topical Meeting on Glasses for Optoelectronics
35. Pope EJA, Mackenzie JD (1986) In: Better ceramics through chemistry II, MRS Symp Vol 73, pp 809–814
36. Pope EJA, Mackenzie JD (1986) Mat Sci Res Vol. 20, pp 187–192, Plenum, NY
37. Dunn B, Mackenzie JD, Zink JI, Stafsudd OM (1990) Solid-state tunable lasers based on dye-doped sol–gel materials, SPIE Proc [cf. ref. 7] paper 1328-18
38. Sanchez C (1990) Optical properties of transition metal oxide gels, SPIE Proc [cf. ref. 7] paper 1328-05
39. Levy D, Reisfeld R, Avnir D (1984) Chem Phys Lett 109: 593; Chernyak V, Reisfeld R, under preparation
40. Brusilovsky D, Reisfeld R (1987) Chem Phys Lett 141: 119
41. Berlman IB (1971) Handbook of fluorescence spectra of aromatic molecules, Academic Press, New York
42. Yoruzo T, Hoshino M, Imamura M (1982) J Phys Chem 86: 4426
43. Chen K, Susiya T, Mackenzie JD (1986) J Non-Cryst Solids 81: 227
44. Weyl WA (1959) Coloured glasses, Dawson's of Pall Mall, London
45. Brode WR (1940) J Am Ceram Soc 23: 301
46. Bjerrum J (1989) Coord Chem Rev 94: 1
47. Ballhausen CJ, Jørgensen CK (1955) Acta Chem Scand 9: 397
48. Cotton FA, Goodgame DML, Goodgame M (1961) J Am Chem Soc 83: 4690
49. Reisfeld R, Jørgensen CK (1988) Structure and Bonding 69: 63
50. Cotton FA, Goodgame M (1961) J Am Chem Soc 83: 1777
51. Cotton FA, Goodgame DML, Goodgame M, Sacco A (1961) J Am Chem Soc 83: 4157
52. Griffith JS (1961) Theory of transition-metal ions, Cambridge University Press
53. Schmitz-Dumont O, Brokopf H, Burkhardt K (1958) Z Anorg Allg Chem 295: 7
54. Cotton FA, Faut OD, Mague JT (1964) Inorg Chem 3: 17
55. Carlin RL, Holt SL (1963) Inorg Chem 2: 849
56. Jørgensen CK (1962) Acta Chem Scand 16: 2017
57. Fine DA (1962) J Am Chem Soc 84: 1139
58. Bjerrum J (1957) Metal Ammine formation in aqueous solution, 2. P. Haase (ed) Copenhagen
59. Cotton FA, Faut OD, Goodgame DML, Holm RH (1961) J Am Chem Soc 83: 1780
60. Genet M, Brandel V (1990) Transparent gel and xerogel of thorium phosphate: optical spectroscopy with Nd^{+3}, Er^{+3}, Eu^{+3}, Cr^{+3} and Rhodamine 6G, SPIE Proc [cf. ref. 7] paper 1328–20
61. Simona MG, Balsenc LR, Jørgensen CK (1978) Helv Chim Acta 61: 1984
62. Reisfeld R, Jørgensen CK (1987) In: Gschneidner KA, Eyring L (eds) Handbook on the physics and chemistry of rare earths, North-Holland, Amsterdam, Vol 9, pp 1–90
63. Yamanaka T, Takahashi Y, Kitamuri T, Uchida K (1990) Chem Phys Lett 172: 29
64. Yamanaka T, Takahashi Y, Uchida K (1990) Chem Phys Lett 172: 405
65. Horrocks WDeW, Burlone DA (1976) J Am Chem Soc 98: 6512
66. Jørgensen CK (1971) Modern aspects of ligand field theory, North-Holland, Amsterdam
67. Glerup J, Mønsted O, Schäffer CE (1975) Inorg Chem 15: 1399
68. Jørgensen CK (1988) Chem Phys Lett 153: 185
69. Cotton FA, Meyers MD (1960) J Am Chem Soc 82: 5023
70. Reisfeld R, Eyal M, Jørgensen CK (1987) Chimia 41: 117
71. Cotton FA, Bergman JG (1964) J Am Chem Soc 86: 2941
72. Meyer TJ (1989) Acc Chem Res 22: 163
73. Kalyanasundaram K (1982) Coord Chem Rev 46: 159
74. Krausz E (1988) J Luminescence 42: 283
75. Van Houten J, Watts RJ (1976) J Am Chem Soc 98: 4853
76. Demas JN, Crosby GA (1971) J Am Chem Soc 93: 2841
77. Elfring WH, Crosby GA (1981) J Am Chem Soc 103: 2683
78. Belser P, Von Zelewsky A, Juris A, Barigelletti F, Tucci A, Balzani V (1984) Chem Phys Lett 89: 101
79. Hashimoto K, Hiramoto M, Lever ABP, Sakata T (1988) J Phys Chem 92: 1016
80. Reisfeld R, Manor N, Avnir D (1983) Solar Energy Mat 8: 399
81. Reisfeld R, Eyal M, Gvishi R (1987) Chem Phys Lett 138: 377

82. Eyal M, Reisfeld R (1988) J Luminescence 40: 539
83. Reisfeld R, Jørgensen CK (1982) Structure and Bonding 49: 1
84. Williams RJP (1955) J Chem Soc 137
85. Jørgensen CK (1957) Acta Chem Scand 11: 151 and 166
86. Gafney HD (1990) Coord Chem Rev 104: 113
87. Jørgensen CK (1989) Topics Current Chem 150: 1
88. Jørgensen CK (1962) Acta Chem Scand 16: 2406
89. Juris A, Balzani V, Barigelletti F, Campagna S, Belser P, Von Zelewsky A (1988) Coord Chem Rev 84: 85
90. Jørgensen CK (1963) Adv Chem Phys 5: 33
91. Skibsted H (1989) Coord Chem Rev 94: 151
92. Joensen F, Schäffer CE (1984) Acta Chem Scand A 38: 819
93. Jørgensen CK (1969) Oxidation numbers and oxidation states, Springer, Berlin/New York
94. Schröder KA (1962) J. Chem Phys 37: 2553
95. Reisfeld R, Brusilovsky D, Eyal M, Jørgensen CK (1989) Chimia 43: 385
96. Gvishi R (1989) MSc thesis, Spectroscopic and physical properties of laser dye oxazine 170 in sol–gel glass as compared to solutions, and applications, Hebrew University, Jerusalem
97. Dürr H, Dörr G, Zengerle K, Reis B, Braun AM (1983) Chimia 37: 245
98. Ulrich DR (1988) Overview: Non-linear optical organics and devices, in: Hann RA, Bloor D (eds) Organic materials for non-linear optics, Proc Conf Applied Solid State Chemistry Group of the Dalton Division of the Royal Society of Chemistry, London, pp 241–263
99. Zyss J, Tsoucaris G (1990) Chemistry, symmetry and optics, in: Pierrot M (ed) Structure and optics of molecular crystals, Elsevier, Amsterdam
100. Shkunov VV, Zeldovich BYa (1985) Scientific American 253: 40 (December)
101. Tompkin WR, Boyd RW, Hall DW, Tick PA (1987) J Opt Soc Am B 4: 1030
102. Kramer MA, Tompkin WR, Boyd RW (1986) Phys Rev A 34: 2026
103. Kramer MA, Tompkin WR, Boyd RW (1986) Phys Rev A 34: 3005
104. Martin MM, Lindquist L (1975) J Luminescence 10: 381
105. Lewis GN, Lipkin D, Magel TT (1941) J Am Chem Soc 63: 3005
106. Todorov T, Nikolova L, Tomova N, Dragostinova V (1981) Opt Quant Electron 13: 209
107. Todorov T, Nikolova L, Tomova N, Dragostinova V (1986) IEEE J Quantum Electronics QE-22: 1262
108. Graham S, Renner R, Klingshirn C, Schnepp W, Reisfeld R, Brusilovsky D, Eyal M (1989) Pump- and probe beam measurements in organic materials, Int Conf Materials for Non-linear and Electro-optics, Cambridge 1989, Inst Phys Conf Series no. 103, section 2.2, pp 157–162, Bristol and New York
109. Graham S, Eyal M, Thoma M, Brusilovsky D, Reisfeld R, Klingshirn C (1991) J Luminescence 48: 325
110. Graham S, Thoma, M, Eyal M, Brusilovsky D, Reisfeld R, Gaponenko S, Zimin LG, Klingshirn C (1990) Laser induced gratings and nonlinear optics in organic materials; presented at conference, Oxford, 4–6 September 1990
111. Eichler HJ, Gunter P, Pohl DW (1986) Laser-induced dynamic gratings. Springer, Berlin/Heidelberg
112. Lambsdorff M, Dornfield C, Klingshirn C (1986) Z Physik B 64: 409
113. Thoma N (1990) Diplomarbeit: Dynamische, raum-zeitliche Strukturbildung durch thermische laserinduzierte Gitter in CdS. Kaiserslautern
114. Boothroyd SA, Chrostkowski J, O'Sullivan MS (1989) J Opt Soc Am B 6: 766
115. Eichler H, Enterlein G, Munschau J, Stahl H (1971) Z Angew Physik 31: 1
116. Burgdorff C, Kircher T, Löhmannsröben (1988) Spectrochim Acta 44 A: 1137
117. Altshuler GB, Bakhanov VA, Dulneva EG, Erofeev AV, Mazurin OV, Roskova GFP, Teskhomskaya TS (1988) Opt Spectrosc 62: 707
118. Knobbe ET, Dunn B, Fuqua P, Nishida F, Zink JI (in press) Laser behaviour and nonlinear optical properties in organic dye doped sol–gel materials, in: Uhlmann DR et al (eds) Ultrastructure processing of ceramics
119. Seybold G, Wagenblast G (1989) Dyes and Pigments 11: 303
120. Reisfeld R, Brusilovsky D, Eyal M, Miron E, Burstein Z, Ivri J, Perylene dye in a composite sol–gel glass; a new solid-state tunable laser in the visible range, Proc Binational French–Israeli Workshop on Solid State Lasers (1989) SPIE Proc 1182: 230
121. Reisfeld R, Chernyak V, New solid-state laser configuration based on thin dye-doped glass films, in preparation

122. Reisfeld R, Seybold G, Stable solid-state tunable lasers in the visible, Lecture presented at the International Conference on Luminescence, Lisboa, 16–20 July 1990, (1991) J Luminescence 49: 898
123. Langhals H, Demmig S, Huber H (1988) Spectrochim Acta 44 A: 1189
124. Burgdorff C, Reisfeld R, Löhmannsröben HG, Spectroscopy and laser characteristics of N,N′-bis(2,5-di-tert-butylphenyl)-3, 4; 9, 10-perylenebis(carboximide) DBPI in sol–gel glass, in preparation
125. Bulska H, Grabowska A, Grabowski ZR (1986) J Luminescence 35: 189
126. Sepiol J, Bulska H, Grabowska A (1987) Chem Phys Lett 140: 607
127. Sepiol J, Grabowska A, Bulska H, Mordzinski A, Perez Salgado F, Rettschnick RPH (1989) Chem Phys Lett 163: 443
128. Eyal M, Reisfeld R, Chernyak V, Kaczmarek L, Grabowska A (1991) Chem Phys Lett 176: 531
129. Jørgensen CK (1988) Chimia 42: 21
130. Reisfeld R (1987) Inorg Chim Acta 140: 345
131. Reisfeld R (1984) Inorg Chim Acta 95: 69
132. Reisfeld R (1983) J Less-Common Metals 93: 243
133. Reisfeld R (1985) J Less-Common Metals 112: 9
134. Neuroth N, Hasper R (1986) Int Soc Opt Eng Optical Mater Technol, Energy Efficiency and Solar Energy Conversion, Proc SPIE 653: 88
135. Zastrow A (1985) Phys Unsere Zeit 16: 167
136. Wittwer V, Heidler K, Zastrow A, Goetzberger A (1981) J Luminescence 24: 873
137. Rapp CF, Boling L (1978) Proc 13th IEEE Photovoltaics Specialists Conference, p 690
138. Reisfeld R, Eyal M, Chernyak V, Zusman R (1988) Solar Energy Mater 17: 439
139. Reisfeld R, Chernyak V, Eyal M, Weitz A (1988) Optical materials technology for energy efficiency and solar energy conversion VII, Solar collecting devices, Hamburg, Sept 19–23, SPIE Proc 1016: 240
140. Ghosh BK, Chakravorty A (1989) Coord Chem Rev 95: 239
141. DeArmond MK, Myrick ML (1989) Acc Chem Res 22: 364
142. Krausz E, Ferguson J (1989) Progress Inorg Chem 37: 193
143. DeArmond MK, Myrick ML, Pittman R, Des Enfants RE (1990) Coord Chem Rev 97: 261
144. Krause RA (1987) Structure and Bonding 67: 1
145. Lumpkin RS, Kober EM, Worl LA, Murtaza Z, Meyer TJ (1990) J Phys Chem 94: 239
146. Slama-Schwok A, Avnir D, Ottolenghi M (1989) J Phys Chem 93: 7544
147. Nord G, Hazell AC, Hazell RG, Farver O (1983) Inorg Chem 22: 3429
148. Newkome GR, Puckett WE, Gupta VK, Kiefer GE (1986) Chem Rev 86: 451
149. Evans DW, Baker GR, Newkome GR (1989) Coord Chem Rev 93: 155
150. Kozik M, Sutin N, Winkler JR (1990) Coord Chem Rev 97: 23
151. Buncel E, Rajagopal S (1990) Acc Chem Res 23: 226
152. Gibson HW, Bailey FC (1974) Tetrahedron 30: 2043
153. Jørgensen CK (1989) Proceedings of the OSA Topical Meeting, North Falmouth MA, May 1–3, 1989, printed as: Optical Society of America Proceedings on Tunable Solid State Lasers 5: 252
154. Jørgensen CK (1990) Proceedings of the OSA Topical Meeting, Advanced Solid State Lasers, Salt Lake City, March 5–7, 1990, in press
155. Lecomte M, Viana B, Sanchez C (1991) J Chim Phys (Paris) 88: 39
156. Schmidt H, Popall M (1990) Inorganic–organic composites (ORMOCER) for optical application, SPIE Proc [cf. ref. 7] paper 1328-25
157. Nass R, Schmidt H (1990) Synthesis and properties of transparent ZrO_2 containing SiO_2 polymethacrylate polymers, SPIE Proc [cf. ref. 7] paper 1328-26
158. Hench LL, West JK (1990) Chem Rev 90: 33
159. Novoplanski A, Sachs T, Cohen D, Bar R, Bodenheimer J, Reisfeld R (1990) Solar Energy Mat 21: 17
160. Chernyak V, Reisfeld R, Gvishi R, Venezky D (1990) Oxazine 170 in sol–gel glass and PMMA films as reversible optical waveguide sensor for ammonia and acids, Sensors and Materials 2: 117
161. Peyghambrian N, Koch SW (1987) Revue Physique Appl 22: 1711
162. Remillard JT, Steel DG (1988) Opt Lett 13: 30
163. Jerominek H, Pigeon M, Patela S, Jakubczyk Z, Delisle C, Trembley R (1988) J Appl Phys 63: 957
164. Borelli NF, Hall DW, Halland HJ, Smith DW (1987) J Appl Phys 61: 5399

165. Steigerwald ML, Brus LE (1990) Acc Chem Res 23: 183
166. Rosetti R, Nakahara S, Brus LE (1983) J Chem Phys 79: 1086
167. Rosetti R, Ellison JL, Gibson JM, Brus LE (1984) J Chem Phys 80: 4464
168. Rosetti R, Hull R, Gibson JM, Brus LE (1985) J Chem Phys 82: 552
169. Fojtik A, Weller H, Koch U, Henglein A (1984) Ber Bunsenges Physik Chem 88: 969
170. Lianos P, Thomas JK (1986) Chem Phys Lett 125: 299
171. Modes S, Lianos P (1988) Chem Phys Lett 153: 351
172. Stramel RD, Nakamura T, Thomas JK (1988) JCS Faraday Trans I 84: 969
173. Rajh T, Vucemilovich MI, Dimitrijevic NM, Micic OI, Nozik AJ (1988) Chem Phys Lett 143: 305
174. Reisfeld R, Minti H (1990) Preparations and optical properties of thin glass films prepared by sol–gel method, including CdS crystallites, paper under preparation
175. Bagnall CM, Zarzycki J (1990) Obtention of $CdSe_xS_{1-x}$ crystallites, SPIE Proc [cf. ref. 7] paper 1328-11
176. Henglein A (1989) Chem Rev 89: 1861
177. Ruller JA, Dutt DA, Williams GM, Justus BL, Campillo AJ, Friebele EJ (1990) Characterization of quantum-confined CuCl microcrystals in a glassy matrix, SPIE Proc [cf. ref. 7] paper 1327-13
178. Hutter F, Schmidt H, Scholze H (1986) J Non-Cryst Solids 82: 373
179. Robin MB, Day P (1967) Adv Inorg Radiochem 10: 248
180. Jørgensen CK (1970) Progress Inorg Chem 12: 101
181. Blasse G (in press) Structure and Bonding
182. Potember RS, Speck KR (1990) Optical switches based on semiconducting vanadium dioxide films prepared by the sol–gel process, SPIE Proc [cf. ref. 7] paper 1328-36
183. Aegerter MA, Baudry PE, Martins-Rodrigues AC, LaSerra ER (1990) Optical and electrochemical properties of an all solid state electrochemical window, SPIE Proc [cf. ref. 7] paper 1328-37
184. Judeinstein P, Livage J (1990) electrochromic properties of WO_3 thin films obtained via the sol–gel process, SPIE Proc [cf. ref. 7] paper 1328-34
185. Jørgensen CK (1991) Quantum background for parametrization of *J*-levels, Eur J Solid State Inorg Chem 28: 139
186. Jørgensen CK (in preparation) Comments Inorg Chem
187. Brusilovsky D, Eyal M, Reisfeld R (1988) Chem Phys Lett 153: 203 & 210
188. Turkevich J, Stevenson PC, Hillier J (1951) Discuss Faraday Soc 11: 55
189. Wiesner J, Wokaun A (1989) Chem Phys Lett 157: 569
190. Reisfeld R, Seybold G (1990) Chimia 44: 295
191. Catania P, Hovnanian N, Cot L (1990) Eur J Solid State Inorg Chem 27: 659

Note added in proof

Very recent results on thin glass bulks prepared from tetraethoxysilane (TEOS) doped by 10^{-4} M/L of 2,2′-Bipyridyl-3,3′-Diol (in the starting solution) prepared by M. Eyal and measured in Warsaw Institute of Organic and Physical Chemistry of Polish Academy of Sciences by P. Borowicz and A. Grabowska, revealed much higher quantum yields than the films reported in Table 3. The lifetimes now obtained were about 5.6 nanoseconds (as compared to 0.7 ns) and the quantum yield 0.6, four times higher than in Table 3. We attribute this remarkable increase of quantum yield to the aging of the glass samples, and removal of water.

Author Index Volumes 1–77

GPSR Compliance

The European Union's (EU) General Product Safety Regulation (GPSR) is a set of rules that requires consumer products to be safe and our obligations to ensure this.

If you have any concerns about our products, you can contact us on ProductSafety@springernature.com

In case Publisher is established outside the EU, the EU authorized representative is:

Springer Nature Customer Service Center GmbH
Europaplatz 3
69115 Heidelberg, Germany

Batch number: 10620141

Printed by Printforce, the Netherlands